Maximum Likelihood Estimation with Stata

Fifth Edition

Maximum Likelihood Estimation with Stata

Fifth Edition

JEFFREY PITBLADO
StataCorp LLC

BRIAN POI
Poi Consulting LLC

WILLIAM GOULD
StataCorp LLC

A Stata Press Publication
StataCorp LLC
College Station, Texas

Copyright © 1999, 2003, 2006, 2010, 2024 StataCorp LLC
All rights reserved. First edition 1999
Second edition 2003
Third edition 2006
Fourth edition 2010
Fifth edition 2024

Published by Stata Press, 4905 Lakeway Drive, College Station, Texas 77845
Typeset in LaTeX 2_ε
Printed in the United States of America
10 9 8 7 6 5 4 3 2 1

Print ISBN-10: 1-59718-411-X
Print ISBN-13: 978-1-59718-411-3
ePub ISBN-10: 1-59718-412-8
ePub ISBN-13: 978-1-59718-412-0

Library of Congress Control Number: 2023947523

No part of this book may be reproduced, stored in a retrieval system, or transcribed, in any
form or by any means—electronic, mechanical, photocopy, recording, or otherwise—without
the prior written permission of StataCorp LLC.

Stata, **stata**, Stata Press, Mata, **mata**, and NetCourse are registered trademarks of
StataCorp LLC.

Stata and Stata Press are registered trademarks with the World Intellectual Property Organi-
zation of the United Nations.

NetCourseNow is a trademark of StataCorp LLC.

LaTeX 2_ε is a trademark of the American Mathematical Society.

Other brand and product names are registered trademarks or trademarks of their respective
companies.

Contents

Tables

Figures

Preface to the fifth edition

Maximum Likelihood Estimation with Stata, Fifth Edition is written for researchers in all disciplines who need to compute maximum likelihood estimators that are not available as prepackaged routines. To get the most from this book, you should be familiar with Stata, but you will not need any special programming skills, except in chapters 13 and 14, which detail how to take an estimation technique you have written and add it as a new *command* to Stata. No special theoretical knowledge is needed either, other than an understanding of the likelihood function that will be maximized.

Like the rest of Stata, the tools one uses to implement maximum likelihood estimators in Stata have undergone many enhancements over the years, and a new version of this book reflecting those changes is warranted. The core of the book continues to focus on the `ml` suite of commands. We have also added a new chapter for the `mlexp` command, which is useful not only for pedagogical and prototyping purposes but also for implementing relatively simple estimators with zero programming. For those who are familiar with Mata, Stata's matrix programming language, we have also included a new chapter describing the `moptimize()` suite of functions for implementing maximum likelihood estimators entirely within Mata.

Chapter 1 provides a general overview of maximum likelihood estimation theory and numerical optimization methods, with an emphasis on the practical implications of each for applied work. Chapter 2 covers the `mlexp` command for implementing relatively simple estimators with no programming skills required. Chapter 3 is an introduction to the `ml` command, which provides substantially more flexibility than `mlexp` and can be used to implement arbitrarily complex maximum-likelihood estimators. Chapter 4 is an overview of the `ml` command and the notation used throughout the rest of the book. Chapters 5–11 detail, step by step, how to use Stata to maximize user-written likelihood functions. Chapter 12 describes how to package all the user-written code in a do-file so that it can be conveniently reapplied to different datasets and model specifications. Chapter 13 details how to structure the code in an ado-file to create a new Stata estimation command. Chapter 14 shows how to add survey estimation features to existing `ml`-based estimation commands.

Chapters 15 and 16 are more advanced and show how to use Mata to implement maximum likelihood estimators. Chapter 15 shows how to write your likelihood evaluator in Mata while continuing to use the `ml` command to specify your model, maximize the likelihood function, and report results. Chapter 16 shows how to implement an estimator using Mata's `moptimize()` function and bypass `ml` altogether.

Chapter 17, the final chapter, provides examples. For a set of estimation problems, we derive the log-likelihood function, show the derivatives that make up the gradient and Hessian, write one or more likelihood-evaluation programs, and so provide a fully functional estimation command. We use the estimation command to fit the model to a dataset. An estimation command is developed for each of the following:

- Logit and probit models
- Linear regression
- Weibull regression
- Cox proportional hazards model
- Random-effects linear regression for panel data
- Seemingly unrelated regression
- Bivariate Poisson regression

Appendices contain full syntax diagrams for all the `ml` subroutines, useful checklists for implementing each maximization method, and program listings of each estimation command covered in chapter 17.

We acknowledge William Sribney as one of the original developers of `ml` and the principal author of the first edition of this book.

College Station, TX   Jeffrey Pitblado
October 2023   Brian Poi
                                                                    William Gould

Versions of Stata

This book was written for Stata 18. Regardless of what version of Stata you are using, verify that your copy of Stata is up to date and obtain any free updates; to do this, enter Stata, type

```
. update query
```

and follow the instructions.

Having done that, if you are still using a version older than 18—such as Stata 16—you will likely run into compatibility issues with some of the code and examples in this book. In that case, you should purchase an upgrade to Stata before continuing. We will assume that you are running Stata 18 or perhaps an even newer version.

All the programs in this book follow the outline

```
program myprog
        version 18
        ...
end
```

Because Stata 18 is the current release of Stata at the time this book was written, we write **version 18** at the top of our programs. You could omit the line, but we recommend that you include it because Stata is continually being developed and sometimes details of syntax change. Placing **version 18** at the top of your program tells Stata that, if anything has changed, you want the version 18 interpretation.

Coding **version 18** at the top of your programs ensures they will continue to work in the future.

What about programs you write in the future? Perhaps the here and now for you is Stata 19 or Stata 20. Using this book, should you put **version 18** at the top of your programs, or should you put **version 19** or **version 20**? Probably, you should substitute the more modern version number. The only reason you would not want to make the substitution is that the syntax of **ml** itself has changed and you want to use the version of syntax described in this book.

Anyway, if you are using a version more recent than 18, type **help whatsnew** to see a complete listing of what has changed. That will help you decide what to code at the top of your programs: unless the listing clearly states that **ml**'s syntax has changed, substitute the more recent version number.

Notation and typography

In this book, we assume that you are somewhat familiar with Stata. You should know how to input data and to use previously created datasets, create new variables, run regressions, and the like.

We designed this book for you to learn by doing, so we expect you to read this book while sitting at a computer and trying to use the sequences of commands contained in the book to replicate our results. In this way, you will be able to generalize these sequences to suit your own needs.

Generally, we use the `typewriter font` to refer to Stata commands, syntax, and variables. A "dot" prompt (.) followed by a command indicates that you can type verbatim what is displayed after the dot (in context) to replicate the results in the book.

Except for some very small expository datasets, all the data we use in this book are freely available for you to download, using a net-aware Stata, from the Stata Press web site, https://www.stata-press.com. In fact, when we introduce new datasets, we load them into Stata the same way that you would. For example,

```
. use https://www.stata-press.com/data/ml5/tablef7-1
```

Try it. Also, the ado-files (and a limited number of do-files) used may be obtained by typing

```
. net from https://www.stata-press.com/data/ml5
  (output omitted )
. net install ml5_ado
  (output omitted )
. ml5_ado
  (output omitted )
```

The `ml5_ado` command will load the ado-files in your current directory, where you can look at them and use them.

This text complements but does not replace the material in the Stata manuals, so we often refer to the Stata manuals using [R], [P], etc. For example, [R] **logit** refers to the *Stata Base Reference Manual* entry for `logit`, and [P] **syntax** refers to the entry for `syntax` in the *Stata Programming Reference Manual*.

The following mathematical notation is used throughout this book:

- $F(\cdot)$ is a cumulative probability distribution function.
- $f(\cdot)$ is a probability density function.
- $L(\cdot)$ is the likelihood function.
- ℓ_j is the likelihood function for the jth observation or group.
- g_{ij} is the gradient for the ith parameter and jth observation or group (i is suppressed in single-parameter models).
- H_{ikj} is the Hessian with respect to the ith and kth parameters and the jth observation or group (i and k are suppressed in single-parameter models).
- μ, σ, η, γ, and π denote parameters for specific probability models (we generically refer to the ith parameter as θ_i).
- $\boldsymbol{\beta}_i$ is the coefficient vector for the ith `ml` equation.

When we show the derivatives of the log-likelihood function for a model, we will use one of two forms. For models that meet the linear-form restrictions (see section 5.1), we will take derivatives with respect to (possibly functions of) the parameters of the probability model

$$H_{ikj} = \frac{\partial^2 \ln \ell_j}{\partial \theta_{ij} \partial \theta_{kj}} = \frac{\partial^2 \ln \ell_j}{\partial \mathbf{x}_j \boldsymbol{\beta}_i \partial \mathbf{x}_j \boldsymbol{\beta}_k} = \cdots$$

We will then use some programming utilities to perform the chain-rule calculations that result in the gradient vector and Hessian matrix. For the other models, we may have to take derivatives with respect to the coefficient vector of an `ml` equation

$$\mathbf{H}_{ikj} = \frac{\partial^2 \ln \ell_j}{\partial \boldsymbol{\beta}'_i \partial \boldsymbol{\beta}_k} = \cdots$$

In either case, we will always identify what it is we are taking derivatives with respect to.

1 Theory and practice

Stata can fit user-defined models using the method of maximum likelihood (ML) through Stata's `mlexp` and `ml` commands. `mlexp` has a simple syntax that allows you to obtain estimates with minimal effort, but it is restricted in the types of models that it can handle. `ml` has a formidable syntax diagram (see appendix B) but is nevertheless relatively easy to use and can fit just about any model you can imagine.

In the classical linear regression model, we have

$$y_j = \mathbf{x}_j' \boldsymbol{\beta} + \epsilon_j \qquad \epsilon_j \sim N(0, \sigma^2)$$

or, equivalently,

$$y_j \sim N(\mathbf{x}_j' \boldsymbol{\beta}, \sigma^2)$$

Here we use `mlexp` to fit a classical linear regression:

```
. sysuse auto
(1978 automobile data)
. mlexp (ln(normalden(mpg, {xb:weight displacement _cons}, {sigma})))
Initial:     Log likelihood =     -<inf>  (could not be evaluated)
Feasible:    Log likelihood = -10383.274
Rescale:     Log likelihood = -292.89564
Rescale eq:  Log likelihood = -238.45986
Iteration 0: Log likelihood = -238.45986  (not concave)
Iteration 1: Log likelihood = -225.48871  (not concave)
Iteration 2: Log likelihood = -215.66781
Iteration 3: Log likelihood = -199.18065
Iteration 4: Log likelihood = -195.29858
Iteration 5: Log likelihood =    -195.24
Iteration 6: Log likelihood =  -195.2398
Iteration 7: Log likelihood =  -195.2398
Maximum likelihood estimation
Log likelihood = -195.2398                          Number of obs = 74
```

	Coefficient	Std. err.	z	P>|z|	[95% conf.	interval]
xb						
weight	-.0065671	.0011424	-5.75	0.000	-.0088061	-.0043281
displacement	.0052808	.0096674	0.55	0.585	-.0136671	.0242286
_cons	40.08452	1.978738	20.26	0.000	36.20627	43.96278
/sigma	3.385282	.2782684	12.17	0.000	2.839886	3.930678

In our command, we specified the dependent variable, mpg, as a function of weight, displacement, and a constant term _cons. We explain this command more fully in chapter 2 after we go over some theory, but this example shows how easy it can be to fit maximum likelihood models in Stata.

In probit regression, the dependent variable y_i is either 0 or 1, and we assume that

$$\Pr[y_j = 1 | \mathbf{x}_j; \boldsymbol{\beta}] = \Phi(\mathbf{x}_j' \boldsymbol{\beta})$$

where $\Phi(\cdot)$ is the standard normal cumulative distribution function. Here we use ml to implement probit regression and to fit a particular model:

```
                                            ─────── begin myprobit_lf.ado ───────
program myprobit_lf
        version 18
        args lnfj xb
        quietly replace 'lnfj' = ln(normal( 'xb')) if $ML_y1 == 1
        quietly replace 'lnfj' = ln(normal(-'xb')) if $ML_y1 == 0
end
                                            ─────────── end myprobit_lf.ado ───────────
. sysuse cancer
(Patient survival in drug trial)

. ml model lf myprobit_lf (died = i.drug age)

. ml maximize

Initial:        Log likelihood = -33.271065
Alternative:    Log likelihood = -31.427839
Rescale:        Log likelihood = -31.424556
Iteration 0:    Log likelihood = -31.424556
Iteration 1:    Log likelihood = -21.883453
Iteration 2:    Log likelihood = -21.710901
Iteration 3:    Log likelihood = -21.710799
Iteration 4:    Log likelihood = -21.710799
```

					Number of obs =	48
					Wald chi2(3) =	13.58
Log likelihood = -21.710799					Prob > chi2 =	0.0035

died	Coefficient	Std. err.	z	P>|z|	[95% conf. interval]	
drug						
Other	-1.941275	.597057	-3.25	0.001	-3.111485	-.7710644
NA	-1.792924	.5845829	-3.07	0.002	-2.938686	-.6471626
age	.0666733	.0410666	1.62	0.104	-.0138159	.1471624
_cons	-2.044876	2.284283	-0.90	0.371	-6.521988	2.432235

For our example, we entered a four-line program into an ado-file (myprobit_lf.ado) and then typed two more lines of code after reading some data into memory. That was all that was required to program and fit a probit model. Moreover, our four-line program is good enough to fit any probit model. This program reports conventional (inverse, negative second-derivative) variance estimates, but by specifying an option to ml model, we could obtain the outer product of the gradients or Huber/White/sandwich robust variance estimates, all without changing our simple four-line program.

We will discuss `ml` and how to use it in chapter 4, but first we discuss the theory and practice of maximizing likelihood functions.

Also, we will discuss theory so that we can use terms such as *conventional (inverse, negative second-derivative) variance estimates, outer product of the gradients variance estimates,* and *Huber/White/sandwich robust variance estimates,* and you can understand not only what the terms mean but also some of the theory behind them.

As for practice, a little understanding of how numerical optimizers work goes a long way toward reducing the frustration of programming ML estimators. A knowledgeable person can glance at output and conclude when better starting values are needed, when more iterations are needed, or when—even though the software reported convergence—the process has not converged.

1.1   The likelihood-maximization problem

The foundation for the theory and practice of ML estimation is a probability model

$$\Pr(Z \leq z) = F(z; \boldsymbol{\theta})$$

where Z is the random variable distributed according to a cumulative probability distribution function $F(\cdot)$ with parameter vector $\boldsymbol{\theta}' = (\theta_1, \theta_2, \ldots, \theta_E)$ from $\boldsymbol{\Theta}$, which is the parameter space for $F(\cdot)$. Typically, there is more than one variable of interest, so the model

$$\Pr(Z_1 \leq z_1, Z_2 \leq z_2, \ldots, Z_k \leq z_k) = F(\mathbf{z}; \boldsymbol{\theta}) \qquad (1.1)$$

describes the joint distribution of the random variables, with $\mathbf{z} = (z_1, z_2, \ldots, z_k)$. Using $F(\cdot)$, we can compute probabilities for values of the Zs given values of the parameters $\boldsymbol{\theta}$.

In likelihood theory, we turn things around. Given observed values $\mathbf{z}$ of the variables, the likelihood function is

$$\ell(\boldsymbol{\theta}; \mathbf{z}) = f(\mathbf{z}; \boldsymbol{\theta})$$

where $f(\cdot)$ is the probability density function corresponding to $F(\cdot)$. The point is that we are interested in the element (vector) of $\boldsymbol{\Theta}$ that was used to generate $\mathbf{z}$. We denote this vector by $\boldsymbol{\theta}_T$.

Data typically consist of multiple observations on relevant variables, so we will denote a dataset with the matrix $\mathbf{Z}$. Each of N rows, $\mathbf{z}_j$, of $\mathbf{Z}$ consists of jointly observed values of the relevant variables. In this case, we write

$$L(\boldsymbol{\theta}; \mathbf{Z}) = f(\mathbf{Z}; \boldsymbol{\theta}) \qquad (1.2)$$

and acknowledge that $f(\cdot)$ is now the joint-distribution function of the data-generating process. This means that the method by which the data were collected now plays a role in the functional form of $f(\cdot)$; at this point, if this were a textbook, we would introduce the assumption that "observations" are independent and identically distributed (i.i.d.) and rewrite the likelihood as

$$L(\boldsymbol{\theta}; \mathbf{Z}) = \ell(\boldsymbol{\theta}; \mathbf{z}_1) \times \ell(\boldsymbol{\theta}; \mathbf{z}_2) \times \cdots \times \ell(\boldsymbol{\theta}; \mathbf{z}_N)$$

The ML estimates for $\boldsymbol{\theta}$ are the values $\widehat{\boldsymbol{\theta}}$ such that

$$L(\widehat{\boldsymbol{\theta}}; \mathbf{Z}) = \max_{\mathbf{t} \in \boldsymbol{\Theta}} \ L(\mathbf{t}; \mathbf{Z})$$

Most texts will note that the above is equivalent to finding $\widehat{\boldsymbol{\theta}}$ such that

$$\ln L(\widehat{\boldsymbol{\theta}}; \mathbf{Z}) = \max_{\mathbf{t} \in \boldsymbol{\Theta}} \ \ln L(\mathbf{t}; \mathbf{Z})$$

This is true because $L(\cdot)$ is a positive function and $\ln(\cdot)$ is a monotone increasing transformation. Under the i.i.d. assumption, we can rewrite the log likelihood as

$$\ln L(\boldsymbol{\theta}; \mathbf{Z}) = \ln \ell(\boldsymbol{\theta}; \mathbf{z}_1) + \ln \ell(\boldsymbol{\theta}; \mathbf{z}_2) + \cdots + \ln \ell(\boldsymbol{\theta}; \mathbf{z}_N)$$

Why do we take logarithms?

1. Speaking statistically, we know how to take expectations (and variances) of sums, and it is particularly easy when the individual terms are independent.

2. Speaking numerically, some models would be impossible to fit if we did not take logs. That is, we would want to take logs even if logs were not, in the statistical sense, convenient.

To better understand the second point, consider a likelihood function for discrete data, meaning that the likelihoods correspond to probabilities; logit and probit models are examples. In such cases,

$$\ell(\boldsymbol{\theta}; \mathbf{z}_j) = \Pr(\text{we would observe } \mathbf{z}_j)$$

where $\mathbf{z}_j$ is a vector of observed values of one or more response (dependent) variables. All predictor (independent) variables, $\mathbf{x} = (x_1, \ldots, x_p)$, along with their coefficients, $\boldsymbol{\beta}' = (\beta_1, \ldots, \beta_p)$, are part of the model parameterization, so we can refer to the parameter values for the jth observation as $\boldsymbol{\theta}_j$. For instance, $\ell(\boldsymbol{\theta}_j; \mathbf{z}_j)$ might be the probability that $y_j = 1$, conditional on $\mathbf{x}_j$; thus $\mathbf{z}_j = y_j$ and $\boldsymbol{\theta}_j = \mathbf{x}_j\boldsymbol{\beta}$. The overall likelihood function is then the probability that we would observe the y values given the $\mathbf{x}$ values and

$$L(\boldsymbol{\theta}; \mathbf{Z}) = \ell(\boldsymbol{\theta}; \mathbf{z}_1) \times \ell(\boldsymbol{\theta}; \mathbf{z}_2) \times \cdots \times \ell(\boldsymbol{\theta}; \mathbf{z}_N)$$

because the N observations are assumed to be independent. Said differently,

$$\Pr(\text{dataset}) = \Pr(\text{datum } 1) \times \Pr(\text{datum } 2) \times \cdots \times \Pr(\text{datum } N)$$

Probabilities are bound by 0 and 1. In the simple probit or logit case, we can hope that $\ell(\boldsymbol{\theta}; \mathbf{z}_j) > 0.5$ for almost all j, but that may not be true. If there were many possible outcomes, such as multinomial logit, it is unlikely that $\ell(\boldsymbol{\theta}; \mathbf{z}_j)$ would be greater than 0.5. Anyway, suppose that we are lucky and $\ell(\boldsymbol{\theta}; \mathbf{z}_j)$ is right around 0.5 for all N observations. What would be the value of $L(\cdot)$ if we had, say, 500 observations? It would be

$$0.5^{500} \approx 3 \times 10^{-151}$$

That is a very small number. What if we had 1,000 observations? The likelihood would be

$$0.5^{1000} \approx 9 \times 10^{-302}$$

What if we had 2,000 observations? The likelihood would be

$$0.5^{2000} \approx \texttt{<COMPUTER UNDERFLOW>}$$

Mathematically, we can calculate it to be roughly 2×10^{-603}, but that number is too small for most digital computers. Modern computers can process a range of roughly 10^{-301} to 10^{301}.

Therefore, if we were considering ML estimators for the logit or probit models and if we implemented our likelihood function in natural units, we could not deal with more than about 1,000 observations! Taking logs is how programmers solve such problems because logs remap small positive numbers to the entire range of negative numbers. In logs,

$$\ln\left(0.5^{1000}\right) = 1000 \times \ln(0.5) \approx 1000 \times -0.6931 = -693.1$$

a number well within computational range. Similarly, 2,000 observations is not a problem:

$$\ln\left(0.5^{2000}\right) = 2000 \times \ln(0.5) \approx 2000 \times -0.6931 = -1386.2$$

Many datasets have millions of observations. Taking logs makes maximum likelihood estimation feasible with these datasets.

1.2   Likelihood theory

In this section, we examine the asymptotic properties of $\widehat{\boldsymbol{\theta}}$, namely, its probability limit, variance estimators, and distribution. We will reference several theoretical results—the Mean Value Theorem, the Central Limit Theorem, Slutsky's Theorem, the Law of Large Numbers—and provide some details on how they relate to ML estimation. We refer you to Stuart and Ord (1991, 649–706) and Welsh (1996, chap. 4) for more thorough reviews of this and related topics.

Because we are dealing with likelihood functions that are continuous in their parameters, let's define some differential operators to simplify the notation. For any real valued function $a(\mathbf{t})$, we define D by

$$D\,a(\boldsymbol{\theta}) = \left.\frac{\partial a(\mathbf{t})}{\partial \mathbf{t}}\right|_{\mathbf{t}=\boldsymbol{\theta}}$$

and D^2 by

$$D^2\, a(\boldsymbol{\theta}) = \left.\frac{\partial^2 a(\mathbf{t})}{\partial \mathbf{t}\partial \mathbf{t}'}\right|_{\mathbf{t}=\boldsymbol{\theta}}$$

The first derivative of the log-likelihood function with respect to its parameters is commonly referred to as the *gradient vector*, or *score vector*. We denote the gradient vector by

$$\mathbf{g}(\boldsymbol{\theta}) = D\, \ln L(\boldsymbol{\theta}) = \left.\frac{\partial \ln L(\mathbf{t};\mathbf{Z})}{\partial \mathbf{t}}\right|_{\mathbf{t}=\boldsymbol{\theta}}$$

The second derivative of the log-likelihood function with respect to its parameters is commonly referred to as the *Hessian matrix*. We denote the Hessian matrix by

$$\mathbf{H}(\boldsymbol{\theta}) = D\, \mathbf{g}(\boldsymbol{\theta}) = D^2\, \ln L(\boldsymbol{\theta}) = \left.\frac{\partial^2 \ln L(\mathbf{t};\mathbf{Z})}{\partial \mathbf{t}\partial \mathbf{t}'}\right|_{\mathbf{t}=\boldsymbol{\theta}} \qquad (1.3)$$

Let's assume that through some miracle, we obtain $\widehat{\boldsymbol{\theta}}$ satisfying

$$\ln L(\widehat{\boldsymbol{\theta}};\mathbf{Z}) = \max_{\mathbf{t}\in\Theta}\ \{\ln \ell(\mathbf{t};\mathbf{z}_1) + \ln \ell(\mathbf{t};\mathbf{z}_2) + \cdots + \ln \ell(\mathbf{t};\mathbf{z}_N)\}$$

Now $\widehat{\boldsymbol{\theta}}$ identifies the peak of a continuous surface; thus the slope at this point is zero:

$$\mathbf{g}(\widehat{\boldsymbol{\theta}}) = \mathbf{0} \qquad (1.4)$$

By the Mean Value Theorem, that is, a Taylor expansion with first-order remainder, there exists a $\boldsymbol{\theta}^*$ between $\widehat{\boldsymbol{\theta}}$ (the estimated values) and $\boldsymbol{\theta}_T$ (the true values) such that

$$\mathbf{g}(\widehat{\boldsymbol{\theta}}) - \mathbf{g}(\boldsymbol{\theta}_T) = D\mathbf{g}(\boldsymbol{\theta}^*)\,(\widehat{\boldsymbol{\theta}} - \boldsymbol{\theta}_T)$$

From (1.3) and (1.4), we have

$$-\mathbf{g}(\boldsymbol{\theta}_T) = \mathbf{H}(\boldsymbol{\theta}^*)\,(\widehat{\boldsymbol{\theta}} - \boldsymbol{\theta}_T)$$

and assuming that $\mathbf{H}(\boldsymbol{\theta}^*)$ is nonsingular, we can rewrite this as

$$\widehat{\boldsymbol{\theta}} = \boldsymbol{\theta}_T + \{-\mathbf{H}(\boldsymbol{\theta}^*)\}^{-1}\,\mathbf{g}(\boldsymbol{\theta}_T)$$

which, as we will see in section 1.3.1, is motivation for the update step in the Newton–Raphson algorithm.

We are assuming that the $\mathbf{z}_j$ are i.i.d., so

$$\mathbf{g}(\boldsymbol{\theta}) = D\ln \ell(\boldsymbol{\theta};\mathbf{z}_1) + \cdots + D\ln \ell(\boldsymbol{\theta};\mathbf{z}_N)$$

and $\mathbf{g}(\boldsymbol{\theta})$ is the sum of N i.i.d. random variables. By the Central Limit Theorem, the asymptotic distribution of $\mathbf{g}(\boldsymbol{\theta})$ is multivariate normal with mean vector $E\{\mathbf{g}(\boldsymbol{\theta})\}$ and variance matrix $\mathrm{Var}\{\mathbf{g}(\boldsymbol{\theta})\}$. Note that, for clustered observations, we can use similar arguments by identifying the independent cluster groups.

Lemma 1. *Let $E(\cdot)$ denote expectation with respect to the probability measure defined by $F(\cdot)$ in (1.1). Then, given the previous notation (and under the usual regularity conditions),*

$$E\{\mathbf{g}(\boldsymbol{\theta}_T)\} = \mathbf{0} \tag{1.5}$$

$$\mathrm{Var}\{\mathbf{g}(\boldsymbol{\theta}_T)\} = E[\{\mathbf{g}(\boldsymbol{\theta}_T)\}\{\mathbf{g}(\boldsymbol{\theta}_T)\}'] \tag{1.6}$$

❏ Proof

Let $\mathcal{Z}$ denote the sample space associated with the model defined in (1.2). Note that

$$\int_{\mathcal{Z}} L(\boldsymbol{\theta}_T; \mathbf{Z}) \, d\mathbf{Z} = \int_{\mathcal{Z}} f(\mathbf{Z}; \boldsymbol{\theta}_T) \, d\mathbf{Z} = \int_{\mathcal{Z}} dF = 1$$

because $f(\cdot)$ is the density function corresponding to $F(\cdot)$. The standard line at this point being "under appropriate regularity conditions", we can move the derivative under the integral sign to get

$$\mathbf{0} = D\left\{\int_{\mathcal{Z}} L(\boldsymbol{\theta}_T; \mathbf{Z}) \, d\mathbf{Z}\right\} = \int_{\mathcal{Z}} \{D\, L(\boldsymbol{\theta}_T; \mathbf{Z})\} \, d\mathbf{Z}$$

You might think that these regularity conditions are inconsequential for practical problems, but one of the conditions is that the sample space $\mathcal{Z}$ does not depend on $\boldsymbol{\theta}_T$. If it does, all the following likelihood theory falls apart and the following estimation techniques will not work. Thus if the range of the values in the data $\mathbf{Z}$ depends on $\boldsymbol{\theta}_T$, you have to start from scratch.

In any case,

$$\mathbf{0} = \int_{\mathcal{Z}} \{D\, L(\boldsymbol{\theta}_T; \mathbf{Z})\} \, d\mathbf{Z} = \int_{\mathcal{Z}} \{1/L(\boldsymbol{\theta}_T; \mathbf{Z})\} \{D\, L(\boldsymbol{\theta}_T; \mathbf{Z})\} \, L(\boldsymbol{\theta}_T; \mathbf{Z}) \, d\mathbf{Z}$$

$$= \int_{\mathcal{Z}} \{D \ln L(\boldsymbol{\theta}_T; \mathbf{Z})\} \, f(\mathbf{Z}; \boldsymbol{\theta}_T) \, d\mathbf{Z}$$

$$= \int_{\mathcal{Z}} \mathbf{g}(\boldsymbol{\theta}_T) \, f(\mathbf{Z}; \boldsymbol{\theta}_T) \, d\mathbf{Z}$$

$$= E\{\mathbf{g}(\boldsymbol{\theta}_T)\}$$

which concludes the proof of (1.5).

Note that (1.6) follows from (1.5) and the definition of the variance.

❏

The following large-sample arguments may be made once it is established that $\widehat{\boldsymbol{\theta}}$ is consistent for $\boldsymbol{\theta}_T$. By consistent, we mean that $\widehat{\boldsymbol{\theta}}$ converges to $\boldsymbol{\theta}_T$ in probability. Formally, $\widehat{\boldsymbol{\theta}}$ converges to $\boldsymbol{\theta}_T$ in probability if for every $\varepsilon > 0$

$$\lim_{N \to \infty} \mathrm{Pr}(|\widehat{\boldsymbol{\theta}} - \boldsymbol{\theta}_T| > \varepsilon) = 0$$

We will write this as

$$\text{plim}_{N\to\infty} \widehat{\boldsymbol{\theta}} = \boldsymbol{\theta}_T$$

or

$$\widehat{\boldsymbol{\theta}} \overset{p}{\to} \boldsymbol{\theta}_T \quad \text{as} \quad N \to \infty$$

There are multiple papers in the statistics literature that prove this—some are referenced in Welsh (1996). We will accept this without providing the outline for a proof.

Given that

$$\mathbf{H}(\boldsymbol{\theta}) = D^2 \ln \ell(\boldsymbol{\theta}; \mathbf{z}_1) + \cdots D^2 \ln \ell(\boldsymbol{\theta}; \mathbf{z}_N)$$

we now note that consistency of $\widehat{\boldsymbol{\theta}}$ and the Law of Large Numbers imply

$$\text{plim}_{N\to\infty} \{-\mathbf{H}(\widehat{\boldsymbol{\theta}})\} = \text{plim}_{N\to\infty} \{-\mathbf{H}(\boldsymbol{\theta}_T)\} = -E\{\mathbf{H}(\boldsymbol{\theta}_T)\} \tag{1.7}$$

which implies

$$\text{plim}_{N\to\infty} \{-\mathbf{H}(\boldsymbol{\theta}^*)\} = -E\{\mathbf{H}(\boldsymbol{\theta}_T)\}$$

If we can assume that $-E\{\mathbf{H}(\boldsymbol{\theta}_T)\}$ is a nonsingular matrix, then by Slutsky's Theorem, the asymptotic distribution of $\widehat{\boldsymbol{\theta}}$ is multivariate normal with mean vector $\boldsymbol{\theta}_T$ and variance matrix

$$\mathbf{V}_1 = [-E\{\mathbf{H}(\boldsymbol{\theta}_T)\}]^{-1} \text{Var}\{\mathbf{g}(\boldsymbol{\theta}_T)\} [-E\{\mathbf{H}(\boldsymbol{\theta}_T)\}]^{-1} \tag{1.8}$$

which you may recognize as the form of the sandwich (robust variance) estimator; see section 1.2.4. Note that although we defined $\mathbf{g}(\boldsymbol{\theta})$ and $\mathbf{H}(\boldsymbol{\theta})$ in terms of a log-likelihood function, it was only in proving lemma 1 that we used this distributional assumption. There are other, more technical proofs that do not require this assumption.

If, as in the proof of lemma 1, we can take the derivative under the integral a second time, we are left with a simpler formula for the asymptotic variance of $\widehat{\boldsymbol{\theta}}$.

Lemma 2. *Given the assumptions of lemma 1,*

$$\text{Var}\{\mathbf{g}(\boldsymbol{\theta}_T)\} = -E\{\mathbf{H}(\boldsymbol{\theta}_T)\} \tag{1.9}$$

❏ Proof

Taking the derivative of $E\{\mathbf{g}(\boldsymbol{\theta}_T)\} = \mathbf{0}$ yields

$$
\begin{aligned}
\mathbf{0} &= D\left[E\{\mathbf{g}(\boldsymbol{\theta}_T)\}\right] \\
&= D\int_{\mathcal{X}} \{D\ln L(\boldsymbol{\theta}_T; \mathbf{Z})\}\, L(\boldsymbol{\theta}_T; \mathbf{Z})\, d\mathbf{Z} \\
&= \int_{\mathcal{X}} D\left[\{D\ln L(\boldsymbol{\theta}_T; \mathbf{Z})\}\, L(\boldsymbol{\theta}_T; \mathbf{Z})\right] d\mathbf{Z} \\
&= \int_{\mathcal{X}} \left[\{D^2\ln L(\boldsymbol{\theta}_T; \mathbf{Z})\} L(\boldsymbol{\theta}_T; \mathbf{Z}) + \{D\ln L(\boldsymbol{\theta}_T; \mathbf{Z})\}\{D L(\boldsymbol{\theta}_T; \mathbf{Z})\}'\right] d\mathbf{Z} \\
&= \int_{\mathcal{X}} \left[\{D^2\ln L(\boldsymbol{\theta}_T; \mathbf{Z})\} + \{D\ln L(\boldsymbol{\theta}_T; \mathbf{Z})\}\{D\ln L(\boldsymbol{\theta}_T; \mathbf{Z})\}'\right] L(\boldsymbol{\theta}_T; \mathbf{Z})\, d\mathbf{Z} \\
&= E\{D^2\ln L(\boldsymbol{\theta}_T; \mathbf{Z})\} + E[\{D\ln L(\boldsymbol{\theta}_T; \mathbf{Z})\}\{D\ln L(\boldsymbol{\theta}_T; \mathbf{Z})\}'] \\
&= E\{\mathbf{H}(\boldsymbol{\theta}_T)\} + E\{\mathbf{g}(\boldsymbol{\theta}_T)\mathbf{g}(\boldsymbol{\theta}_T)'\}
\end{aligned}
$$

Thus

$$
\mathrm{Var}\{\mathbf{g}(\boldsymbol{\theta}_T)\} = E\{\mathbf{g}(\boldsymbol{\theta}_T)\mathbf{g}(\boldsymbol{\theta}_T)'\} = -E\{\mathbf{H}(\boldsymbol{\theta}_T)\}
$$

which concludes the proof.

❏

Thus $\widehat{\boldsymbol{\theta}}$ is asymptotically multivariate normal with mean $\boldsymbol{\theta}_T$ and variance

$$
\mathbf{V}_2 = [-E\{\mathbf{H}(\boldsymbol{\theta}_T)\}]^{-1}
$$

which follows from (1.8) and (1.9).

It is common to use

$$
\widehat{\mathbf{V}}_2 = \{-\mathbf{H}(\widehat{\boldsymbol{\theta}})\}^{-1}
$$

as a variance estimator for $\widehat{\boldsymbol{\theta}}$. This is justified by (1.7).

1.2.1 All results are asymptotic

The first important consequence of the above discussion is that all results are asymptotic.

1. $\widehat{\boldsymbol{\theta}} \xrightarrow{p} \boldsymbol{\theta}_T$ is guaranteed only as $N \to \infty$.
2. It is not true in general that $E(\widehat{\boldsymbol{\theta}}) = \boldsymbol{\theta}_T$ for finite N. $\widehat{\boldsymbol{\theta}}$ may be biased, and the bias may be substantial for small N.
3. The variance of $\widehat{\boldsymbol{\theta}}$ is $\{-\mathbf{H}(\widehat{\boldsymbol{\theta}})\}^{-1}$ asymptotically. For finite N, we are not sure how good this variance estimator is.

For these reasons, you should not fit an ML model with only a handful of observations.

1.2.2   Likelihood-ratio tests and Wald tests

Stata's `test` command performs a Wald test, which is a statistical test of the coefficients based on the estimated variance $\{-\mathbf{H}(\widehat{\boldsymbol{\theta}})\}^{-1}$. Likelihood-ratio (LR) tests, on the other hand, compare the heights of the likelihood function at $\widehat{\boldsymbol{\theta}}$ and $\boldsymbol{\theta}_0$, where $\boldsymbol{\theta}_0$ is the vector of hypothesized values. Stata's `lrtest` command performs LR tests.

The LR is defined as

$$\text{LR} = \frac{\max_{\mathbf{t}=\boldsymbol{\theta}_0} L(\mathbf{t};\mathbf{Z})}{\max_{\mathbf{t}\in\Theta} L(\mathbf{t};\mathbf{Z})} = \frac{\max_{\boldsymbol{\theta}=\boldsymbol{\theta}_0} L(\boldsymbol{\theta};\mathbf{Z})}{L(\widehat{\boldsymbol{\theta}};\mathbf{Z})}$$

The null hypothesis, $\text{H}_0\colon \boldsymbol{\theta}=\boldsymbol{\theta}_0$, may be simple—all values of $\boldsymbol{\theta}$ are hypothesized to be some set of values, such as $\boldsymbol{\theta}_0' = (0,0,\ldots,0)$—or it may be a composite hypothesis— only some values of $\boldsymbol{\theta}$ are hypothesized, such as $\boldsymbol{\theta}_0' = (0,?,?,\ldots,?)$, where ? means that the value can be anything feasible.

In general, we can write $\boldsymbol{\theta}_0' = (\boldsymbol{\theta}_r', \boldsymbol{\theta}_u')$, where $\boldsymbol{\theta}_r$ is fixed and $\boldsymbol{\theta}_u$ is not. Thus

$$\max_{\boldsymbol{\theta}=\boldsymbol{\theta}_0} L(\boldsymbol{\theta};\mathbf{Z}) = \max_{\boldsymbol{\theta}_u} L(\boldsymbol{\theta}_r,\boldsymbol{\theta}_u;\mathbf{Z}) = L(\boldsymbol{\theta}_r,\widehat{\boldsymbol{\theta}}_u;\mathbf{Z})$$

and the LR becomes

$$\text{LR} = \frac{L(\boldsymbol{\theta}_r,\widehat{\boldsymbol{\theta}}_u;\mathbf{Z})}{L(\widehat{\boldsymbol{\theta}};\mathbf{Z})}$$

It can be shown that $-2\ln(\text{LR})$ is asymptotically distributed as chi-squared with r degrees of freedom, where r is the dimension of $\boldsymbol{\theta}_r$. This theory holds only for small to moderate values of r. If r is very large (say, $r > 100$), the chi-squared distribution result becomes questionable.

To compute LR, we must do two maximizations: one to get $\widehat{\boldsymbol{\theta}}$ and another to compute $\widehat{\boldsymbol{\theta}}_u$. Hence, LR tests can be time consuming because you must do an "extra" maximization for each hypothesis test.

The Wald test, on the other hand, simply uses $\text{Var}(\widehat{\boldsymbol{\theta}})$, which is estimated assuming the true values of $\boldsymbol{\theta}$; that is, it uses $\{-\mathbf{H}(\widehat{\boldsymbol{\theta}})\}^{-1}$ for the variance. For a linear hypothesis $\mathbf{R}\boldsymbol{\theta} = \mathbf{r}$, the Wald test statistic is

$$W = (\mathbf{R}\widehat{\boldsymbol{\theta}} - \mathbf{r})'\{\mathbf{R}\text{Var}(\widehat{\boldsymbol{\theta}})\mathbf{R}'\}^{-1}(\mathbf{R}\widehat{\boldsymbol{\theta}} - \mathbf{r})$$

If $\widehat{\boldsymbol{\theta}}$ is normally distributed, W has a chi-squared distribution with $r = \text{rank}(\mathbf{R}\mathbf{R}')$ degrees of freedom. Obviously, this is an easy computation to make once we have $\text{Var}(\widehat{\boldsymbol{\theta}})$.

As Davidson and MacKinnon (1993, 278) discuss, when the sample size is not large, these tests may have very different finite-sample properties, so in general, we cannot claim that one test is better than the other in all cases.

One advantage that the LR test does have over the Wald test is the so-called invariance property. Suppose we want to test the hypothesis $\text{H}_0\colon \beta_2 = 2$. Saying $\beta_2 = 2$ is clearly equivalent to saying $1/\beta_2 = 1/2$. With the LR test, whether we state our null hypothesis as $H_0 : \beta_2 = 2$ or $H_0 : 1/\beta_2 = 1/2$ makes absolutely no difference; we will reach the same conclusion either way. On the other hand, you may be surprised to know that we will obtain different Wald test statistics for tests of these two hypotheses. In fact, in some cases, the Wald test may lead to the rejection of one null hypothesis but does not allow the rejection of a mathematically equivalent null hypothesis formulated differently. The LR test is said to be "invariant to nonlinear transformations", and the Wald test is said to be "manipulable."

1.2.3   The outer product of gradients variance estimator

In discussing the large-sample properties of $\widehat{\boldsymbol{\theta}}$, we established

$$\mathrm{Var}(\widehat{\boldsymbol{\theta}}) \approx [-E\{\mathbf{H}(\boldsymbol{\theta}_T)\}]^{-1}$$

and indicated that $\{-\mathbf{H}(\widehat{\boldsymbol{\theta}})\}^{-1}$ is consistent for the variance of $\widehat{\boldsymbol{\theta}}$.

There are other ways we can obtain estimates of the variance. From (1.9), we may rewrite the above as

$$\mathrm{Var}(\widehat{\boldsymbol{\theta}}) \approx [\mathrm{Var}\{\mathbf{g}(\boldsymbol{\theta}_T)\}]^{-1}$$

Note that

$$\mathbf{g}(\boldsymbol{\theta}) = \mathbf{g}_1(\boldsymbol{\theta}) + \cdots + \mathbf{g}_N(\boldsymbol{\theta})$$

where

$$\mathbf{g}_j(\boldsymbol{\theta}) = D \ln \ell(\boldsymbol{\theta}; \mathbf{z}_j)$$

which is to say, $\mathbf{g}(\boldsymbol{\theta})$ is the sum of N random variables. If the observations are independent, we can further say that $\mathbf{g}(\boldsymbol{\theta})$ is the sum of N i.i.d. random variables. What is a good estimator for the variance of the sum of N i.i.d. random variables? Do not let the fact that we are summing evaluations of functions fool you. A good estimator for the variance of the mean of N i.i.d. random variates $y_1, y_2, \ldots, y_N$ is

$$\frac{s^2}{N} = \frac{1}{N(N-1)} \sum_{j=1}^{N} (y_j - \overline{y})^2$$

Therefore, a good estimator for the variance of the total is simply N^2 times the variance estimator of the mean:

$$N s^2 = \frac{N}{N-1} \sum_{j=1}^{N} (y_j - \overline{y})^2 \tag{1.10}$$

We estimate $\mathrm{Var}\{\mathbf{g}(\boldsymbol{\theta}_T)\}$ by simply substituting $y_j = \mathbf{g}_j(\boldsymbol{\theta}_T)$ in (1.10), using the respective matrix operators. Let $\overline{\mathbf{g}}(\boldsymbol{\theta}) = \mathbf{g}(\boldsymbol{\theta})/N$, and then we have

$$\mathrm{Var}\{\mathbf{g}(\boldsymbol{\theta}_T)\} \approx \frac{N}{N-1} \sum_{j=1}^{N} \{\mathbf{g}_j(\boldsymbol{\theta}_T) - \overline{\mathbf{g}}(\boldsymbol{\theta}_T)\} \{\mathbf{g}_j(\boldsymbol{\theta}_T) - \overline{\mathbf{g}}(\boldsymbol{\theta}_T)\}' \tag{1.11}$$

and thus another variance estimator for $\widehat{\boldsymbol{\theta}}$ is

$$\text{Var}(\widehat{\boldsymbol{\theta}}) \approx \left\{ \sum_{j=1}^{N} \mathbf{g}_j(\widehat{\boldsymbol{\theta}}) \mathbf{g}_j(\widehat{\boldsymbol{\theta}})' \right\}^{-1} \tag{1.12}$$

where we substituted $\widehat{\boldsymbol{\theta}}$ for $\boldsymbol{\theta}_T$ and simplified by remembering that $\mathbf{g}(\widehat{\boldsymbol{\theta}}) = \mathbf{0}$ (implying $\overline{\mathbf{g}}(\widehat{\boldsymbol{\theta}})$ is also $\mathbf{0}$) and by substituting 1 for $N/(N-1)$ (which it is for large N). In any case, (1.12) is known as the outer product of the gradients variance estimator—often called the OPG—and, in fact, the OPG has the same asymptotic properties as $\{-\mathbf{H}(\widehat{\boldsymbol{\theta}})\}^{-1}$, the "conventional variance estimator".

We refer to $\{-\mathbf{H}(\widehat{\boldsymbol{\theta}})\}^{-1}$ as the conventional variance estimator not because it is better but because it is more commonly reported. As we will discuss later, if we use an optimization method like Newton–Raphson for maximizing the likelihood function, we need to compute the Hessian matrix anyway, and thus at the end of the process, we will have $\{-\mathbf{H}(\widehat{\boldsymbol{\theta}})\}^{-1}$ at our fingertips.

We will also discuss ways in which functions can be maximized that do not require calculation of $\{-\mathbf{H}(\widehat{\boldsymbol{\theta}})\}^{-1}$, such as the Davidon–Fletcher–Powell algorithm. These algorithms are often used on functions for which calculating the second derivatives would be computationally expensive, and given that expense, it is common to report the OPG variance estimate once the maximum is found.

The OPG variance estimator has much to recommend it. In fact, it is more empirical than the conventional calculation because it is like using the sample variance of a random variable instead of plugging the ML parameter estimates into the variance expression for the assumed probability distribution. Note, however, that the above development requires that the data come from a simple random sample.

1.2.4 Robust variance estimates

Our result that $\text{Var}(\widehat{\boldsymbol{\theta}})$ is asymptotically $\{-\mathbf{H}(\widehat{\boldsymbol{\theta}})\}^{-1}$ hinges on lemma 1 and lemma 2. In proving these lemmas, and only in proving them, we assumed the likelihood function $L(\boldsymbol{\theta}_T; \mathbf{Z})$ was the density function for $\mathbf{Z}$. If $L(\boldsymbol{\theta}_T; \mathbf{Z})$ is not the true density function, our lemmas do not apply and all our subsequent results do not necessarily hold. In practice, the ML estimator and this variance estimator still work reasonably well if $L(\boldsymbol{\theta}_T; \mathbf{Z})$ is a little off from the true density function. For example, if the true density is probit and you fit a logit model (or vice versa), the results will still be accurate.

You can also derive a variance estimator that does not require $L(\boldsymbol{\theta}_T; \mathbf{Z})$ to be the density function for $\mathbf{Z}$. This is the *robust variance estimator*, which is implemented in many Stata estimation commands and in Stata's survey (svy) commands.

The robust variance estimator was first published, we believe, by Peter Huber, a mathematical statistician, in 1967 in conference proceedings (Huber 1967). Survey statisticians were thinking about the same things around this time, at least for linear regression. In the 1970s, the survey statisticians wrote up their work, including

Kish and Frankel (1974), Fuller (1975), and others, all of which was summarized and generalized in an excellent paper by Binder (1983). White, an economist, independently derived the estimator and published it in 1980 for linear regression and in 1982 for ML estimates, both in economics literature. Many others have extended its development, including Kent (1982); Royall (1986); Gail, Tan, and Piantadosi (1988); and Lin and Wei (1989).

The robust variance estimator is called different things by different people. At Stata, we originally called it the Huber variance estimator (Bill Rogers, who first implemented it here, was a student of Huber). Some people call it the sandwich estimator. Survey statisticians call it the Taylor-series linearization method, linearization method, or design-based variance estimate. Economists often call it the White estimator. Statisticians often refer to it as the empirical variance estimator. In any case, they all mean the same variance estimator. We will sketch the derivation here.

The starting point is

$$\text{Var}(\widehat{\boldsymbol{\theta}}) \approx \{-\mathbf{H}(\widehat{\boldsymbol{\theta}})\}^{-1} \text{Var}\{\mathbf{g}(\widehat{\boldsymbol{\theta}})\}\{-\mathbf{H}(\widehat{\boldsymbol{\theta}})\}^{-1} \tag{1.13}$$

Because of this starting point, some people refer to robust variance estimates as the Taylor-series linearization method; we obtained (1.13) from the delta method, which is based on a first-order (linear term only) Taylor series.

The next step causes other people to refer to this as the empirical variance estimator. Using the empirical variance estimator of $\text{Var}\{\mathbf{g}(\boldsymbol{\theta}_T)\}$ from (1.11) for $\text{Var}\{\mathbf{g}(\widehat{\boldsymbol{\theta}})\}$, we have that

$$\text{Var}(\widehat{\boldsymbol{\theta}}) \approx \{-\mathbf{H}(\widehat{\boldsymbol{\theta}})\}^{-1} \left\{ \frac{N}{N-1} \sum_{j=1}^{N} \mathbf{g}_j(\widehat{\boldsymbol{\theta}}) \mathbf{g}_j(\widehat{\boldsymbol{\theta}})' \right\} \{-\mathbf{H}(\widehat{\boldsymbol{\theta}})\}^{-1}$$

This is the robust variance estimator. It is robust for the same reasons that the estimator for the variance of a mean is robust; note our lack of assumptions about $\ln \ell(\cdot)$ and hence $L(\cdot)$. That is, (1.13) arises from the fact that $\widehat{\boldsymbol{\theta}}$ solves $\mathbf{g}(\widehat{\boldsymbol{\theta}}) = \mathbf{0}$, regardless of whether $\mathbf{g}(\cdot)$ is obtained from the true data-generating process.

The estimator for the variance of the total of the $\mathbf{g}_j(\widehat{\boldsymbol{\theta}})$ values relies only on our data's coming from simple random sampling (the observations are i.i.d.). Thus the essential assumption of the robust variance estimator is that the observations are independent selections from the same population.

For cluster sampling, we merely change our estimator for the variance of the total of the $\mathbf{g}_j(\widehat{\boldsymbol{\theta}})$ values to reflect this sampling scheme. Consider superobservations made up of the sum of $\mathbf{g}_j(\widehat{\boldsymbol{\theta}})$ for a cluster: these superobservations are independent, and the above formulas hold with $\mathbf{g}_j(\widehat{\boldsymbol{\theta}})$ replaced by the cluster sums (and N replaced by the number of clusters). See [P] _robust, [SVY] **Variance estimation**, Rogers (1993), Williams (2000), and Wooldridge (2010) for more details.

1.3   The maximization problem

Given the problem

$$L(\widehat{\boldsymbol{\theta}}; \mathbf{Z}) = \max_{\mathbf{t} \in \Theta} \; L(\mathbf{t}; \mathbf{Z})$$

how do we obtain the solution? One way we might solve this analytically is by taking derivatives and setting them to zero:

$$\text{Solve for } \widehat{\boldsymbol{\theta}}: \qquad \left. \frac{\partial \ln L(\mathbf{t}; \mathbf{Z})}{\partial \mathbf{t}} \right|_{\mathbf{t} = \widehat{\boldsymbol{\theta}}} = \mathbf{0} \qquad\qquad (1.14)$$

Let's use a simplified example. If $\ln L(\theta) = -\theta^2 + \theta$, then

$$\frac{d \ln L(\theta)}{d\theta} = -2\theta + 1 = 0 \qquad \Rightarrow \qquad \theta = 1/2$$

In general, however, $\partial \ln L(\mathbf{t}; \mathbf{Z})/\partial \mathbf{t} = \mathbf{0}$ is too complicated to admit an analytic solution, and we are forced to seek a numerical one. We start with (1.14) but use the gradient vector, so our problem is

$$\text{Solve for } \widehat{\boldsymbol{\theta}}: \qquad \mathbf{g}(\widehat{\boldsymbol{\theta}}; \mathbf{Z}) = \mathbf{0}$$

We have converted the maximization problem into a root-finding problem. Thus ML estimates $\widehat{\boldsymbol{\theta}}$ are obtained by finding the roots of $\mathbf{g}(\widehat{\boldsymbol{\theta}}; \mathbf{Z})$. Many computer optimizers (the generic name for maximizers and minimizers) are, at heart, root finders.

1.3.1   Numerical root finding

For one dimension, so that θ and $g(\theta)$ are both scalars, one method of finding roots is Newton's method. This is an iterative technique. You have a guess θ_i (called the initial value), and you update it as follows.

Newton's method

1. The current guess is θ_i.
2. Calculate $g(\theta_i)$.
3. Calculate $g'(\theta_i)$, the slope of $g(\cdot)$ at the current guess.
4. Draw the tangent line at the current guess: $y = g'(\theta_i)(\theta - \theta_i) + g(\theta_i)$.
5. The new guess is θ_{i+1}, the root of the above tangent line:

$$0 = g'(\theta_i)(\theta_{i+1} - \theta_i) + g(\theta_i) \qquad \Rightarrow \qquad \theta_{i+1} = \theta_i - \frac{g(\theta_i)}{g'(\theta_i)}$$

6. Repeat.

The above sequence is not guaranteed to converge, but it generally does. Figure 1.1 illustrates a step in Newton's method.

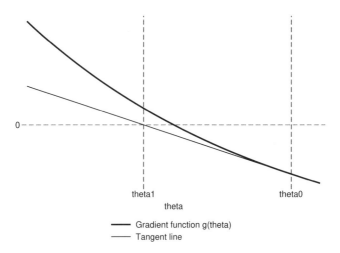

Figure 1.1. Newton's method. The next guess, `theta1`, is the point where the tangent line crosses the horizontal axis.

Newton's method for finding roots can be converted to finding minimums and maximums of a function—say, $f(\cdot)$—by substituting $f'(\cdot)$ for $g(\cdot)$ and $f''(\cdot)$ for $g'(\cdot)$:

To find θ such that $f(\theta)$ is maximized,

1. Start with guess θ_i.
2. Calculate new guess $\theta_{i+1} = \theta_i - f'(\theta_i)/f''(\theta_i)$.
3. Repeat.

We can generalize this to allow θ to be a vector, $\boldsymbol{\theta}$.

To find vector $\widehat{\boldsymbol{\theta}}$ such that $f(\widehat{\boldsymbol{\theta}})$ is maximized,

1. Start with guess $\boldsymbol{\theta}_i$.
2. Calculate new guess $\boldsymbol{\theta}_{i+1} = \boldsymbol{\theta}_i - \{\mathbf{H}(\boldsymbol{\theta}_i)\}^{-1}\mathbf{g}(\boldsymbol{\theta}_i)$, where $\mathbf{g}(\boldsymbol{\theta}_i)$ is the gradient vector and $\mathbf{H}(\boldsymbol{\theta}_i)$ is the matrix of second derivatives. (In the future, we will write this as $\boldsymbol{\theta}_{i+1} = \boldsymbol{\theta}_i + \{-\mathbf{H}(\boldsymbol{\theta}_i)\}^{-1}\mathbf{g}(\boldsymbol{\theta}_i)$ because the matrix $-\mathbf{H}(\boldsymbol{\theta}_i)$ is positive definite and calculating the inverse of a positive-definite matrix is easier.)
3. Repeat.

This is the Newton–Raphson algorithm, which is the method Stata and most other statistical packages use by default, although a few details are commonly changed. In fact, any package that claims to use Newton–Raphson is probably oversimplifying the story. These details are changed because

1. Calculating $f(\boldsymbol{\theta}_i)$ is computationally expensive.
2. Calculating $\mathbf{g}(\boldsymbol{\theta}_i)$ is more computationally expensive.
3. Calculating $\mathbf{H}(\boldsymbol{\theta}_i)$ is even more computationally expensive.

The Newton–Raphson algorithm

Relatively speaking, we can calculate $f(\boldsymbol{\theta}_i)$ many times for the same computer time required to calculate $\mathbf{H}(\boldsymbol{\theta}_i)$ just once. This leads to separating the direction calculation from the step size:

To find vector $\widehat{\boldsymbol{\theta}}$ such that $f(\widehat{\boldsymbol{\theta}})$ is maximized,

1. Start with a guess $\boldsymbol{\theta}_i$.
2. Calculate a direction vector $\mathbf{d} = \{-\mathbf{H}(\boldsymbol{\theta}_i)\}^{-1}\mathbf{g}(\boldsymbol{\theta}_i)$.
3. Calculate a new guess $\boldsymbol{\theta}_{i+1} = \boldsymbol{\theta}_i + s\mathbf{d}$, where s is a scalar, for instance:
   a. Start with $s = 1$.
   b. If $f(\boldsymbol{\theta}_i + \mathbf{d}) > f(\boldsymbol{\theta}_i)$, try $s = 2$. If $f(\boldsymbol{\theta}_i + 2\mathbf{d}) > f(\boldsymbol{\theta}_i + \mathbf{d})$, try $s = 3$ or even $s = 4$, and so on.
   c. If $f(\boldsymbol{\theta}_i + \mathbf{d}) \leq f(\boldsymbol{\theta}_i)$, back up and try $s = 0.5$ or $s = 0.25$, etc.
4. Repeat.

The way that s is obtained varies from package to package, but the idea is simple enough: go in the direction $\mathbf{d} = \{-\mathbf{H}(\boldsymbol{\theta}_i)\}^{-1}\mathbf{g}(\boldsymbol{\theta}_i)$ as far as you possibly can before spending computer cycles to compute a new direction. This all goes under the rubric "step size calculation".

The other important detail goes under the rubric "nonconcavity", meaning that if $\{-\mathbf{H}(\boldsymbol{\theta}_i)\}^{-1}$ does not exist—if $\mathbf{H}(\boldsymbol{\theta}_i)$ is singular—then the direction calculation $\mathbf{d} = \{-\mathbf{H}(\boldsymbol{\theta}_i)\}^{-1}\mathbf{g}(\boldsymbol{\theta}_i)$ cannot be made. It is worth thinking about this in the single-dimension case: in the maximization problem, it means that the second derivative of $f(\cdot)$ is zero. In the root-finding problem, it means that the derivative of $g(\cdot)$ is zero. The gradient is flat, so there is no root, as illustrated in figure 1.2.

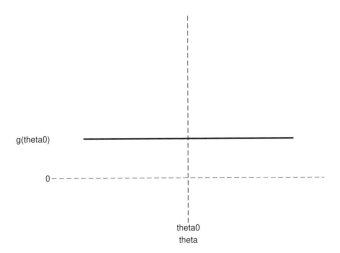

Figure 1.2. Nonconcavity. The slope of the tangent line at `theta` is zero, so where is the root?

Remember, all you really know is that $g'(\cdot)$ evaluated at θ_i is 0. Even though we have drawn a line, you do not really know that $g'(\theta) = 0$ for all θ. In fact, you are reasonably certain that "$g'(\theta) = 0$ for all θ" is not true because $g(\cdot)$ is the gradient of a likelihood function and the likelihood certainly cannot be increasing or decreasing without bound. Probably, θ_i is just a poor guess and far from the heart of the function.

The tangent line in figure 1.1 is how you would usually project to find the new guess θ_{i+1}—it goes through the point $\{\theta_i, g(\theta_i)\}$ and has slope $g'(\theta_i)$. Figure 1.2 emphasizes why this rule is not helpful in regions of nonconcavity.

So, pretend that you are the optimizer charged with finding the maximum of $f(\theta)$; θ_i is the current guess and $g'(\theta_i) = 0$. What is your next guess? Put that aside. Is your next guess θ_{i+1} to the left or right of θ_i?

This is where knowing that this is a maximization problem rather than merely a root-finding problem is useful. If you told us just to find the roots, we would have no reason to select $\theta_{i+1} > \theta_i$ or $\theta_{i+1} < \theta_i$; we would have to try both. But if you tell us this is a maximization problem, we observe that in the graph drawn, $g(\theta_i) > 0$, so increasing θ_i increases the value of $f(\cdot)$. We would try something to the right of θ_i. If $g(\theta_i) < 0$, we would try something to the left of θ_i. And if this were a minimization rather than a maximization problem, we would reverse the rules.

The important point is that at least we have some idea of which direction to go.

When we generalize to allow θ to be a vector, our choices proliferate. Corresponding to our scalar go-with-the-sign-of-$g(\cdot)$ rule, the vector equivalent is the go-with-the-vector-$g(\cdot)$ rule, which amounts to treating the noninvertible matrix $\mathbf{H}(\theta_i)$ as if $\{-\mathbf{H}(\theta_i)\}^{-1} = \mathbf{I}$ because our general rule is direction $\mathbf{d} = \{-\mathbf{H}(\theta_i)\}^{-1}\mathbf{g}(\theta_i)$. That is called *steepest ascent*. It is called ascent because, just as in the scalar case, we choose the direction based on the sign of the gradient and the knowledge that we want to maximize $f(\cdot)$. It is called steepest because, mathematically, it works out that among all directions, we could go with $f(\cdot)$ increasing, the direction $\mathbf{g}(\boldsymbol{\theta}_i)$ is steepest. By the way, the steepest-ascent rule is often called such because most of the numerical optimization literature is written in terms of function minimization rather than maximization.

This rule can be improved upon. Rather than treating $\{-\mathbf{H}(\boldsymbol{\theta}_i)\}^{-1}$ as $\mathbf{I}$, we can add a constant to the diagonal elements of $\{-\mathbf{H}(\boldsymbol{\theta}_i)\}^{-1}$ until it is invertible. This is like mixing in a little steepest ascent to the otherwise "best" direction and is called the *modified Marquardt algorithm* (Marquardt 1963).

Stata follows one of two variations on this scheme. In the first variation, rather than adding a constant, c, to the diagonal terms, `ml` adds a proportional term $c|h_{ii}|$, where h_{ii} is the ith diagonal element of $\{-\mathbf{H}(\boldsymbol{\theta})\}^{-1}$. `ml` then renormalizes the resulting matrix so that it has the original trace, thus attempting to keep the scale of the matrix the same.

If you specify `ml`'s `difficult` option, it does more work, computing the eigenvalues of $\{-\mathbf{H}(\boldsymbol{\theta})\}^{-1}$. Then for the part of the orthogonal subspace where the eigenvalues are negative or small positive numbers, `difficult` uses steepest ascent; in the other subspace, it uses a regular Newton–Raphson step.

1.3.2 Quasi-Newton methods

The Newton–Raphson algorithm is not the only method for finding maximums. Related techniques are known as quasi-Newton methods. These methods follow the same steps as the Newton–Raphson algorithm (see page 16) but make a substitution for the Hessian matrix $\mathbf{H}(\boldsymbol{\theta}_i)$ in computing the direction to go, $\mathbf{d} = \{-\mathbf{H}(\boldsymbol{\theta}_i)\}^{-1}\mathbf{g}(\boldsymbol{\theta}_i)$.

Often calculation of $\mathbf{H}(\boldsymbol{\theta}_i)$ can be computationally expensive—very computationally expensive—and these quasi-Newton methods seek to avoid that expense by substituting something cheaper to calculate and (it is advertised) nearly as good.

You may be bothered by having choices with no criterion with which to choose among them. Certainly, you will say to yourself, there must be a best way to maximize a function. Were computation free and computers infinitely fast, then we could seek a method certain to find the maximum regardless of the amount of computation it took to reach our goal. That is not a difficult problem, and you can probably think of a few solutions yourself. Perhaps the easiest would be just to calculate the function at every point in $\boldsymbol{\Theta}$ and choose the maximum, not worrying that an infinite amount of calculations would be required.

In the real world, the problem is to find the maximum with as little calculation as possible. What you must remember is that there are many paths to the top of a hill and the particular path you choose does not matter, assuming you reach the top. The paths differ only in the amount of computation you expend in arriving at the solution.

These quasi-Newton methods in general choose poorer paths, but they expend less computation on the path. They are of particular interest when the calculation of $\mathbf{H}(\boldsymbol{\theta}_i)$ is known to be exorbitantly expensive.

Even Newton–Raphson is not guaranteed to find the maximum. In general, Newton–Raphson works well, but if it does fail, it is often because of some very fine detail in the likelihood function that interacts with the Newton–Raphson rules, and the substitution of another set of rules—any other method—will be sufficient to work around the problem.

The BHHH algorithm

BHHH stands for Berndt, Hall, Hall, and Hausman—its authors—and is often pronounced B-H-cubed.

The BHHH algorithm substitutes the outer product of the gradients for the observed negative Hessian matrix. Thus if $\mathbf{g}_j(\boldsymbol{\theta}_i)$ is the gradient vector from the jth observation using the current guess $\boldsymbol{\theta}_i$, then the BHHH algorithm uses

$$-\mathbf{H}_{\mathrm{BHHH}}(\boldsymbol{\theta}_i) = \sum_{j=1}^{N} \mathbf{g}_j(\boldsymbol{\theta}_i)\mathbf{g}_j(\boldsymbol{\theta}_i)'$$

instead of the observed negative Hessian matrix. For further details about this algorithm, see Berndt et al. (1974).

The DFP and BFGS algorithms

The DFP and BFGS methods substitute an ever-improving estimate for $\mathbf{H}(\boldsymbol{\theta}_i)$ at every step. DFP stands for Davidon (1959), Fletcher (1970), and Powell (Fletcher and Powell 1963), and BFGS stands for Broyden (1967), Fletcher (1970), Goldfarb (1970), and Shanno (1970).

As above, let i be the index for the current iteration; $\boldsymbol{\theta}_i$ be the guess for the ML estimate of $\boldsymbol{\theta}_T$ at the ith iteration; $\mathbf{g}_i = \mathbf{g}(\boldsymbol{\theta}_i)$, the associated gradient vector; and $\mathbf{H}_i = \mathbf{H}(\boldsymbol{\theta}_i)$, the associated Hessian matrix. According to the Newton–Raphson algorithm, the next guess is

$$\boldsymbol{\theta}_{i+1} = \boldsymbol{\theta}_i + (-\mathbf{H}_i)^{-1}\mathbf{g}_i$$

Suppose that $\mathbf{A}_i$ is a substitute for $(-\mathbf{H}_i)^{-1}$. For both DFP and BFGS, $\mathbf{A}_{i+1}$ is calculated by adding in a correction to $\mathbf{A}_i$ so that

$$\Delta\boldsymbol{\theta}_{i+1} = \mathbf{A}_{i+1}\Delta\mathbf{g}_{i+1}$$

where $\Delta\boldsymbol{\theta}_{i+1} = \boldsymbol{\theta}_{i+1} - \boldsymbol{\theta}_i$ and $\Delta\mathbf{g}_{i+1} = \mathbf{g}_{i+1} - \mathbf{g}_i$.

In the DFP algorithm, $\mathbf{A}_{i+1}$ is updated by

$$\mathbf{A}_{i+1} = \mathbf{A}_i + \frac{\Delta\boldsymbol{\theta}_{i+1}\Delta\boldsymbol{\theta}_{i+1}'}{\Delta\boldsymbol{\theta}_{i+1}'\Delta\mathbf{g}_{i+1}} - \frac{\mathbf{A}_i\Delta\mathbf{g}_{i+1}\Delta\mathbf{g}_{i+1}'\mathbf{A}_i}{\Delta\mathbf{g}_{i+1}'\mathbf{A}_i\Delta\mathbf{g}_{i+1}}$$

In the BFGS algorithm, $\mathbf{A}_{i+1}$ is updated by

$$\mathbf{A}_{i+1} = \mathbf{A}_i + \frac{\Delta\boldsymbol{\theta}_{i+1}\Delta\boldsymbol{\theta}'_{i+1}}{\Delta\boldsymbol{\theta}'_{i+1}\Delta\mathbf{g}_{i+1}} - \frac{\mathbf{A}_i\Delta\mathbf{g}_{i+1}\Delta\mathbf{g}'_{i+1}\mathbf{A}_i}{\Delta\mathbf{g}'_{i+1}\mathbf{A}_i\Delta\mathbf{g}_{i+1}} + \Delta\mathbf{g}'_{i+1}\mathbf{A}_i\Delta\mathbf{g}_{i+1}\mathbf{u}\mathbf{u}'$$

where

$$\mathbf{u} = \frac{\Delta\boldsymbol{\theta}_{i+1}}{\Delta\boldsymbol{\theta}'_{i+1}\Delta\mathbf{g}_{i+1}} - \frac{\mathbf{A}_i\Delta\mathbf{g}_{i+1}}{\Delta\mathbf{g}'_{i+1}\mathbf{A}_i\Delta\mathbf{g}_{i+1}}$$

Thus the BFGS algorithm is essentially the same as DFP with an added correction to $\mathbf{A}_i$ when calculating $\mathbf{A}_{i+1}$. For both algorithms, `ml` uses $\mathbf{g}_0 = \mathbf{0}$ (vector of zeros) and $\mathbf{A}_0 = I$ (identity matrix) as starting values. For further details about these algorithms, see Fletcher (1987) and Press et al. (2007).

1.3.3 Numerical maximization

To summarize, to find $\widehat{\boldsymbol{\theta}}$ such that $f(\widehat{\boldsymbol{\theta}})$ is maximized,

1. Start with a guess $\boldsymbol{\theta}_i$.

2. Calculate a direction vector $\mathbf{d} = \{-\mathbf{H}(\boldsymbol{\theta}_i)\}^{-1}\mathbf{g}(\boldsymbol{\theta}_i)$, if you can. $\mathbf{H}(\boldsymbol{\theta}_i)$ can be calculated several ways: by direct calculation of the second derivatives, by the outer product of the gradients, or by an iterative technique. However you get $\mathbf{H}(\boldsymbol{\theta}_i)$, if $-\mathbf{H}(\boldsymbol{\theta}_i)$ is not invertible, then substitute for $-\mathbf{H}(\boldsymbol{\theta}_i)$ the matrix $\mathbf{I}$, or $-\mathbf{H}(\boldsymbol{\theta}_i) + c\mathbf{I}$ for some small, positive scalar c, or some other matrix related to $-\mathbf{H}(\boldsymbol{\theta}_i)$.

3. Calculate a new guess $\boldsymbol{\theta}_{i+1} = \boldsymbol{\theta}_i + s\mathbf{d}$, where s is a scalar.

   a. For instance, start with $s = 1$.

   b. If $f(\boldsymbol{\theta}_i + \mathbf{d}) > f(\boldsymbol{\theta}_i)$, try $s = 2$. If $f(\boldsymbol{\theta}_i + 2\mathbf{d}) > f(\boldsymbol{\theta}_i + \mathbf{d})$, try $s = 3$ or even $s = 4$, and so on.

   c. If $f(\boldsymbol{\theta}_i + \mathbf{d}) \leq f(\boldsymbol{\theta}_i)$, back up and try $s = 0.5$ or $s = 0.25$, etc.

4. Repeat.

If you have ever used Stata's `ml` command or any of Stata's preprogrammed ML estimators, you have probably seen remnants of these rules.

When Stata says

```
Iteration #:  Log likelihood = ...
```

that means it is at step 2 and calculating a direction. The log likelihood reported is $f(\boldsymbol{\theta}_i)$.

You may have seen

```
Iteration #:  Log likelihood = ...  (not concave)
```

That arose during step 2. $-\mathbf{H}(\boldsymbol{\theta}_i)$ was not invertible, and Stata had to use something else to choose a direction. So what does it mean? If the message occurs in early iterations, you can safely ignore it. It just means that the current guess for $\widehat{\boldsymbol{\theta}}$ is rather poor and that the matrix of second derivatives calculated at that point is noninvertible—corresponding in the scalar case to the gradient being flat.

In the multidimensional case, a noninvertible $-\mathbf{H}(\boldsymbol{\theta}_i)$ means that the likelihood function $f(\cdot)$ locally has a ridge, a flat section, or a saddle point. Real likelihood functions have few points like this, and when such arise, it is typically because the gradient is almost, but not perfectly, flat. Roundoff error makes the nearly flat gradient appear absolutely flat. In any case, the optimizer encountered the problem and worked around it.

If, on the other hand, this message were to arise at the final iteration, just before results are presented, it would be of great concern.

If you have seen Stata say

```
Iteration #:  Log likelihood = ...  (backed up)
```

that arose in step 3. Stata started with $s = 1$. With $s = 1$, Stata expected $f(\boldsymbol{\theta}_i + s\mathbf{d}) = f(\boldsymbol{\theta}_i + \mathbf{d}) > f(\boldsymbol{\theta}_i)$, but that did not happen. Actually, Stata mentions that it backed up only if it halved the original step size six or more times. As with the not-concave message, this should not concern you unless it occurs at the last iteration, in which case the declaration of convergence is questionable. (You will learn to specify ml's `gradient` option to see the gradient vector in cases such as this so that you can determine whether its elements really are close to zero.)

1.3.4   Numerical derivatives

The maximizer that we have sketched has substantial analytic requirements. In addition to specifying $f(\mathbf{t}) = \ln L(\mathbf{t}; \mathbf{Z})$, the routine must be able to calculate $\mathbf{g}(\mathbf{t}) = \partial f(\mathbf{t})/\partial \mathbf{t}$ and, in the Newton–Raphson case, $\mathbf{H}(\mathbf{t}) = \partial^2 f(\mathbf{t})/\partial \mathbf{t}\partial \mathbf{t}'$. It needs the function, its first derivatives, and its second derivatives.

Stata provides facilities so that you do not actually have to program the derivative calculations, but really, Stata has only one optimizer. When you do not provide the derivatives, Stata calculates them numerically.

The definition of an analytic derivative is

$$f'(z) = \frac{df(z)}{dz} = \lim_{h \to 0} \frac{f(z + h) - f(z)}{h}$$

and that leads to a simple approximation formula,

$$f'(z) \approx \frac{f(z+h) - f(z)}{h}$$

for a suitably small but large enough h.

We have a lot to say about how Stata finds a suitably small but large enough h. Stata's use of a formula like $\{f(z+h) - f(z)\}/h$ has an important implication for your program that calculates $f(\cdot)$, the log-likelihood function.

If Stata chooses h just right, in general this formula will be only about half as accurate as the routine that calculates $f(\cdot)$. Because of the way Stata chooses h, it will not be less accurate than that, and it might be more accurate. The loss of accuracy comes about because of the subtraction in the numerator.

Let's show this by calculating the derivative of $\exp(z)$ at $z = 2$. The true answer is $\exp(2)$ because $d\{\exp(z)\}/dz = \exp(z)$. Let's try this formula with $h = 10^{-8}$ ($h = 1e-8$ in computer speak). We will carefully do this calculation with 16 digits of accuracy:

```
exp(2 + 1e-8)        =  7.389056172821211   (accuracy is 16 digits)
exp(2)               =  7.389056098930650   (accuracy is 16 digits)
─────────────────────────────────────────
difference           =   .000000073890561

exp(2)               =  7.389056098930650   (true answer)
difference / 1e-8    =  7.389056033701991   (approximation formula)
─────────────────────────────────────────
error                    .000000065228659   (approximation is correct to
                       1 2345678             8 digits)
```

The major source of error was introduced when we calculated the difference $\exp(2 + 1e-8) - \exp(2)$.

```
exp(2 + 1e-8)        =  7.389056172821211   (accuracy is 16 digits)
exp(2)               =  7.389056098930650   (accuracy is 16 digits)
─────────────────────────────────────────
difference           =   .000000073890561
                          12345678          (accuracy is 8 digits)
```

Our full 16 digits of accuracy were lost because half the digits of $\exp(2 + 1e-8)$ and $\exp(2)$ were in common and so canceled each other.

This is an unpleasant feature of numerical derivatives. Given how Stata chooses h, if you start with k digits of accuracy, you will fall to $k/2$ digits at worst. You can get lucky, the best case being when $f(z) = 0$, in which case you can get all k digits back, but that is ignoring the inaccuracy introduced by the algorithm itself.

Thus as a programmer of likelihood functions, if you are going to use numerical derivatives, it is vitally important that you supply $\ln L(\cdot)$ as accurately as you can. This means that all your **generate** statements should explicitly specify **double**:

```
. generate double ... = ...
```

If you do not specify `double`, Stata will store the result as `float`, which has about 7 digits of accuracy, meaning that the numerical first derivatives will be accurate to only 3.5 digits. (Even if you specify `double`, your routine is probably not accurate to 16 digits; it is merely carrying around 16 digits, and the roundoff error inherent in your calculation probably means you will supply only 13 or 14 accurate digits and possibly fewer. The minute you include one `generate` without `double`, however, accuracy falls to 7 digits, and then the roundoff error further reduces it from there.)

The issue in calculating numerical derivatives is choosing h, which Stata does for you. Just so that you are properly appreciative, let's consider that problem:

1. If Stata chooses a value for h that is too small, $f(z+h)$ will numerically equal $f(z)$, so the approximation to $f'(z)$ will be zero.

2. Values that are larger than that but still too small result in poor accuracy because of the numerical roundoff error associated with subtraction that we have just seen. If $f(z+h)$ is very nearly equal to $f(z)$, nearly all the digits are in common, and the difference has few significant digits. For instance, say $f(z)$ and $f(z+h)$ differ only in the 15th and 16th digits:

```
f(x0+h)          =  y.xxxxxxxxxxxxxab
f(x0)            =  y.xxxxxxxxxxxxxcd
─────────────────────────────────────────────
difference       =  0.0000000000000ef   (2 digits of accuracy)
```

Even worse, the 16-digit numbers shown might be accurate to only 14 digits themselves. In that case, the difference $f(z+h) - f(z)$ would be purely noise.

3. On the other hand, if Stata chooses a value for h that is too large, the approximation formula will not estimate the derivative accurately.

Choosing the right value for h is important. Above we considered numerically evaluating the derivative of $\exp(z)$ at $z = 2$. Below we calculate the error for a variety of h's:

```
. drop _all
. set obs 21
Number of observations (_N) was 0, now 21.
. generate double h = 10^(-(_N-_n))
. generate double approx = (exp(2+h)-exp(2))/h
. generate double error = exp(2) - approx
. generate double relerr = abs(error/exp(2))
```

```
. list, sep(0) divider
```

	h	approx	error	relerr
1.	1.000e-20	0	7.3890561	1
2.	1.000e-19	0	7.3890561	1
3.	1.000e-18	0	7.3890561	1
4.	1.000e-17	0	7.3890561	1
5.	1.000e-16	0	7.3890561	1
6.	1.000e-15	6.2172489	1.1718072	.15858685
7.	1.000e-14	7.5495166	-.16046047	.02171596
8.	1.000e-13	7.3807627	.00829343	.00112239
9.	1.000e-12	7.3896445	-.00058835	.00007962
10.	1.000e-11	7.3890227	.00003337	4.516e-06
11.	1.000e-10	7.3890583	-2.155e-06	2.917e-07
12.	1.000e-09	7.3890565	-3.789e-07	5.127e-08
13.	1.000e-08	7.389056	6.523e-08	8.828e-09
14.	1.000e-07	7.3890565	-3.522e-07	4.767e-08
15.	1.000e-06	7.3890598	-3.695e-06	5.001e-07
16.	.00001	7.389093	-.00003695	5.000e-06
17.	.0001	7.3894256	-.00036947	.00005
18.	.001	7.3927519	-.00369576	.00050017
19.	.01	7.4261248	-.03706874	.00501671
20.	.1	7.7711381	-.38208204	.05170918
21.	1	12.696481	-5.3074247	.71828183

In comparing these alternative h's, you should focus on the **relerr** column. The minimum relative error in this experiment is recorded in observation 13, $h = $ 1e–8.

So should h be 1e–8? Not in general: 1e–8 is a good choice for calculating the derivative of $\exp(\cdot)$ at 2, but at some other location, a different value of h would be best. Even if you map out the entire $\exp(\cdot)$ function, you will not have solved the problem. Change the function, and you change the best value of h.

Nash (1990, 219) suggested that we choose h so that $z + h$ differs from z in at least half its digits (the least significant ones). In particular, Nash suggested the formula

$$h(z) = \epsilon \left(|z| + \epsilon \right)$$

where ϵ is the square root of machine precision. Machine precision is roughly 1e–16 for double precision, and therefore, $h(2) \approx$ 2e–8. This way of dynamically adjusting h to the value of z works pretty well.

Nash's suggestion can be improved upon if we are willing to spend computer time. Two issues require balancing:

1. If the calculation were carried out in infinite precision, then the smaller h is, the more accurately $\{f(z + h) - f(z)\}/h$ would approximate the derivative.

2. Taking into consideration finite precision, the closer the values $f(z + h)$ and $f(z)$ are, the fewer the significant digits in the calculation $f(z + h) - f(z)$ are.

Thus we use a variation on Nash's suggestion and control the numerator:

$$f(z + h) \text{ and } f(z) \text{ should differ in about half their digits.}$$

That is, we adopt the rule of setting h as small as possible, subject to the constraint that $f(z + h) - f(z)$ will be calculated to at least half accuracy.

This is a computationally expensive rule to implement because, each time a numerical derivative is calculated, the program must search for the optimal value of h, meaning that $f(z + h)$ must be calculated for trial values. The payoff, however, is that ml is remarkably robust.

ml also uses a centered derivative calculation. Rather than using

$$f'(z) \approx \frac{f(z + h) - f(z)}{h}$$

ml uses

$$f'(z) \approx \frac{f(z + h/2) - f(z - h/2)}{h}$$

This also increases the computational time required but results in an important improvement, reducing the order of magnitude of the error (ignoring numerically induced error) from $O(h)$ to $O(h^2)$.

1.3.5 Numerical second derivatives

Newton–Raphson requires not only $f'(z)$ but also $f''(z)$, the second derivatives. We use the same method to calculate the second derivatives that we use to calculate the first derivatives:

$$f''(z) \approx \frac{f'(z + h/2) - f'(z - h/2)}{h}$$

If you make the substitution for $f'(z)$, you obtain

$$f''(z) \approx \frac{f(z + h) - 2f(z) + f(z - h)}{h^2}$$

Think about the accuracy of this. You might first be tempted to reason that if $f(\cdot)$ is calculated to 16 decimal places, $f'(\cdot)$ will be accurate to 16/2 decimal places at worst, assuming optimal h. If $f'(\cdot)$ is accurate to 8 decimal places, $f''(\cdot)$ will be accurate to at least 8/2 decimal places, at worst. Thus $f''(\cdot)$ is accurate to at least 4 decimal places, at worst.

That is not quite correct: you must look at the numerical calculation formula. The numerator can be rewritten as

$$\{f(z + h) - f(z)\} + \{f(z - h) - f(z)\}$$

The first term, $f(z+h) - f(z)$, can be calculated to half accuracy. The second term can similarly be calculated to the same accuracy if we assume that $f(z + h)$ and $f(z - h)$

are of the same order of magnitude (binary magnitude for modern digital computers). Given that roughly half the digits have already been lost, the numerical summation of the two half-accurate results can be done without further loss of precision, assuming that they are of the same order of (binary) magnitude. The net result is that the numerator can be calculated nearly as accurately as the numerator for the first derivatives.

1.4   Monitoring convergence

When you perform ML estimation, before presenting the estimates, Stata displays something like this:

```
Iteration 0:   Log likelihood =   -45.03321
Iteration 1:   Log likelihood = -27.990675
Iteration 2:   Log likelihood = -23.529093
Iteration 3:   Log likelihood = -21.692004
Iteration 4:   Log likelihood = -20.785625
Iteration 5:   Log likelihood = -20.510315
Iteration 6:   Log likelihood = -20.483776
Iteration 7:   Log likelihood = -20.483495
```

Stata displays this partly for the entertainment value. Placing something on the screen periodically convinces you that your computer is still working.

The iteration log, however, contains reassuring information if you know how to read it. Now that you know how Stata's modified Newton–Raphson works—it is based on Newton's method for finding roots—it should be obvious to you that, for smooth functions, it may jump around at first; however, once it gets close to the maximum (root), it should move smoothly toward it, taking smaller and smaller steps. That is what happened above, and the first graph in figure 1.3 plots these log-likelihood values.

Figure 1.3. Monitoring convergence. Left: the log-likelihood values displayed earlier; right: the log-likelihood values jumping around early on but smoothing out as we get closer to convergence.

When we see this, we know that the numerical routines have really converged to a maximum. Every step changes the log likelihood by less than the previous step. It would not bother us too much if the early iterations jumped around, as long as the later iterations followed the expected pattern. An example of this is the second graph in figure 1.3. All this means is that, wherever we were at the second through fourth iterations, the likelihood function flattened out.

The graphs in figure 1.4, however, would concern us because we are not seeing the expected slowing down of increments in the likelihood function as we get close to the solution.

 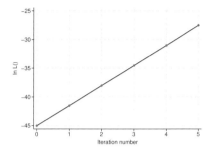

Figure 1.4. Disconcerting "convergence". Left: larger than the previous, indicating that there is still a long way to go to get to the maximum; right: the steps have the same amount of progress, also indicating that there is still a long way to the maximum.

Looking at the second graph of figure 1.4, after thinking about the possibility of a programming bug and dismissing it, we would force more iterations and hope to see something like the first graph in figure 1.5; however, we would not be shocked to see something like the second graph. Either way, it would mean that we just had not done enough iterations previously.

Figure 1.5. Achieving convergence: two possible results from forcing more iterations from the problem identified by the graph on the right in figure 1.4.

We suggest that, after estimation, you type `ml graph` to obtain graphs like the ones above.

Stata is reasonably conservative about declaring convergence, and in fact, it is not uncommon to see irritating evidence of that:

```
Iteration 0:  Log likelihood =   -45.03321
Iteration 1:  Log likelihood = -27.990675
Iteration 2:  Log likelihood = -23.529093
Iteration 3:  Log likelihood = -21.692004
Iteration 4:  Log likelihood = -20.785625
Iteration 5:  Log likelihood = -20.510315
Iteration 6:  Log likelihood = -20.483776
Iteration 7:  Log likelihood = -20.483495
Iteration 8:  Log likelihood = -20.483495
```

Iterations 7 and 8 both present the same value (to the sixth decimal place) of the log-likelihood function. This extra step can be irritating when iterations take 10 or 15 seconds, but the equal value at the end is very reassuring. Stata does not always do that, but we like to see it the first time we code a likelihood function, and whatever Stata does, we invariably force a few extra iterations just for the reassurance.

2 Estimation with mlexp

mlexp is a Stata command that allows you to specify a likelihood function directly in a dialog box or at the command line and obtain maximum likelihood (ML) estimates with a minimum of effort. mlexp works only with certain likelihood functions, but for problems where mlexp is applicable, its simplicity cannot be matched.

In chapter 1, we stated that if the observations are i.i.d., then we can write the log-likelihood function as

$$\ln L(\boldsymbol{\theta}; \mathbf{Z}) = \ln \ell(\boldsymbol{\theta}; \mathbf{z}_1) + \ln \ell(\boldsymbol{\theta}; \mathbf{z}_2) + \cdots + \ln \ell(\boldsymbol{\theta}; \mathbf{z}_N) \tag{2.1}$$

In this case, we can write the log-likelihood function for each observation individually, and the overall log likelihood for the full sample is simply the sum of each observation's log likelihood. In chapter 4, where we define terminology for ml, we say that log likelihoods satisfying this condition meet the "linear form restriction".

To use mlexp, you must have a log-likelihood function that satisfies the functional form in (2.1). Most models for cross-sectional and survey data do meet this restriction, as do many models for survival data. Models for panel data do not meet this restriction because while we can write the log-likelihood function for an individual panel, we cannot write the log-likelihood function for an individual observation. Some time-series models will meet this linear form restriction, while others will not.

2.1 Syntax

The full syntax for mlexp, including a description of all options, is given in appendix A. Here we give just the most basic syntax and discuss some of the available options as we go along in this chapter.

mlexp (lexp) [if] [in] [weight] [, options]

where lexp is how we specify our observation-level likelihood function. Like other Stata estimation commands, mlexp allows us to specify a subset of data using if and in conditions; and we can perform weighted estimation using aweights, fweights, iweights, or pweights.

We specify the likelihood function using a substitutable expression, a Stata expression that also contains parameters to be estimated. The parameters are enclosed in curly braces and must satisfy the naming requirements of Stata variables. {beta} represents a scalar parameter β.

Most regression like models depend on a linear combination of variables $\mathbf{x}_j'\boldsymbol{\beta}$. Rather than typing out an expression like

```
{b1}*weight + {b2}*displacement + {b0}
```

you can use the shorthand

```
{xb:weight displacement _cons}
```

The notation `_cons` indicates that the linear combination is to include a constant term; the constant term is not included automatically. If your likelihood function refers to the same linear combination multiple times, on the second and subsequent times, you can type `{xb:}` to refer to its value. If for some reason you want to refer to the value of a particular variable's parameter within the linear combination, you can type, for example, `{xb:weight}` to get the value of the parameter on the variable `weight`.

By default, parameters are initialized to zero. Later, we will see how to override that in the context of `mlexp`.

2.2   Normal linear regression

We began chapter 1 showing how easy it can be to fit a normal linear regression model. The model can be written succinctly as

$$y_j \sim N(\mathbf{x}_j'\boldsymbol{\beta}, \sigma^2)$$

If y_j follows a normal distribution with mean $\mathbf{x}_j'\boldsymbol{\beta}$ and variance σ^2, then we can write its probability density function as

$$f(y_j|\mathbf{x}_j; \boldsymbol{\beta}, \sigma^2) = \frac{1}{\sigma\sqrt{2\pi}} \exp\left\{-\frac{1}{2}\left(\frac{y_j - \mathbf{x}_j'\boldsymbol{\beta}}{\sigma}\right)^2\right\} \tag{2.2}$$

We could write $f(y_j|\mathbf{x}_j; \boldsymbol{\beta}, \sigma^2)$ in terms of $\phi(z)$, the probability density function of the standard normal distribution; however, there is no need to do so here: Stata's `normalden()` function, which returns the probability density function of a normally distributed variable, allows us to pass in the mean and standard deviation. If we did write $f(y_j|\mathbf{x}_j; \boldsymbol{\beta}, \sigma^2)$ in terms of $\phi(z)$, the expression we use with `mlexp` would be slightly more complicated.

The likelihood function for observation j is simply a restatement of the probability density function where we treat the data as fixed and the parameters as random variables, so

$$\ell(\boldsymbol{\beta}, \sigma^2|y_j, \mathbf{x}_j) = f(y_j|\mathbf{x}_j; \boldsymbol{\beta}, \sigma^2)$$

Remember the discussion from section 1.1. We want to maximize the natural logarithm of this likelihood function. In fact, with `mlexp` you must specify the natural logarithm of

the likelihood function because `mlexp` maximizes the sum of the individual observations'
log likelihoods.

Using the venerable `auto.dta` dataset, we fit the classical normal regression model

$$\text{mpg}_j = \beta_1 \cdot \text{weight}_j + \beta_2 \cdot \text{displacement}_j + \beta_0 + \epsilon_j \qquad \epsilon_j \sim N(0, \sigma^2)$$

by typing

```
. sysuse auto
(1978 automobile data)

. mlexp (ln(normalden(mpg, {xb:weight displacement _cons}, {sigma})))
Initial:       Log likelihood =      -<inf>  (could not be evaluated)
Feasible:      Log likelihood = -10383.274
Rescale:       Log likelihood = -292.89564
Rescale eq:    Log likelihood = -238.45986
Iteration 0:   Log likelihood = -238.45986  (not concave)
Iteration 1:   Log likelihood = -225.48871  (not concave)
Iteration 2:   Log likelihood = -215.66781
Iteration 3:   Log likelihood = -199.18065
Iteration 4:   Log likelihood = -195.29858
Iteration 5:   Log likelihood =    -195.24
Iteration 6:   Log likelihood = -195.2398
Iteration 7:   Log likelihood = -195.2398

Maximum likelihood estimation

Log likelihood = -195.2398                              Number of obs = 74
```

	Coefficient	Std. err.	z	P>|z|	[95% conf. interval]	
xb						
weight	-.0065671	.0011424	-5.75	0.000	-.0088061	-.0043281
displacement	.0052808	.0096674	0.55	0.585	-.0136671	.0242286
_cons	40.08452	1.978738	20.26	0.000	36.20627	43.96278
/sigma	3.385282	.2782684	12.17	0.000	2.839886	3.930678

Here we used the shorthand notation described above to specify our linear combination
of variables `weight`, `displacement`, and a constant term. Doing this makes changing
the model specification easy. If we want to add another regressor, we just add it to the
list. If we want to change some of or all the variables, we just change them. The linear
combination notation takes care of the corresponding parameters for us.

2.3 Initial values

The first line of the iteration log of the previous example says

```
Initial:       Log likelihood =      -<inf>  (could not be evaluated)
```

All parameters by default are initialized to zero. That includes `sigma`, our parameter
for σ, the standard deviation of the normal distribution. Looking at (2.2), we see that
if $\sigma = 0$, then the probability density function is not defined. Saying the log likelihood

is -<inf> is equivalent to saying the function is either not defined or else cannot be
calculated with the current parameter values.

Both mlexp and ml are good at seeking out alternative starting values when the
provided (or default) starting values result in an undefined log likelihood. On the other
hand, the better your starting values, the fewer iterations mlexp or ml will take, and
the faster you will obtain results.

The standard deviation of mpg is about 5.8, so suppose we think $\sigma = 5.8$ is a good
initial value. Then when we declare our parameter sigma, we just include an equals
sign and 5.8 to override the default of 0.

```
. sysuse auto
(1978 automobile data)
. mlexp (ln(normalden(mpg, {xb:weight displacement _cons}, {sigma=5.8})))
Initial:         Log likelihood = -733.27913
Rescale:         Log likelihood = -334.11848
Rescale eq:      Log likelihood = -334.11848
Iteration 0:     Log likelihood = -334.11848  (not concave)
Iteration 1:     Log likelihood = -279.14531  (not concave)
Iteration 2:     Log likelihood = -262.55907  (not concave)
Iteration 3:     Log likelihood = -237.27484  (not concave)
Iteration 4:     Log likelihood = -221.79481  (not concave)
Iteration 5:     Log likelihood =  -217.5619
Iteration 6:     Log likelihood = -198.95135
Iteration 7:     Log likelihood = -195.40835
Iteration 8:     Log likelihood =  -195.2398
Iteration 9:     Log likelihood =  -195.2398

Maximum likelihood estimation

Log likelihood = -195.2398                              Number of obs = 74
```

	Coefficient	Std. err.	z	P>|z|	[95% conf. interval]	
xb						
weight	-.0065671	.0011424	-5.75	0.000	-.0088061	-.0043281
displacement	.0052808	.0096675	0.55	0.585	-.0136671	.0242286
_cons	40.08452	1.978739	20.26	0.000	36.20627	43.96278
/sigma	3.385282	.2782685	12.17	0.000	2.839886	3.930678

The iteration log no longer indicates that log likelihood could not be calculated
at the initial values. However, the iteration log does show that mlexp required nine
iterations, while the previous example required only seven plus an extra step to find
feasible initial values. Whether the search for feasible initial values or a maximization
iteration takes longer depends on the complexity of the likelihood function and the
number of parameters in the model.

You cannot specify initial values for the parameters in a linear combination. In some
applications, you might find that being able to specify an initial value for the constant
term aids in obtaining convergence. The solution is to not specify a constant in the
linear combination but rather include it in the log-likelihood function as a separate
parameter. Here we refit the previous model and initialize the constant term we call b0

to equal 21.3, the mean of `mpg`. We will also initialize `sigma` as we did in the previous example. We type

```
. sysuse auto
(1978 automobile data)

. mlexp (ln(normalden(mpg, {xb:weight displacement} + {b0=21.3}, {sigma=5.8})))
Initial:       Log likelihood = -234.40071
Rescale:       Log likelihood = -234.40071
Rescale eq:    Log likelihood = -234.40071
Iteration 0:   Log likelihood = -234.40071  (not concave)
Iteration 1:   Log likelihood = -227.51818  (not concave)
Iteration 2:   Log likelihood = -220.61578  (not concave)
Iteration 3:   Log likelihood = -216.48798
Iteration 4:   Log likelihood = -202.00319
Iteration 5:   Log likelihood = -195.24569
Iteration 6:   Log likelihood =  -195.2398
Iteration 7:   Log likelihood =  -195.2398

Maximum likelihood estimation

Log likelihood = -195.2398                            Number of obs = 74
```

	Coefficient	Std. err.	z	P>|z|	[95% conf. interval]	
xb						
weight	-.0065671	.0011424	-5.75	0.000	-.0088061	-.0043281
displacement	.0052808	.0096674	0.55	0.585	-.0136671	.0242286
/b0	40.08452	1.978738	20.26	0.000	36.20627	43.96278
/sigma	3.385282	.2782684	12.17	0.000	2.839886	3.930678

2.4   Restricted parameters

`mlexp` and `ml` assume that the parameters in your likelihood function can take on any value. Parameters like the slope coefficients and constant term in our regression model are not restricted. In contrast, the parameter σ in our model must be strictly positive; otherwise, the normal distribution is not defined. In many cases, `mlexp` and `ml` will work around invalid parameter values for you. In our first classical normal regression example, parameter `sigma` was initialized to zero, an invalid value; yet `mlexp` was able to achieve convergence anyway.

For parameters that must be positive, a solution is to define an auxiliary parameter equal to the natural logarithm of the desired parameter. For example, say we let $\theta = \ln \sigma$. The properties of the natural logarithm imply that θ can take on any value. Now, when we maximize over unconstrained parameter θ, we have $\sigma = \exp(\theta) > 0$ as required by our model.

When we have a parameter in our model like σ that must be positive, we name the auxiliary parameter θ as, say, `lnsigma` rather than `theta` to make clear that the parameter is actually the natural logarithm of a parameter we care about. Here we refit our model forcing σ to be positive, and we use `nlcom` to recover our estimate of σ:

```
. sysuse auto
(1978 automobile data)

. mlexp (ln(normalden(mpg,
>   {xb:weight displacement} + {b0=21.3},
>   exp({lnsigma})))) 
Initial:       Log likelihood = -1289.7315
Rescale:       Log likelihood = -1289.7315
Rescale eq:    Log likelihood = -1289.7315
Iteration 0:   Log likelihood = -1289.7315
Iteration 1:   Log likelihood = -842.25864
Iteration 2:   Log likelihood = -200.52701
Iteration 3:   Log likelihood = -195.25082
Iteration 4:   Log likelihood =  -195.2398
Iteration 5:   Log likelihood =  -195.2398

Maximum likelihood estimation

Log likelihood = -195.2398                        Number of obs = 74
```

	Coefficient	Std. err.	z	P>|z|	[95% conf. interval]	
xb						
weight	-.0065671	.0011424	-5.75	0.000	-.0088061	-.0043281
displacement	.0052808	.0096674	0.55	0.585	-.0136671	.0242286
/b0	40.08452	1.978738	20.26	0.000	36.20627	43.96278
/lnsigma	1.219437	.0821995	14.84	0.000	1.058329	1.380545

```
. nlcom exp(_b[/lnsigma])

       _nl_1: exp(_b[/lnsigma])
```

	Coefficient	Std. err.	z	P>|z|	[95% conf. interval]	
_nl_1	3.385282	.2782684	12.17	0.000	2.839886	3.930678

In other cases, you might want to restrict a parameter π to be between zero and one; for example, π could be a probability. In that case, we would maximize our log-likelihood function with respect to unrestricted parameter θ and use the relationship $\pi = \exp(\theta)/\{1 + \exp(\theta)\}$. Then for any value of θ, we will have $\pi \in (0,1)$. The previous function is known as the inverse logit, calculated by Stata's `invlogit()` function. We would name the auxiliary parameter `logitpi` and use the Stata expression `invlogit({logitpi})` to obtain values of π in our substitutable expression.

Correlation coefficients must be bounded between -1 and 1. Here we would take unrestricted parameter $\theta = \operatorname{atanh}(\rho)$, where atanh is the area hyperbolic tangent function. Then for any value of θ, we have $-1 < \rho = \tanh(\theta) < 1$, where tanh is the hyperbolic tangent function. We would call θ `athrho` and use the Stata expression `tanh({athrho})` to obtain ρ in our substitutable expression.

2.5   Robust standard errors

In section 3.3, we derived the robust variance estimator that in some cases provides correct inference even if we misspecify the log-likelihood function. The classical normal regression model assumes that $\epsilon_j \sim N(0, \sigma^2)$ for all j. The assumption of a constant error variance, known as homoskedasticity, is often violated in practice. Even if the errors are heteroskedastic, meaning the variance is not constant across observations, one can still prove that the estimated parameter vector $\widehat{\boldsymbol{\beta}}$ converges in probability to its true value $\boldsymbol{\beta}$. Thus, we can still use the maximum likelihood estimator; however, because the likelihood function is misspecified, we must use the robust variance estimator. This is easy to do with `mlexp`.

```
. sysuse auto
(1978 automobile data)
. mlexp (ln(normalden(mpg, {xb:weight displacement _cons}, {sigma}))),
> vce(robust)
Initial:       Log pseudolikelihood =     -<inf>  (could not be evaluated)
Feasible:      Log pseudolikelihood = -10383.274
Rescale:       Log pseudolikelihood = -292.89564
Rescale eq:    Log pseudolikelihood = -238.45986
Iteration 0:   Log pseudolikelihood = -238.45986  (not concave)
Iteration 1:   Log pseudolikelihood = -225.48871  (not concave)
Iteration 2:   Log pseudolikelihood = -215.66781
Iteration 3:   Log pseudolikelihood = -199.18065
Iteration 4:   Log pseudolikelihood = -195.29858
Iteration 5:   Log pseudolikelihood =    -195.24
Iteration 6:   Log pseudolikelihood =  -195.2398
Iteration 7:   Log pseudolikelihood =  -195.2398

Maximum likelihood estimation

Log pseudolikelihood = -195.2398                      Number of obs = 74
```

	Coefficient	Robust std. err.	z	P>|z|	[95% conf. interval]	
xb						
weight	-.0065671	.0009195	-7.14	0.000	-.0083692	-.004765
displacement	.0052808	.0073381	0.72	0.472	-.0091017	.0196633
_cons	40.08452	2.045516	19.60	0.000	36.07538	44.09366
/sigma	3.385282	.4544255	7.45	0.000	2.494624	4.275939

The header of the coefficient table now indicates that we have robust standard errors, just as we requested. These standard errors are smaller than those we would obtain by typing

```
. regress mpg weight displacement, vce(robust)
```

by a factor of $\sqrt{71/74}$ because `regress` makes a degree-of-freedom adjustment while `mlexp` and `ml` do not. As we explained in chapter 1, maximum likelihood estimators are justified asymptotically, and so they should not be used with very small datasets, where degree-of-freedom adjustments make the most difference.

In many datasets, observations are clustered. For example, you may have a dataset of student test scores, and you suspect that the scores for students within the same classroom are correlated because those students had the same teacher. The classical normal regression model assumes that all observations are independent, but here that is not the case. Although more sophisticated modeling approaches exist for hierarchical data like these, a simple alternative is to fit a classical normal regression model and obtain standard errors that allow for correlation among observations within each cluster.

Here we use the NHANES II dataset to fit a classical normal regression model of systolic blood pressure as a function of cholesterol and body mass index. We suspect that measurements taken within the same location may be correlated because the same machine was used to measure blood pressure. We type

```
. webuse nhanes2

. mlexp (ln(normalden(bpsystol, {xb:tcresult age bmi _cons}, {sigma}))),
>   vce(cluster location)
Initial:       Log pseudolikelihood =      -<inf>  (could not be evaluated)
Feasible:      Log pseudolikelihood = -817782.77
Rescale:       Log pseudolikelihood = -57404.104
Rescale eq:    Log pseudolikelihood = -52198.979
Iteration 0:   Log pseudolikelihood = -52198.979  (not concave)
Iteration 1:   Log pseudolikelihood = -48574.414  (not concave)
Iteration 2:   Log pseudolikelihood = -47874.684
Iteration 3:   Log pseudolikelihood =  -45835.99
Iteration 4:   Log pseudolikelihood = -45418.078
Iteration 5:   Log pseudolikelihood = -45399.273
Iteration 6:   Log pseudolikelihood = -45399.267
Iteration 7:   Log pseudolikelihood = -45399.267

Maximum likelihood estimation

Log pseudolikelihood = -45399.267                       Number of obs = 10,351
```

(Std. err. adjusted for 62 clusters in location)

	Coefficient	Robust std. err.	z	P>|z|	[95% conf. interval]	
xb						
tcresult	.0197722	.0046124	4.29	0.000	.0107321	.0288122
age	.5668307	.0200393	28.29	0.000	.5275544	.606107
bmi	1.283265	.0559505	22.94	0.000	1.173604	1.392926
_cons	66.83673	1.719312	38.87	0.000	63.46694	70.20652
/sigma	19.43432	.2953367	65.80	0.000	18.85547	20.01317

2.6   The probit model

Again using `auto.dta`, say that we want to fit a probit model to predict whether a car is foreign or domestic based on its weight and price. Our statistical model is

$$\pi_j = \Pr(\texttt{foreign}_j \mid \texttt{weight}_j, \texttt{price}_j)$$
$$= \Phi(\beta_1 \texttt{weight}_j + \beta_2 \texttt{price}_j + \beta_0)$$

where $\Phi(\cdot)$ denotes the standard normal distribution function. The log likelihood for the jth observation is

$$\ln \ell_j = \begin{cases} \ln \Phi(\beta_1 \text{weight}_j + \beta_2 \text{price}_j + \beta_0) & \text{if } \text{foreign}_j = 1 \\ 1 - \ln \Phi(\beta_1 \text{weight}_j + \beta_2 \text{price}_j + \beta_0) & \text{if } \text{foreign}_j = 0 \end{cases}$$

Because the normal density function is symmetric about zero, $1 - \Phi(w) = \Phi(-w)$, and computers can more accurately calculate the latter than the former. Therefore, we are better off writing the log likelihood as

$$\ln \ell_j = \begin{cases} \ln \Phi(\beta_1 \text{weight}_j + \beta_2 \text{price}_j + \beta_0) & \text{if } \text{foreign}_j = 1 \\ \ln \Phi(-\beta_1 \text{weight}_j - \beta_2 \text{price}_j - \beta_0) & \text{if } \text{foreign}_j = 0 \end{cases}$$

The log-likelihood function looks a bit more difficult because we must deal with the case where $\text{foreign}_j = 1$ differently from where $\text{foreign}_j = 0$. Stata's cond(x, a, b) function makes that easy, though. If the condition x is true, then the function returns a; otherwise, it returns b. The condition x in our application is foreign==1. We type

```
. sysuse auto
(1978 automobile data)
. mlexp (cond(foreign==1, ln(normal({xb:weight price _cons})),
>                        ln(normal(-1*{xb:})))))
Initial:       Log likelihood = -51.292891
Alternative:   Log likelihood = -45.055272
Rescale:       Log likelihood = -45.055272
Iteration 0:   Log likelihood = -45.055272
Iteration 1:   Log likelihood = -20.769042
Iteration 2:   Log likelihood = -18.201831
Iteration 3:   Log likelihood = -18.007514
Iteration 4:   Log likelihood = -18.006571
Iteration 5:   Log likelihood = -18.006571

Maximum likelihood estimation

Log likelihood = -18.006571                          Number of obs = 74
```

xb	Coefficient	Std. err.	z	P>|z|	[95% conf. interval]	
weight	-.003238	.0008643	-3.75	0.000	-.004932	-.0015441
price	.000517	.0001591	3.25	0.001	.0002052	.0008287
_cons	4.921935	1.330065	3.70	0.000	2.315055	7.528816

The log likelihood for the case where foreign is zero includes the linear combination of variables multiplied by negative one. Because we already defined the linear combination where foreign is one, we use the syntax {xb:} to get its value again.

A minor quibble about our implementation of probit is that we do not do any sanity checking to make sure that foreign is either zero or one. Instead, as long as foreign is not one, we assume it is zero. If we were writing a formal estimation command, we would implement a test to make sure our left-hand-side variable satisfied the requirements of the model. Typically, we use mlexp for less formal purposes and do not bother with those types of details.

2.7   Specifying derivatives

`mlexp` allows you to specify analytic first derivatives of the log-likelihood function, and this almost always speeds up convergence. You specify the derivatives using substitutable expressions just as you specify the log-likelihood function. The derivatives for the probit model are relatively easy, so we will use that as our example.

We are going to use the same statistical model as the previous example, but to illustrate how to specify derivatives of both individual parameters and of linear combinations, imagine that we specified the `mlexp` command as

```
. mlexp (cond(foreign==1, ln(normal({xb:weight price} + {b0})), ///
                          ln(normal(-1*{xb:} - {b0}))))
```

Now we have both linear combination `xb:` and parameter `b0` in our model. `mlexp` requires derivatives for both of those.

We first need to do a bit of math. The log-likelihood function with the constant term made explicit is

$$
\ln \ell_j = \begin{cases} \ln \Phi(\mathbf{x}'_j \boldsymbol{\beta} + \beta_0) & \text{if foreign}_j = 1 \\ \ln \Phi(-\mathbf{x}'_j \boldsymbol{\beta} - \beta_0) & \text{if foreign}_j = 0 \end{cases}
$$

The derivative of the standard normal distribution function $\Phi(z)$ with respect to z is simply the standard normal density function $\phi(z)$. Moreover, $\phi(z)$ is symmetric about zero: $\phi(-z) = \phi(z)$. Thus, the derivative with respect to the linear combination $\mathbf{x}'_j \boldsymbol{\beta}$ is

$$
\frac{\partial \ln \ell_j}{\partial \mathbf{x}'_j \boldsymbol{\beta}} = \begin{cases} \phi(\mathbf{x}'_j \boldsymbol{\beta} + \beta_0)/\Phi(\mathbf{x}'_j \boldsymbol{\beta} + \beta_0) & \text{if foreign}_j = 1 \\ -\phi(\mathbf{x}'_j \boldsymbol{\beta} + \beta_0)/\Phi(-\mathbf{x}'_j \boldsymbol{\beta} - \beta_0) & \text{if foreign}_j = 0 \end{cases}
$$

and the derivative with respect to the constant term is also

$$
\frac{\partial \ln \ell_j}{\partial \beta_0} = \begin{cases} \phi(\mathbf{x}'_j \boldsymbol{\beta} + \beta_0)/\Phi(\mathbf{x}'_j \boldsymbol{\beta} + \beta_0) & \text{if foreign}_j = 1 \\ -\phi(\mathbf{x}'_j \boldsymbol{\beta} + \beta_0)/\Phi(-\mathbf{x}'_j \boldsymbol{\beta} - \beta_0) & \text{if foreign}_j = 0 \end{cases}
$$

Looking at the derivatives, we can see that we are going to be using the expression `{xb:}+{b0}` quite a few times. To save ourselves a bit of typing, we are going to define local macro `xb0` to equal that expression. We type

```
. sysuse auto
(1978 automobile data)
. local xb0 ({xb:} + {b0})
. mlexp (cond(foreign==1, ln(normal({xb:weight price} + {b0}))),
>                         ln(normal(-1*`xb0')))),
>       deriv(/xb = cond(foreign==1,
>                        normalden(`xb0')/normal(`xb0'),
>                        -1*normalden(`xb0')/normal(-1*`xb0')))
>       deriv(/b0 = cond(foreign==1,
>                        normalden(`xb0')/normal(`xb0'),
>                        -1*normalden(`xb0')/normal(-1*`xb0')))
Initial:        Log likelihood = -51.292891
Alternative:    Log likelihood = -44.872898
Rescale:        Log likelihood = -43.177049
Rescale eq:     Log likelihood = -41.349514
Iteration 0:    Log likelihood = -41.349514
Iteration 1:    Log likelihood = -20.508893
Iteration 2:    Log likelihood = -18.168331
Iteration 3:    Log likelihood = -18.007368
Iteration 4:    Log likelihood = -18.006572
Iteration 5:    Log likelihood = -18.006571

Maximum likelihood estimation

Log likelihood = -18.006571                    Number of obs = 74
```

	Coefficient	Std. err.	z	P>|z|	[95% conf. interval]	
xb						
weight	-.003238	.0008643	-3.75	0.000	-.004932	-.0015441
price	.000517	.0001591	3.25	0.001	.0002052	.0008287
/b0	4.921935	1.330065	3.70	0.000	2.315055	7.528815

We have one linear combination and one regular parameter, so we specify the option `deriv()` twice. The syntax for regular parameters is to specify a forward slash, the parameter name, and the derivative as a substitutable expression. For linear combinations of variables, we specify the linear combination's name (without the colon) rather than a parameter name. We specified the variables included in the linear combination when we specified the log-likelihood function, so the shorthand `{xb:}` is available to use in our derivative expressions.

We pulled one subtle trick out of our hat when we defined the local macro `xb0`: we bound our expression `{xb:}+{b0}` in parentheses. That allowed us to specify `-{xb:}-{b0}` as `-1*`xb0'`. Without the parentheses multiplying the local macro by negative one would expand to `-1*{xb:}+{b0}`, which is not what appears in our derivatives. We often use local macros to define repetitive terms in substitutable expressions, and we almost always bind the definition in parentheses.

We maximize the log-likelihood function with respect to the parameters in our model. How, then, are we able to specify the derivative with respect to a linear combination of variables? In short, `mlexp` understands the chain rule. That is,

$$\frac{\partial \ln \ell_j(\mathbf{x}_j'\boldsymbol{\beta})}{\partial \beta_i} = \frac{\partial \ln \ell_j(\mathbf{x}_j'\boldsymbol{\beta})}{\partial \mathbf{x}_j'\boldsymbol{\beta}} \frac{\partial \mathbf{x}_j'\boldsymbol{\beta}}{\partial \beta_i} = \frac{\partial \ln \ell_j(\mathbf{x}_j'\boldsymbol{\beta})}{\partial \mathbf{x}_j'\boldsymbol{\beta}} x_{ij}$$

`mlexp` knows the variables in the linear combination, so if it knows $\partial \ln \ell(\mathbf{x}_j'\boldsymbol{\beta})/\partial \mathbf{x}_j'\boldsymbol{\beta}$, it can work out the derivative of each parameter in the linear combination. An advantage of this is that if we change the variables in our linear combination when we specify the substitutable expression for the log-likelihood function, we do not need to do anything to our derivatives.

❑ **Technical note**

Another way to specify the log-likelihood function for the probit model accounting for the $\texttt{foreign}_j = 0$ and $\texttt{foreign}_j = 1$ cases is to define the auxiliary variable $q_j = 2 \cdot \texttt{foreign}_j - 1$. Then we could write the log-likelihood function as

$$\ln \ell_j = \ln \Phi \left\{ q_j \cdot (\beta_1 \texttt{weight}_j + \beta_2 \texttt{price}_j + \beta_0) \right\}$$

We illustrate how to specify this log-likelihood function and its derivatives in [R] **mlexp**.

❑

2.8   Additional estimation features

Like most estimation commands in Stata, `mlexp` can perform weighted estimation, and it accepts several useful prefix commands, including `bootstrap` and `svy`. You can thus use `mlexp` to implement custom maximum-likelihood estimators and take advantage of many Stata features without any additional programming.

Suppose we want to obtain bootstrap standard errors of our linear regression model. We can simply specify the `bootstrap` prefix command, specify the number of replications, set the random-number seed, and invoke `mlexp`. Here we refit our linear regression model and obtain normal-approximation bootstrap confidence intervals based on 1,000 replications. We set the random-number seed so that the results are reproducible. We suppress the bootstrap replication dots here to save space, but you may wish to omit the `nodots` option to monitor progress:

```
. sysuse auto
(1978 automobile data)

. bootstrap, reps(1000) seed(1) nodots:
>  mlexp (ln(normalden(mpg, {xb:weight displacement _cons}, {sigma})))

Maximum likelihood estimation

Bootstrap results                                Number of obs =      74
Log likelihood = -195.2398                       Replications  = 1,000
```

	Observed coefficient	Bootstrap std. err.	z	P>|z|	Normal-based [95% conf. interval]	
xb						
weight	-.0065671	.0010272	-6.39	0.000	-.0085804	-.0045538
displacement	.0052808	.0083257	0.63	0.526	-.0110372	.0215988
_cons	40.08452	2.21428	18.10	0.000	35.74461	44.42443
/sigma	3.385282	.4474113	7.57	0.000	2.508372	4.262192

The NHANES II dataset we used earlier is based on a survey. To refit that model and account for the survey design, we just need to `svyset` our data and use the `svy` prefix with `mlexp`:

```
. webuse nhanes2

. svyset psu [pweight=finalwgt], strata(strata)

Sampling weights: finalwgt
             VCE: linearized
     Single unit: missing
        Strata 1: strata
  Sampling unit 1: psu
           FPC 1: <zero>

. svy: mlexp (ln(normalden(bpsystol, {xb:tcresult age i.race _cons}, {sigma})))
(running mlexp on estimation sample)

Survey data analysis

Number of strata = 31                Number of obs   =        10,351
Number of PSUs   = 62                Population size = 117,157,513
                                     Design df       =            31
```

	Coefficient	Linearized std. err.	t	P>|t|	[95% conf. interval]	
tcresult	.039044	.00551	7.09	0.000	.0278063	.0502818
age	.5922628	.0173295	34.18	0.000	.5569191	.6276066
race						
Black	3.317389	.8004878	4.14	0.000	1.684783	4.949995
Other	-1.405489	1.195789	-1.18	0.249	-3.844317	1.033339
_cons	93.31971	1.092803	85.39	0.000	91.09092	95.54849
/sigma	18.85556	.2697697	69.90	0.000	18.30537	19.40576

2.9   Wrapping up

mlexp is the easiest way to implement a maximum likelihood estimator in Stata. No programming is necessary, and you can take advantage of many useful estimation features, including weighted estimation, robust standard errors, and bootstrapping. You can also use the collect prefix command to retain results for inclusion in a table, and you can use the by prefix to run mlexp on subsets of data.

However, mlexp has several limitations. While the substitutable expressions used by mlexp allow you to avoid programming, more complicated log-likelihood functions are difficult if not impossible to code using them. In addition, mlexp requires that you be able to write down the log likelihood for each observation, which is not possible for many important models. Providing the analytic gradient speeds up convergence, but also providing the Hessian matrix greatly increases the speed; mlexp has no facilities to provide the Hessian matrix, but ml does.

The remainder of this book focuses on the ml command, which gives you much greater flexibility to implement your maximum likelihood estimators. In exchange for a bit of programming, ml lets you overcome all the limitations of mlexp. You can implement arbitrarily complex log-likelihood functions, and you can (but are not required to) provide both the analytic gradient and the analytic Hessian matrix of your log-likelihood function. You can program estimators for which the log-likelihood function cannot be written for a single observation but rather can be written only for groups of observations as in panel data or for the entire sample of observations. Indeed, many of Stata's commands that implement maximum likelihood estimators are written atop ml.

3 Introduction to ml

`ml` is the Stata command to implement maximum likelihood (ML) estimators that cannot be handled by `mlexp` and to write your own commands that perform ML estimation. Obtaining ML estimates requires the following steps:

1. Derive the log-likelihood function from your probability model.

2. Write a program that calculates the log-likelihood values and, optionally, its derivatives. This program is known as a likelihood evaluator.

3. Identify a particular model to fit using your data variables and the `ml model` statement.

4. Fit the model using `ml maximize`.

This chapter illustrates steps 2, 3, and 4 using the probit model for dichotomous (0/1) variables and the linear regression model assuming normally distributed errors.

In this chapter, we fit our models explicitly, handling each coefficient and variable individually. New users of `ml` will appreciate this approach because it closely reflects how you would write down the model you wish to fit on paper; and it allows us to focus on some of the basic features of `ml` without becoming overly encumbered with programming details. We will also illustrate this strategy's shortcomings so that once you become familiar with the basics of `ml` by reading this chapter, you will want to think of your model in a slightly more abstract form, providing much more flexibility.

In the next chapter, we discuss `ml`'s probability model parameter notation, which is particularly useful when, as is inevitably the case, you decide to change some of the variables appearing in your model. If you are already familiar with `ml`'s θ-parameter notation, you can skip this chapter with virtually no loss of continuity with the rest of the book.

Chapter 17 contains the derivations of log-likelihood functions (step 1) for models discussed in this book.

3.1 The probit model

In section 2.6, we fit a probit model to predict whether a car is foreign or domestic based on its weight and price using `auto.dta`. Here we illustrate how to fit that model using `ml`. Recall that our statistical model is

$$\begin{aligned} \pi_j &= \Pr(\texttt{foreign}_j \mid \texttt{weight}_j, \texttt{price}_j) \\ &= \Phi(\beta_1 \texttt{weight}_j + \beta_2 \texttt{price}_j + \beta_0) \end{aligned}$$

where we use the subscript j to denote observations and $\Phi(\cdot)$ denotes the standard normal distribution function. The log likelihood for the jth observation is

$$\ln \ell_j = \begin{cases} \ln \Phi(\beta_1 \texttt{weight}_j + \beta_2 \texttt{price}_j + \beta_0) & \text{if } \texttt{foreign}_j = 1 \\ 1 - \ln \Phi(\beta_1 \texttt{weight}_j + \beta_2 \texttt{price}_j + \beta_0) & \text{if } \texttt{foreign}_j = 0 \end{cases}$$

$$= \begin{cases} \ln \Phi(\beta_1 \texttt{weight}_j + \beta_2 \texttt{price}_j + \beta_0) & \text{if } \texttt{foreign}_j = 1 \\ \ln \Phi(-\beta_1 \texttt{weight}_j - \beta_2 \texttt{price}_j - \beta_0) & \text{if } \texttt{foreign}_j = 0 \end{cases} \tag{3.1}$$

where we used the fact that $1 - \Phi(w) = \Phi(-w)$.

With our log-likelihood function in hand, we write a program to evaluate it:

```
───────────────────────────────────── begin myprobit_gf0.ado ─────────
program myprobit_gf0
        version 18
        args todo b lnfj

        tempvar xb
        quietly generate double 'xb' = 'b'[1,1]*weight + 'b'[1,2]*price + ///
                                'b'[1,3]
        quietly replace 'lnfj' = ln(normal('xb')) if foreign == 1
        quietly replace 'lnfj' = ln(normal(-1*'xb')) if foreign == 0
end
────────────────────────────────────── end myprobit_gf0.ado ─────────
```

We named our program `myprobit_gf0.ado`, but you could name it anything you want as long as it has the extension `.ado`. The name without the `.ado` extension is what we use to tell `ml model` about our likelihood function. We added `gf0` to our name to emphasize that our evaluator is a general-form problem and that we are going to specify no (0) derivatives. We will return to this issue when we use the `ml model` statement.

Our program accepts three arguments. The first, `todo`, we can safely ignore for now. In later chapters, when we discuss other types of likelihood-evaluator programs, we will need that argument. The second, `b`, contains a row vector containing the parameters of our model (β_0, β_1, and β_2). The third argument, `lnfj`, is the name of a temporary variable that we are to fill in with the values of the log-likelihood function evaluated at the coefficient vector `b`. Our program then created a temporary variable to hold the values of $\beta_1 \texttt{weight}_j + \beta_2 \texttt{price}_j + \beta_0$. We created that variable to have storage type `double`; we will discuss this point in greater detail in the next chapter, but for now you should remember that when coding your likelihood evaluator, you must create temporary variables as `double`s. The last two lines replace `lnfj` with the values of the log likelihood for $\texttt{foreign}_j$ equal to 0 and 1, respectively. We could have used Stata's `cond()` function to combine those two statements into one, but using two statements is arguably clearer. Because `b` and `lnfj` are arguments passed to our program and `xb` is a temporary variable we created with the `tempvar` commands, their names are local macros that must be dereferenced using left- and right-hand single quote marks to use them; see [U] **18.7 Temporary objects**.

The next step is to identify our model using `ml model`. To that end, we type

```
. sysuse auto
(1978 automobile data)
. ml model gf0 myprobit_gf0 (foreign = weight price)
```

We first loaded in our dataset, because `ml model` will not work without the dataset in memory. Next we told `ml` that we have a method-`gf0` likelihood evaluator named `myprobit_gf0`, our dependent variable is `foreign`, and our independent variables are `weight` and `price`. In subsequent chapters, we examine all the likelihood-evaluator types; method-`gf0` (general form) evaluator programs most closely follow the mathematical notation we used in (3.1) and are therefore perhaps easiest for new users of `ml` to grasp, but we will see that they have disadvantages as well. This general-form evaluator simply receives a vector of parameters and a variable into which the observations' log-likelihood values are to be stored.

The final step is to tell `ml` to maximize the likelihood function and report the coefficients:

```
. ml maximize
Initial:      Log likelihood = -51.292891
Alternative:  Log likelihood = -45.055272
Rescale:      Log likelihood = -45.055272
Iteration 0:  Log likelihood = -45.055272
Iteration 1:  Log likelihood = -20.770386
Iteration 2:  Log likelihood = -18.023563
Iteration 3:  Log likelihood = -18.006584
Iteration 4:  Log likelihood = -18.006571
Iteration 5:  Log likelihood = -18.006571
```

					Number of obs =	74
					Wald chi2(2) =	14.09
Log likelihood = -18.006571					Prob > chi2 =	0.0009

foreign	Coefficient	Std. err.	z	P>|z|	[95% conf. interval]	
weight	-.003238	.0008643	-3.75	0.000	-.004932	-.0015441
price	.000517	.0001591	3.25	0.001	.0002052	.0008287
_cons	4.921935	1.330065	3.70	0.000	2.315055	7.528816

You can verify that we would obtain identical results using `probit`:

```
. probit foreign weight price
```

This example was straightforward because we had only one equation and no auxiliary parameters. Next we consider linear regression with normally distributed errors.

3.2   Normal linear regression

Now suppose we want to fit a linear regression of `turn` on `length` and `headroom`:

$$\text{turn}_j = \beta_1\text{length}_j + \beta_2\text{headroom}_j + \beta_3 + \epsilon_j$$

where ϵ_j is an error term. If we assume that each ϵ_j is independent and identically distributed as a normal random variable with mean zero and variance σ^2, we have what is often called normal linear regression; and we can fit the model by ML. As derived in section 17.3, when we assume homoskedasticity, we can write the log likelihood for the jth observation in terms of $\phi(z)$, the standard normal density function, as

$$\ln \ell_j = \ln \phi \left(\frac{\text{turn}_j - \beta_1\text{length}_j - \beta_2\text{headroom}_j - \beta_3}{\sigma} \right) - \ln \sigma$$

There are four parameters in our model: β_1, β_2, β_3, and σ, so we will specify our `ml model` statement so that our likelihood evaluator receives a vector of coefficients with four columns. As a matter of convention, we will use the four elements of that vector in the order we just listed so that, for example, β_2 is the second element and σ is the fourth element. Our likelihood-evaluator program is

```
─────────────────────────────── begin mynormal1_gf0.ado ───────────
program mynormal1_gf0
        version 18
        args todo b lnfj

        tempvar xb
        quietly generate double 'xb' = 'b'[1,1]*length +               ///
                                'b'[1,2]*headroom + 'b'[1,3]

        quietly replace 'lnfj' = ln(normalden((turn - 'xb')/'b'[1,4])) - ///
                                ln('b'[1,4])
end
─────────────────────────────────── end mynormal1_gf0.ado ───────────
```

In our previous example, when we typed

```
. ml model gf0 myprobit_gf0 (foreign = weight price)
```

`ml` knew to create a coefficient vector with three elements because we specified two right-hand-side variables, and by default `ml` includes a constant term unless we specify the `noconstant` option, which we discuss in the next chapter. How do we get `ml` to include a fourth parameter for σ? Perhaps the most transparent solution is to type

```
. ml model gf0 mynormal1_gf0 (turn = length headroom) /sigma
```

The notation `/sigma` tells `ml` to include a fourth element in our coefficient vector and to label it `sigma` in the output. In chapter 4, we will see other ways to specify parameters like σ and have them labeled slightly differently. Having identified our model, we can now maximize the log-likelihood function:

```
. ml maximize
Initial:     Log likelihood =     -<inf>  (could not be evaluated)
Feasible:    Log likelihood =  -8418.567
Rescale:     Log likelihood = -327.16314
Rescale eq:  Log likelihood = -215.53986
Iteration 0:  Log likelihood = -215.53986  (not concave)
Iteration 1:  Log likelihood = -213.33272  (not concave)
Iteration 2:  Log likelihood = -211.10519  (not concave)
Iteration 3:  Log likelihood = -209.60577  (not concave)
Iteration 4:  Log likelihood = -207.93771  (not concave)
Iteration 5:  Log likelihood = -206.43844  (not concave)
Iteration 6:  Log likelihood = -205.19618  (not concave)
Iteration 7:  Log likelihood = -204.11373  (not concave)
Iteration 8:  Log likelihood = -203.00329  (not concave)
Iteration 9:  Log likelihood =  -202.1822  (not concave)
Iteration 10: Log likelihood = -201.42449  (not concave)
Iteration 11: Log likelihood = -200.64468  (not concave)
Iteration 12: Log likelihood =  -199.9014  (not concave)
Iteration 13: Log likelihood = -199.18937  (not concave)
Iteration 14: Log likelihood = -198.48172  (not concave)
Iteration 15: Log likelihood = -197.78556  (not concave)
Iteration 16: Log likelihood = -197.10597  (not concave)
Iteration 17: Log likelihood = -196.43819  (not concave)
Iteration 18: Log likelihood = -195.78002  (not concave)
Iteration 19: Log likelihood = -195.13253  (not concave)
Iteration 20: Log likelihood =  -194.4956  (not concave)
Iteration 21: Log likelihood = -193.86829  (not concave)
Iteration 22: Log likelihood =  -193.2503  (not concave)
Iteration 23: Log likelihood = -192.64164  (not concave)
Iteration 24: Log likelihood =  -192.0421  (not concave)
Iteration 25: Log likelihood = -191.45135  (not concave)
Iteration 26: Log likelihood = -190.86936  (not concave)
Iteration 27: Log likelihood = -190.29601  (not concave)
Iteration 28: Log likelihood = -189.73109  (not concave)
Iteration 29: Log likelihood = -189.17464  (not concave)
Iteration 30: Log likelihood = -188.62662
Iteration 31: Log likelihood = -167.31438  (backed up)
Iteration 32: Log likelihood = -163.31365
Iteration 33: Log likelihood = -163.18798
Iteration 34: Log likelihood = -163.18765
Iteration 35: Log likelihood = -163.18765
```

```
                                          Number of obs =      74
                                          Wald chi2(2)  = 219.18
Log likelihood = -163.18765               Prob > chi2   = 0.0000
```

turn	Coefficient	Std. err.	z	P>|z|	[95% conf. interval]	
length	.1737846	.0134739	12.90	0.000	.1473762	.2001929
headroom	-.1542078	.3546293	-0.43	0.664	-.8492685	.5408529
_cons	7.450477	2.19735	3.39	0.001	3.143751	11.7572
/sigma	2.195259	.1804492	12.17	0.000	1.841585	2.548933

The point estimates match those we obtain from typing

```
. regress turn length headroom
```

The standard errors differ by a factor of $\sqrt{71/74}$ because of the same degree-of-freedom adjustment we discussed in section 2.5.

3.3   Robust standard errors

Robust standard errors are commonly reported nowadays along with linear regression results because they allow for correct statistical inference even when the tenuous assumption of homoskedasticity is not met. Cluster–robust standard errors can be used when related observations' errors are correlated. Obtaining standard errors with most estimation commands is trivial: you just specify the option vce(robust) or vce(cluster id), where id is the name of a variable identifying groups. Using our previous regression example, you might type

```
. regress turn length headroom, vce(robust)
```

For the evaluator functions we have written so far, both of which have been method gf0, obtaining robust or cluster–robust standard errors is no more difficult than with other estimation commands. To refit our linear regression model, obtaining robust standard errors, we type

```
. ml model gf0 mynormal1_gf0 (turn = length headroom) /sigma, vce(robust)
. ml maximize, nolog
Initial:       Log pseudolikelihood =      -<inf>  (could not be evaluated)
Feasible:      Log pseudolikelihood =  -8418.567
Rescale:       Log pseudolikelihood = -327.16314
Rescale eq:    Log pseudolikelihood = -215.53986
```

```
                                               Number of obs =        74
                                               Wald chi2(2)  =    298.85
Log pseudolikelihood = -163.18765              Prob > chi2   =    0.0000
```

turn	Coefficient	Robust std. err.	z	P>|z|	[95% conf. interval]	
length	.1737846	.0107714	16.13	0.000	.152673	.1948961
headroom	-.1542078	.2955882	-0.52	0.602	-.73355	.4251344
_cons	7.450477	1.858003	4.01	0.000	3.808857	11.0921
/sigma	2.195259	.2886184	7.61	0.000	1.629577	2.760941

ml model accepts vce(cluster id) with method-gf0 evaluators just as readily as it accepts vce(robust).

Being able to obtain robust standard errors just by specifying an option to ml model should titillate you. When we discuss other types of evaluator programs, we will see that in fact there is a lot of work happening behind the scenes to produce robust standard errors. With method-gf0 evaluators (and other linear-form evaluators), ml does all the work for you.

3.4   Weighted estimation

Stata provides four types of weights that the end-user can apply to estimation problems. Frequency weights, known as `fweights` in the Stata vernacular, represent duplicated observations; instead of having five observations that record identical information, `fweights` allow you to record that observation once in your dataset along with a frequency weight of 5, indicating that observation is to be repeated a total of five times. Analytic weights, called `aweights`, are inversely proportional to the variance of an observation and are used with group-mean data. Sampling weights, called `pweights`, denote the inverse of the probability that an observation is sampled and are used with survey data where some people are more likely to be sampled than others. Importance weights, called `iweights`, indicate the relative "importance" of the observation and are intended for use by programmers who want to produce a certain computation.

Obtaining weighted estimates with method-`gf0` likelihood evaluators is the same as with most other estimation commands. Suppose that in `auto.dta`, `rep78` is actually a frequency weight variable. To obtain frequency-weighted estimates of our probit model, we type

```
. ml model gf0 myprobit_gf0 (foreign = weight price) [fweight = rep78]

. ml maximize
Initial:       Log likelihood = -162.88959
Alternative:   Log likelihood = -159.32929
Rescale:       Log likelihood = -156.55825
Iteration 0:   Log likelihood = -156.55825
Iteration 1:   Log likelihood = -72.414357
Iteration 2:   Log likelihood =  -66.82292
Iteration 3:   Log likelihood = -66.426129
Iteration 4:   Log likelihood = -66.424675
Iteration 5:   Log likelihood = -66.424675
```

		Number of obs =	235
		Wald chi2(2) =	58.94
Log likelihood = -66.424675		Prob > chi2 =	0.0000

foreign	Coefficient	Std. err.	z	P>|z|	[95% conf. interval]	
weight	-.0027387	.0003576	-7.66	0.000	-.0034396	-.0020379
price	.0004361	.0000718	6.07	0.000	.0002953	.0005768
_cons	4.386445	.5810931	7.55	0.000	3.247523	5.525366

Just like with obtaining robust standard errors, we did not have to do anything to our likelihood-evaluator program. We just added a weight specification, and `ml` did all the heavy lifting to make that work. You should be impressed. Other evaluator types require you to account for weights yourself, which is not always a trivial task.

3.5   Other features of method-gf0 evaluators

In addition to easily obtaining robust standard errors and weighted estimates, method-`gf0` likelihood evaluators provide several other features. By specifying the `svy` option to `ml model`, you can obtain results that take into account the complex survey design of your data. Before using the `svy` option, you must first `svyset` your data; see [U] **26.19 Survey data**.

You can restrict the estimation sample by using `if` and `in` conditions in your `ml model` statement. Again, method-`gf0` evaluators require you to do nothing special to make them work. See [U] **11 Language syntax** to learn about `if` and `in` qualifiers.

3.6   Limitations

We have introduced `ml` using method-`gf0` evaluators because they align most closely with the way you would write the likelihood function for a specific model. However, writing your likelihood evaluator in terms of a particular model with prespecified variables severely limits your flexibility.

For example, say that we had a binary variable `good` that we wanted to use instead of `foreign` as the dependent variable in our probit model. If we simply change our `ml model` statement to read

```
. ml model gf0 myprobit_gf0 (good = weight price)
```

the output from `ml maximize` will label the dependent variable as `good`, but the output will otherwise be unchanged! When we wrote our likelihood-evaluator program, we hardcoded in the name of the dependent variable. As far as our likelihood-evaluator program is concerned, changing the dependent variable in our `ml model` statement did nothing.

When you specify the dependent variable in your `ml model` statement, `ml` stores the variable name in the global macro `$ML_y1`. Thus a better version of our `myprobit_gf0` program would be

```
——————————————————————— begin myprobit_gf0_good.ado ———————
program myprobit_gf0_good
        version 18
        args todo b lnfj

        tempvar xb
        quietly generate double 'xb' = 'b'[1,1]*weight + 'b'[1,2]*price + ///
                                'b'[1,3]
        quietly replace 'lnfj' = ln(normal('xb')) if $ML_y1 == 1
        quietly replace 'lnfj' = ln(normal(-1*'xb')) if $ML_y1 == 0
end
——————————————————————————— end myprobit_gf0_good.ado ———————
```

With this change, we can specify dependent variables at will.

Adapting our program to accept an arbitrary dependent variable was straightforward. Unfortunately, making it accept an arbitrary set of independent variables is much more difficult. We wrote our likelihood evaluator assuming that the coefficient vector `b` had three elements, and we hardcoded the names of our independent variables in the likelihood-evaluator program. If we were hell-bent on making our method-`gf0` evaluator work with an arbitrary number of independent variables, we could examine the column names of '`b`' and deduce the number of variables, their names, and even the number of equations. In the next chapter, we will learn a better way to approach problems using `ml` that affords us the ability to change regressors without having to modify our evaluator program in any way.

4   Overview of ml

In the previous chapter, we gave an introduction to `ml`, programming our likelihood evaluator much like you would write the likelihood function for a particular model and dataset on paper. In this chapter, we show how `ml`'s syntax allows us to write likelihood evaluators in more general terms, without having to specify variable names or refer to individual coefficients within our evaluator program. By using the methods in this chapter, we will overcome the shortcomings we encountered in the last chapter.

4.1   The terminology of ml

Before we can really discuss using `ml` for model fitting, we must first settle on some notation and terminology.

Any likelihood function can be written as a function of parameters and data variables (both dependent and independent). Thus in the most general case, we can write

$$\ln L = \ln L\{(\theta_{1j}, \theta_{2j}, \dots, \theta_{Ej}; y_{1j}, y_{2j}, \dots, y_{Dj}) : j = 1, 2, \dots, N\}$$
$$= \ln L(\boldsymbol{\theta}_1, \boldsymbol{\theta}_2, \dots, \boldsymbol{\theta}_E; \mathbf{y}_1, \mathbf{y}_2, \dots, \mathbf{y}_D)$$

$$\theta_{ij} = \mathbf{x}_{ij}\boldsymbol{\beta}_i = \beta_{i0} + x_{ij1}\beta_{i1} + \dots + x_{ijk}\beta_{ik}$$
$$\boldsymbol{\theta}_i = \mathbf{X}_i\boldsymbol{\beta}_i$$

where

- j is the observation index and N is the number of observations;
- D is the number of dependent variables ($D \geq 0$);
- $\boldsymbol{\theta}_i' = (\theta_{i1}, \theta_{i2}, \dots, \theta_{iN})$ is the ith model parameter, which we will refer to as the ith equation because it may vary from one observation to another as a linear combination of independent variables;
- i is the equation index and E is the number of equations ($E \geq 1$);
- $\boldsymbol{\beta}_i$ is the vector of coefficients for the ith equation; and
- $\mathbf{X}_i$ is the data matrix of independent variables for the ith equation, with each row representing an observation so that $\mathbf{x}_{ij}$ is the row vector of values from the jth observation of the independent variables in the ith equation.

We use θ to denote probability model parameters and β to denote coefficients when discussing log-likelihood functions in general. By probability model parameters, we mean parameters that appear directly in the density function we use to derive the likelihood function. For example, the normal density function depends on the mean μ and variance σ^2; here we have $\theta_1 = \mu$ and $\theta_2 = \sigma^2$.

The β coefficients are the weights of a linear combination of variables that determines the value of a model parameter. For example, if θ_1 represents the mean of a distribution and the mean depends on two variables that you specify, then $\theta_{1j} = \beta_1 x_{1j} + \beta_2 x_{2j} + \beta_0$. ml includes a constant term (β_0 here) in linear combinations unless you tell it not to.

We will use β to denote coefficients throughout this book; however, we will try to use the conventional parameter symbols when discussing a specific density function, such as μ and σ for the normal density instead of θ_1 and θ_2.

In the special case when

$$\ln \ell_j = \ln \ell(\theta_{1j}, \theta_{2j}, \ldots, \theta_{Ej}; y_{1j}, y_{2j}, \ldots, y_{Dj})$$

$$\ln L = \sum_{j=1}^{N} \ln \ell_j$$

that is, when 1) the log-likelihood contribution can be calculated separately for each observation and 2) the sum of the individual contributions equals the overall log likelihood, the log-likelihood function is said to meet the linear-form restrictions.

As an example, the probit likelihood function meets the linear-form restrictions because we can express the log likelihood for a single observation; and the log likelihood for the entire sample is just the sum of the observations' log-likelihood values. Conditional logistic regression, on the other hand, violates the linear-form requirements because log-likelihood values can be calculated only for groups of observations.

4.2   Equations in ml

The key to understanding ml is getting comfortable with ml's equation notation. To do this, we will continue to focus on the normal linear regression model.

The log likelihood for the normal linear regression model is

$$\ln L = \sum_{j=1}^{N} \left[\ln \phi \{ (y_j - \mathbf{x}_j \boldsymbol{\beta}) / \sigma \} - \ln \sigma \right]$$

where the point is to estimate $\boldsymbol{\beta}$, the vector of coefficients, and σ, the standard deviation of the residuals. Identifying the regression equation

$$\mu_j = E(y_j) = \mathbf{x}_j \boldsymbol{\beta}$$

we can rewrite the above log likelihood as

$$\ln L = \sum_{j=1}^{N} [\ln \phi\{(y_j - \theta_{1j})/\theta_{2j}\} - \ln \theta_{2j}]$$

$$\theta_{1j} = \mu_j = \mathbf{x}_{1j}\boldsymbol{\beta}_{1j}$$

$$\theta_{2j} = \sigma_j = \mathbf{x}_{2j}\boldsymbol{\beta}_{2j}$$

So we see that our log likelihood is a function of y_j, μ_j, and σ_j. Notice that we identify the σ_j parameter as an `ml` equation. We emphasize this so that you will think about and write your likelihood evaluator in terms of the inputs supplied by `ml`. Later, when it comes time to fit a model, you will specify the independent variables that go in each linear equation, and `ml` will produce the corresponding coefficient estimates.

After deriving the log-likelihood function from the probability model, we write a program to evaluate the log-likelihood function

```
program mynormal1_lf
        This program receives data variables for μⱼ and σⱼ as arguments and yⱼ in $ML_y1.
        It returns ln L as output.
end
```

We can then specify the particular model to be fit by typing

```
. ml model lf mynormal1_lf (y = equation for μ) (equation for σ)
```

and finally fit the model by typing

```
. ml maximize
```

Using `auto.dta`, we would fit a linear regression of `mpg` on `weight` and `displacement` by typing

```
. ml model lf mynormal1_lf (mpg = weight displacement) ()
. ml maximize
```

The part of the `ml model` statement that reads (`mpg = weight displacement`) specifies that

$$y_j = \mathtt{mpg}$$

$$\mu_j = \beta_1\mathtt{weight}_j + \beta_2\mathtt{displacement}_j + \beta_3$$

The odd-looking () at the end of the `ml model` statement specifies that $\sigma_j = \beta_4$. Later, we will discuss why we used () instead of `/sigma` as in section 3.2.

Once the `ml model` statement is specified, `ml` now has everything it needs to begin optimizing the log likelihood. It has a program, `mynormal1_lf`, that defines the log likelihood $\ln L(\mu_j, \sigma_j; y_j)$, and it has definitions for y_j, μ_j, and σ_j.

Our likelihood evaluator is

```
——————————————————————————————— begin mynormal1_lf.ado ———————
program mynormal1_lf
        version 18
        args lnfj mu sigma
        quietly replace 'lnfj' = ln(normalden($ML_y1,'mu','sigma'))
end
————————————————————————————————————— end mynormal1_lf.ado ———————
```

In our evaluator, we took advantage of the fact that the `normalden()` function allows us to specify the mean and standard deviation of the normal distribution rather than restricting us to use the "standard normal" variant. Indeed, we could have written the penultimate line of our evaluator program as

```
quietly replace 'lnf' = ln(normalden(($ML_y1 - 'mu')/'sigma')) - ln('sigma')
```

though arguably the version we used in our evaluator program is easier. On the other hand, the properties of $\phi(z)$ are so ingrained in our heads that working through the math is trivial for us.

The result from typing the above `ml` commands is

```
. sysuse auto
(1978 automobile data)

. ml model lf mynormal1_lf (mpg = weight displacement) ()

. ml maximize
Initial:      Log likelihood =       -<inf>  (could not be evaluated)
Feasible:     Log likelihood = -10383.274
Rescale:      Log likelihood = -292.89564
Rescale eq:   Log likelihood = -238.45986
Iteration 0:  Log likelihood = -238.45986  (not concave)
Iteration 1:  Log likelihood = -225.48871  (not concave)
Iteration 2:  Log likelihood = -215.66781
Iteration 3:  Log likelihood = -199.18065
Iteration 4:  Log likelihood = -195.29858
Iteration 5:  Log likelihood =     -195.24
Iteration 6:  Log likelihood =  -195.2398
Iteration 7:  Log likelihood =  -195.2398
```

```
                                              Number of obs =       74
                                              Wald chi2(2)  = 139.21
Log likelihood = -195.2398                    Prob > chi2   = 0.0000
```

mpg	Coefficient	Std. err.	z	P>|z|	[95% conf. interval]	
eq1						
weight	-.0065671	.0011424	-5.75	0.000	-.0088061	-.0043281
displacement	.0052808	.0096674	0.55	0.585	-.0136671	.0242286
_cons	40.08452	1.978738	20.26	0.000	36.20627	43.96278
eq2						
_cons	3.385282	.2782684	12.17	0.000	2.839886	3.930678

In the output above, `eq1` stands for equation 1, corresponding to μ, and `eq2` stands for equation 2, corresponding to σ. We just fit the model defined by

$$\ln L = \sum_{j=1}^{N} [\ln \phi\{(y_j - \theta_{1j})/\theta_{2j}\} - \ln \theta_{2j}]$$

$$y_j = \texttt{mpg}_j$$
$$\theta_{1j} = \mu_j = \beta_{1,1}\texttt{weight}_j + \beta_{1,2}\texttt{displacement}_j + \beta_{1,3}$$
$$\theta_{2j} = \sigma_j = \beta_{2,1}$$

because we wrote a program to calculate

$$\ln L = \sum_{j=1}^{N} [\ln \phi\{(y_j - \theta_{1j})/\theta_{2j}\} - \ln \theta_{2j}]$$

and we defined y_j, θ_{1j}, and θ_{2j} by typing

```
. ml model lf mynormal1_lf (mpg = weight displacement) ()
```

In general, any likelihood problem can be written as

$$\ln L = \ln L\{(\theta_{1j}, \theta_{2j}, \ldots, \theta_{Ej}; y_{1j}, y_{2j}, \ldots, y_{Dj}) : \ j = 1, 2, \ldots, N\}$$
$$y_{1j} = \ldots$$
$$y_{2j} = \ldots$$
$$\vdots$$
$$y_{Dj} = \ldots$$
$$\theta_{1j} = \ldots$$
$$\theta_{2j} = \ldots$$
$$\vdots$$
$$\theta_{Ej} = \ldots$$

The program you write fills in the definition of $\ln L$. The `ml model` statement you issue fills in the dots.

Equations are specified in the `ml model` statement:

```
. ml model ... (equation for θ₁) (equation for θ₂) ...
```

The syntax for specifying an equation is

$(\,[\,equation\_name:\,]\,[\,varlist_y\ =\,]\,[\,varlist_x\,]\,[\,,\ \underline{\text{noconst}}\text{ant}\,]\,)$

equation_name is optional. If you do not specify equation names, the first equation will be named `eq1`, the second `eq2`, etc. Specifying an equation name usually makes the output more readable. Had we, in our linear regression example, specified the equation for θ_{2j} as (`sigma:`) rather than (), the output would have shown the name `sigma` where it now has `eq2`.

varlist$_y$ specifies the dependent variables and must always be followed by =.

varlist$_x$ specifies the independent variables that make up the linear combination for the equation. If no independent variables are specified, the equation is constant across all observations.

`noconstant` determines whether the linear equation includes a constant term. It is a syntax error to specify this option in an equation where no independent variables are specified.

Let's forget the linear regression example for a minute. Without the context of a particular likelihood function, here are some `ml model` specifications with their translations into dependent variables and linear equations:

```
. ml model ... (foreign = mpg weight)
```

$$y_{1j} = \texttt{foreign}_j$$
$$\theta_{1j} = \beta_1 \texttt{mpg}_j + \beta_2 \texttt{weight}_j + \beta_3$$

```
. ml model ... (mpg = weight displacement) ()
```

$$y_{1j} = \texttt{mpg}_j$$
$$\theta_{1j} = \beta_{1,1}\texttt{weight}_j + \beta_{1,2}\texttt{displacement}_j + \beta_{1,3}$$
$$\theta_{2j} = \beta_{2,1}$$

```
. ml model ... (mpg = weight displacement) (price)
```

$$y_{1j} = \texttt{mpg}_j$$
$$\theta_{1j} = \beta_{1,1}\texttt{weight}_j + \beta_{1,2}\texttt{displacement}_j + \beta_{1,3}$$
$$\theta_{2j} = \beta_{2,1}\texttt{price}_j + \beta_{2,2}$$

```
. ml model ... (mpg = weight displacement, nocons) (price)
```

$$y_{1j} = \texttt{mpg}_j$$
$$\theta_{1j} = \beta_{1,1}\texttt{weight}_j + \beta_{1,2}\texttt{displacement}_j$$
$$\theta_{2j} = \beta_{2,1}\texttt{price}_j + \beta_{2,2}$$

```
. ml model ... (mpg = weight displacement) (price, nocons)
```

$$y_{1j} = \texttt{mpg}_j$$
$$\theta_{1j} = \beta_{1,1}\texttt{weight}_j + \beta_{1,2}\texttt{displacement}_j + \beta_{1,3}$$
$$\theta_{2j} = \beta_{2,1}\texttt{price}_j$$

```
. ml model ... (mpg = weight displacement) (price = weight) ()
```

$$y_{1j} = \texttt{mpg}_j$$
$$y_{2j} = \texttt{price}_j$$
$$\theta_{1j} = \beta_{1,1}\texttt{weight}_j + \beta_{1,2}\texttt{displacement}_j + \beta_{1,3}$$
$$\theta_{2j} = \beta_{2,1}\texttt{weight}_j + \beta_{2,2}$$
$$\theta_{3j} = \beta_{3,1}$$

```
. ml model ... (mpg price = weight displacement) (weight) ()
```

$$y_{1j} = \texttt{mpg}_j$$
$$y_{2j} = \texttt{price}_j$$
$$\theta_{1j} = \beta_{1,1}\texttt{weight}_j + \beta_{1,2}\texttt{displacement}_j + \beta_{1,3}$$
$$\theta_{2j} = \beta_{2,1}\texttt{weight}_j + \beta_{2,2}$$
$$\theta_{3j} = \beta_{3,1}$$

This is how `ml` equations work. Note especially the last two examples:

```
. ml model ... (mpg = weight displacement) (price = weight) ()
. ml model ... (mpg price = weight displacement) (weight) ()
```

Both resulted in the same definitions of y_{1j}, y_{2j}, θ_{1j}, θ_{2j}, and θ_{3j}. When you specify the dependent variables, it really does not matter in which equation they appear. The first y variable is y_1, the second is y_2, and so on.

In our example above, we never bothered to specify equation names. Let's go back and show the result of specifying them.

```
. sysuse auto
(1978 automobile data)

. ml model lf mynormal1_lf (mu: mpg = weight displacement) (sigma:)

. ml maximize
Initial:      Log likelihood =      -<inf>  (could not be evaluated)
Feasible:     Log likelihood = -10383.274
Rescale:      Log likelihood = -292.89564
Rescale eq:   Log likelihood = -238.45986
Iteration 0:  Log likelihood = -238.45986  (not concave)
Iteration 1:  Log likelihood = -225.48871  (not concave)
Iteration 2:  Log likelihood = -215.66781
Iteration 3:  Log likelihood = -199.18065
Iteration 4:  Log likelihood = -195.29858
Iteration 5:  Log likelihood =    -195.24
Iteration 6:  Log likelihood = -195.2398
Iteration 7:  Log likelihood = -195.2398
```

				Number of obs = 74
				Wald chi2(2) = 139.21
Log likelihood = -195.2398				Prob > chi2 = 0.0000

mpg	Coefficient	Std. err.	z	P>|z|	[95% conf. interval]
mu					
weight	-.0065671	.0011424	-5.75	0.000	-.0088061 -.0043281
displacement	.0052808	.0096674	0.55	0.585	-.0136671 .0242286
_cons	40.08452	1.978738	20.26	0.000	36.20627 43.96278
sigma					
_cons	3.385282	.2782684	12.17	0.000	2.839886 3.930678

Now, instead of `eq1` and `eq2`, our equation names are labeled `mu` and `sigma`.

Actually, specifying an equation name does more than just change how the output is labeled. If—after we fit a model—we wish to refer to the coefficient on `weight` in an expression, we would refer to it as `[equation_name]_b[weight]`. For example, after the model we just fit, we would type `[mu]_b[weight]`. In fact, we would probably simply type `_b[weight]` because Stata assumes we are referring to the first equation when an equation is not specified. Thus we only need to specify the equation name to refer to a coefficient from an equation other than the first equation.

There is another way `ml model` will let us specify equations that are just scalar parameters. Instead of typing (`equation_name:`), which is one way of specifying a named constant equation, we can type /*equation_name*. Thus in our linear regression example, we could specify the model as

```
. ml model lf mynormal1_lf (mu: mpg = weight displacement) /sigma
```

rather than

```
. ml model lf mynormal1_lf (mu: mpg = weight displacement) (sigma:)
```

or

```
. ml model lf mynormal1_lf (mpg = weight displacement) ()
```

If we use this slash notation to specify σ, we would use the syntax _b[/sigma] to refer to its estimated value. We demonstrate the slash notation in section 4.5. See whether you can spot any differences in the output. In section 4.7, we discuss a few differences among the various ways of specifying σ that more advanced programmers will want to bear in mind.

Now that we have explained ml equations, let's show their real power. In the last chapter, we lamented that because our method-gf0 linear regression likelihood evaluator worked directly with the β coefficients and variables of our model, changing the specification of our regression required us to modify our likelihood evaluator. With a method-lf evaluator, we instead express our likelihood function in terms of the θ parameters. As the previous ml model specifications we considered indicate, all we need to do to change the definition of a θ parameter is change the variables specified in the ml model command.

We have shown that

———————————————————————— begin mynormal1_lf.ado ————————

```
program mynormal1_lf
        version 18
        args lnfj mu sigma
        quietly replace 'lnfj' = ln(normalden($ML_y1,'mu','sigma'))
end
```

———————————————————————————— end mynormal1_lf.ado ————————

```
. ml model lf mynormal1_lf (mu: mpg = weight displacement) (sigma:)
. ml maximize
```

will produce the maximum likelihood estimates of

$$\text{mpg}_j = \beta_1 \text{weight}_j + \beta_2 \text{displacement}_j + \beta_3 + \epsilon_j, \quad \epsilon_j \sim N(0, \sigma^2)$$

Without changing our likelihood evaluator, we can just as easily fit a model in which σ varies with price; that is, we can fit a linear regression model with conditional heteroskedasticity. To illustrate, we will fit $\sigma_j = \beta_{2,1} \text{price}_j + \beta_{2,2}$:

```
. sysuse auto
(1978 automobile data)

. ml model lf mynormal1_lf (mu: mpg = weight displacement) (sigma: price)

. ml maximize, nolog
Initial:       Log likelihood =      -<inf>  (could not be evaluated)
Feasible:      Log likelihood = -10383.274
Rescale:       Log likelihood = -292.89564
Rescale eq:    Log likelihood = -238.45986
                                              Number of obs =       74
                                              Wald chi2(2)  = 126.71
Log likelihood = -195.14254                   Prob > chi2   = 0.0000
```

mpg	Coefficient	Std. err.	z	P>|z|	[95% conf. interval]	
mu						
weight	-.0065258	.0011369	-5.74	0.000	-.008754	-.0042975
displacement	.0041644	.0099923	0.42	0.677	-.01542	.0237489
_cons	40.1942	1.976217	20.34	0.000	36.32089	44.06752
sigma						
price	.0000438	.0001061	0.41	0.680	-.0001641	.0002517
_cons	3.11326	.6915815	4.50	0.000	1.757785	4.468735

4.3   Likelihood-evaluator methods

ml provides a total of 12 types of likelihood-evaluator functions that can be grouped
into four families. Each evaluator has both a short name as well as a longer, more
descriptive name. We tend to use the short names exclusively, though you may specify
either the short or the long name for each evaluator. Table 4.1 classifies the evaluators:

Table 4.1. Likelihood-evaluator family names

Family	Short name	Long name
Linear form	lf	linearform
	lf0	linearform0
	lf1	linearform1
	lf2	linearform2
	lf1debug	linearform1debug
	lf2debug	linearform2debug
Derivative	d0	derivative0
	d1	derivative1
	d2	derivative2
	d1debug	derivative1debug
	d2debug	derivative2debug
General form	gf0	generalform0

Evaluators whose names end in a number indicate the analytic derivatives you are to provide: 0 indicates no derivatives are required; 1 indicates first derivatives are required; and 2 indicates both first and second derivatives are required. The suffix debug indicates the evaluator is used for debugging purposes. These evaluators compute derivatives numerically and provide a report comparing those derivatives to the ones computed using the analytic formulas you provided; see chapter 8 for details and examples.

Among the four families,

1. Use method lf when the likelihood function meets the linear-form restrictions and you do not wish to program first or second derivatives. This is the most popular method among users and is often the only method one ever needs.

2. Also use methods lf0, lf1, and lf2 when the likelihood function meets the linear-form restrictions. In exchange for a slightly more complicated syntax, methods lf1 and lf2 allow you to provide first and second derivatives, respectively. lf0 does not require you to provide any derivatives and is used as a stepping stone toward writing lf1 and lf2. If you have no intention of providing derivatives, use method lf instead because it has a simpler syntax but is otherwise comparable to method lf0.

3. Use d-family evaluators with any likelihood function—not just those meeting the linear-form restrictions. Method d0 does not require you to program any derivatives. Method d1 requires you to program first derivatives of the overall likelihood function with respect to model parameters. Method d2 requires you to program first and second derivatives of the overall likelihood function with respect to the model parameters. Robust standard errors and survey data support are not available with d-family evaluators.

4. Use `gf0` evaluators similarly to methods `lf0`, except that the log-likelihood function for `gf0` evaluators may exist only at the group level rather the observation level. (This also impacts how the gradient and Hessian are computed.) The overall log likelihood is equal to the sum of the group-level log-likelihood functions. Virtually every model you are likely to encounter can be fit using one of the other evaluator families. The only exceptions we can think of are panel-data estimators and the Cox proportional-hazards model, and unless robust standard errors are required, these models can be implemented using d-family evaluators, as we illustrate in chapter 7.

There is a clear tradeoff between speed of programming and speed of execution. Evaluators that do not require you to specify first or second derivatives are easier to write, but `ml` will take longer to execute because it must compute derivatives numerically. Evaluators that require both first and second derivatives execute most quickly, though obtaining the second derivatives of complicated likelihood functions can be difficult. First derivatives are generally easier to obtain than second derivatives, and providing just first derivatives will typically still have a noticeable impact on speed.

Although the numerical derivative engine in `ml` is quite good, analytic derivatives are still more accurate. Therefore, evaluators that require second derivatives are most accurate, followed by those that require first derivatives; evaluators that compute all derivatives numerically tend to be least accurate. An exception to this rule is the method-`lf` evaluator, which we have carefully programmed to produce accurate results even though it does not require you to specify any derivatives.

There is also a clear tradeoff between the ease of use of `mlexp` and the flexibility of `ml`. `mlexp` can fit the same models as `ml` methods `lf`, `lf0`, and `lf1`, and it requires no programming. On the other hand, implementing all but the simplest likelihood functions using substitutable expressions is arguably more tedious than just writing a simple likelihood evaluator program. For a quick one-off solution, we are tempted to use `mlexp`. If our goal is to develop a new estimation command that we, and perhaps others, will use often, then `ml` is the right tool to use. If your model requires a d-family or gf0 evaluator, you must use `ml`.

In general, we heartily recommend method `lf` because it is easy to program, quick to execute, and accurate. Use one of the other methods only when method `lf` does not provide the capabilities you need.

You specify the method and the name of the program you write to evaluate the likelihood function in the `ml model` statement:

ml model *method progname* ...

In the next several chapters, we discuss each of the four families of likelihood evaluators in detail. However, in the previous chapter, we introduced maximum likelihood estimation using method-`gf0` evaluators, and in this chapter, we have used a method-`lf` evaluator for linear regression. What we want to emphasize right now is that this method-`lf` evaluator is short but entirely functional.

──────────────────────────── begin mynormal1_lf.ado ────────────

```
program mynormal1_lf
        version 18
        args lnfj mu sigma
        quietly replace 'lnfj' = ln(normalden($ML_y1,'mu','sigma'))
end
```

──────────────────────────── end mynormal1_lf.ado ────────────

In the rest of this chapter, we will continue using this example, and we want to direct your attention to the many things `ml` will do for you once you have written your evaluator and without your having to write more code.

4.4   Tools for the ml programmer

As you write and improve your likelihood evaluator, the following utilities may be helpful:

- `ml check` helps verify that your program has no syntax errors.
- `ml init` allows you to specify initial values.
- `ml search` allows you to hunt for initial values.
- `ml plot` graphs the log-likelihood function and helps you find initial values.

What you do about initial values is up to you. In most cases, `ml maximize` will succeed even if the log-likelihood function cannot be evaluated at the default initial values, which are a vector of zeros. Providing better initial values usually speeds maximization.

The following utilities can be used after `ml maximize` finishes fitting the model:

- `ml graph` graphs the log-likelihood iteration values.
- `ml display` redisplays the estimation results.

4.5   Common ml options

The syntax and options for `ml` are described in appendix B. The following sections give an overview of the most commonly used options of the `ml model` statement.

4.5.1   Subsamples

If you wish to fit your model on a subsample of the data, specify the restriction in the
ml model statement using if or in. Continuing with our linear regression example,
suppose that we want to fit the model for data in which trunk>8. Using if, we type

```
. sysuse auto
(1978 automobile data)
. ml model lf mynormal1_lf (mpg = weight displacement) /sigma if trunk>8
. ml maximize
Initial:      Log likelihood =      -<inf>  (could not be evaluated)
Feasible:     Log likelihood = -8455.0133
Rescale:      Log likelihood = -252.88251
Rescale eq:   Log likelihood = -203.43609
Iteration 0:  Log likelihood = -203.43609  (not concave)
Iteration 1:  Log likelihood = -194.40235  (not concave)
Iteration 2:  Log likelihood = -189.62404
Iteration 3:  Log likelihood =  -173.4495
Iteration 4:  Log likelihood = -169.91639
Iteration 5:  Log likelihood = -169.85719
Iteration 6:  Log likelihood = -169.85702
Iteration 7:  Log likelihood = -169.85702
```

				Number of obs =	64
				Wald chi2(2) =	109.55
Log likelihood = -169.85702				Prob > chi2 =	0.0000

mpg	Coefficient	Std. err.	z	P>|z|	[95% conf. interval]	
weight	-.0064416	.0011849	-5.44	0.000	-.008764	-.0041191
displacement	.0062196	.0099341	0.63	0.531	-.0132509	.0256901
_cons	39.38041	2.134564	18.45	0.000	35.19674	43.56408
/sigma	3.438645	.3039361	11.31	0.000	2.842941	4.034349

You can specify one or both of if and in in the ml model statement.

4.5.2 Weights

If you wish to perform weighted estimation, specify the weights in the `ml model` statement. For example,

```
. sysuse auto
(1978 automobile data)
. ml model lf mynormal1_lf (mpg = weight displacement) /sigma [aweight=trunk]
. ml maximize, nolog
Initial:       Log likelihood =      -<inf>  (could not be evaluated)
Feasible:      Log likelihood =    -9382.79
Rescale:       Log likelihood = -288.88285
Rescale eq:    Log likelihood = -233.20956
                                                    Number of obs =       74
                                                    Wald chi2(2)  = 138.35
Log likelihood = -193.1554                          Prob > chi2   = 0.0000
```

mpg	Coefficient	Std. err.	z	P>|z|	[95% conf. interval]	
weight	-.0064548	.0011702	-5.52	0.000	-.0087484	-.0041612
displacement	.0060624	.0096706	0.63	0.531	-.0128917	.0250165
_cons	39.48953	2.072494	19.05	0.000	35.42751	43.55154
/sigma	3.291257	.2705397	12.17	0.000	2.761009	3.821505

`ml` allows `fweight`s, `pweight`s, `aweight`s, and `iweight`s; however, you must decide if a given type of weight is meaningful for your probability model. Specifying `pweight`s implies robust estimates of variance. See [U] **11.1.6 weight** for more information on weights.

4.5.3   OPG estimates of variance

The theory and formulas for computing variance estimates using the outer product of gradients (OPG) were explained in section 1.2.3. To obtain OPG variance estimates for the estimated coefficients, specify the vce(opg) option on the ml model statement. For example,

```
. sysuse auto
(1978 automobile data)
. ml model lf mynormal1_lf (mpg = weight displacement) /sigma, vce(opg)
. ml maximize, nolog
Initial:       Log likelihood =     -<inf>  (could not be evaluated)
Feasible:      Log likelihood = -10383.274
Rescale:       Log likelihood = -292.89564
Rescale eq:    Log likelihood = -238.45986
                                                     Number of obs =      74
                                                     Wald chi2(2)  = 128.47
Log likelihood = -195.2398                           Prob > chi2   = 0.0000
```

mpg	Coefficient	OPG std. err.	z	P>|z|	[95% conf. interval]	
weight	-.0065671	.0014932	-4.40	0.000	-.0094938	-.0036404
displacement	.0052808	.0132686	0.40	0.691	-.0207252	.0312868
_cons	40.08452	2.292743	17.48	0.000	35.59083	44.57822
/sigma	3.385282	.2181536	15.52	0.000	2.957709	3.812855

These are virtually the same results as were obtained on page 56, where we did not specify vce(opg). The differences are that the standard errors are different, albeit similar, and that the $\chi^2(2)$ test statistic reported in the header changes from 139.21 to 128.47. This test statistic is for the Wald test that all coefficients in the first equation (except the intercept) are zero, and it is based on the variance estimate chosen.

Method lf allows obtaining OPG variance estimates without your having to make any change to your likelihood-evaluation program—you need only specify the vce(opg) option on the ml model statement. As we will discuss later, you can also specify vce(opg) when you use lf-family evaluators. The OPG variance estimate is not available with d-family evaluators.

4.5.4   Robust estimates of variance

The theory and formulas for computing robust estimates of variance were explained in section 1.2.4. To obtain robust variance estimates for the estimated coefficients, specify the vce(robust) option on the ml model statement. For example,

```
. sysuse auto
(1978 automobile data)

. ml model lf mynormal1_lf (mpg = weight displacement) /sigma, vce(robust)

. ml maximize, nolog
Initial:     Log pseudolikelihood =     -<inf>  (could not be evaluated)
Feasible:    Log pseudolikelihood = -10383.274
Rescale:     Log pseudolikelihood = -292.89564
Rescale eq:  Log pseudolikelihood = -238.45986
                                              Number of obs =       74
                                              Wald chi2(2)  = 111.39
Log pseudolikelihood = -195.2398              Prob > chi2   = 0.0000
```

mpg	Coefficient	Robust std. err.	z	P>|z|	[95% conf. interval]	
weight	−.0065671	.0009195	−7.14	0.000	−.0083692	−.004765
displacement	.0052808	.0073381	0.72	0.472	−.0091017	.0196633
_cons	40.08452	2.045516	19.60	0.000	36.07538	44.09366
/sigma	3.385282	.4544255	7.45	0.000	2.494624	4.275939

These are virtually the same results as reported on pages 56 and 68. The differences are in the standard errors and the $\chi^2(2)$ test—the test that the coefficients on weight and displacement are simultaneously zero—as shown in tables 4.2 and 4.3, respectively.

Table 4.2. Comparison of estimated standard errors

	Conventional	OPG	Robust
weight	0.0011	0.0015	0.0009
displacement	0.0097	0.0133	0.0073
_cons	1.9787	2.2927	2.0455
$\widehat{\sigma}$	0.2783	0.2182	0.4544

Table 4.3. Comparison of the Wald tests

	Conventional	OPG	Robust
Wald $\chi^2(2)$	139.21	128.47	111.39

As with OPG variance estimates, robust estimates are available for all likelihood evaluators except those in the d family. To obtain cluster–robust standard errors instead, just specify vce(cluster *varname*), where *varname* identifies the clusters or groups. Cluster–robust variance estimates relax the assumption that each observation is independently distributed and only assumes that observations in different clusters are independent. Obtaining cluster–robust variance estimates requires no additional programming beyond what is required to obtain robust variance estimates.

If you specify pweights, ml uses vce(robust) automatically.

4.5.5   Survey data

The svy option of ml model causes ml to use the survey design characteristics set by svyset.

```
. use http://www.stata-press.com/data/r11/nhanes2a

. svyset [pweight=leadwt], strata(stratid) psu(psuid)
Sampling weights: leadwt
             VCE: linearized
     Single unit: missing
        Strata 1: stratid
  Sampling unit 1: psuid
           FPC 1: <zero>

. ml model lf mynormal1_lf (loglead = age female) /sigma, svy

. ml maximize
Initial:      Log pseudolikelihood =      -<inf>  (could not be evaluated)
Feasible:     Log pseudolikelihood = -1.040e+09
Rescale:      Log pseudolikelihood = -1.892e+08
Rescale eq:   Log pseudolikelihood = -1.324e+08
Iteration 0:  Log pseudolikelihood = -1.324e+08  (not concave)
Iteration 1:  Log pseudolikelihood =  -84998336  (not concave)
Iteration 2:  Log pseudolikelihood =  -61424477
Iteration 3:  Log pseudolikelihood =  -50149194
Iteration 4:  Log pseudolikelihood =  -48525605
Iteration 5:  Log pseudolikelihood =  -48522256
Iteration 6:  Log pseudolikelihood =  -48522255
```

Number of strata = 31	Number of obs = 4,948
Number of PSUs = 62	Population size = 112,915,016
	Design df = 31
	$F(2, 30)$ = 625.61
	Prob > F = 0.0000

loglead	Coefficient	Linearized std. err.	t	P>|t|	[95% conf. interval]	
age	.0024985	.0004326	5.78	0.000	.0016162	.0033807
female	-.3648611	.0101531	-35.94	0.000	-.3855684	-.3441537
_cons	2.663709	.0295542	90.13	0.000	2.603432	2.723985
/sigma	.3718685	.0109367	34.00	0.000	.349563	.3941741

In addition to the svy option, ml model also accepts the subpop() option, which is not a setting of the svyset command. For more information on svyset, see [SVY] **svyset**, and for information on survey data in general, see the *Survey Data Reference Manual*.

See chapter 14 for details on writing estimation commands with support for the svy prefix.

4.5.6 Constraints

To fit your model with constraints on the coefficients, use the constraints() option on the ml model statement. For example,

```
. sysuse auto
(1978 automobile data)
. constraint 1 weight = -displacement
. ml model lf mynormal1_lf (mpg = weight displacement) /sigma, constraints(1)
. ml maximize, nolog
Initial:     Log likelihood =      -<inf>  (could not be evaluated)
Feasible:    Log likelihood = -10383.274
Rescale:     Log likelihood = -292.89564
Rescale eq:  Log likelihood = -238.45986
```

```
                                              Number of obs =       74
                                              Wald chi2(1)  = 139.15
Log likelihood = -195.25083                   Prob > chi2   = 0.0000
 ( 1)  [eq1]weight + [eq1]displacement = 0
```

mpg	Coefficient	Std. err.	z	P>|z|	[95% conf. interval]	
weight	-.0067142	.0005692	-11.80	0.000	-.0078298	-.0055986
displacement	.0067142	.0005692	11.80	0.000	.0055986	.0078298
_cons	40.24594	1.653851	24.33	0.000	37.00445	43.48743
/sigma	3.385786	.2783099	12.17	0.000	2.840309	3.931264

Constraints are defined using the constraint command; see [R] **constraint**. For more examples of how to define constraints, see [R] **cnsreg**, [R] **mlogit**, and [R] **reg3**.

4.5.7 Choosing among the optimization algorithms

To fit your model using one of the three optimization algorithms described in section 1.3.2, specify the algorithm in the technique() option on the ml model statement.

- technique(nr) specifies the Newton–Raphson algorithm (the default).
- technique(bhhh) specifies the Berndt–Hall–Hall–Hausman (BHHH) algorithm.
- technique(dfp) specifies the Davidon–Fletcher–Powell (DFP) algorithm.
- technique(bfgs) specifies the Broyden–Fletcher–Goldfarb–Shanno (BFGS) algorithm.

Because the BHHH algorithm uses the (observation-level) OPG, it cannot be used with d-family evaluators. All of these algorithms can be used with the other three evaluator families.

Here is an example using the BHHH algorithm:

```
. sysuse auto
(1978 automobile data)
. ml model lf mynormal1_lf (mpg = weight displacement) /sigma, technique(bhhh)
. ml maximize
Initial:      Log likelihood =      -<inf>  (could not be evaluated)
Feasible:     Log likelihood = -10383.274
Rescale:      Log likelihood = -292.89564
Rescale eq:   Log likelihood = -238.45986
Iteration 0:  Log likelihood = -238.45986
(output omitted)
Iteration 22: Log likelihood =  -195.2398
Iteration 23: Log likelihood =  -195.2398
```

			Number of obs =	74
			Wald chi2(2) =	128.52
Log likelihood = -195.2398			Prob > chi2 =	0.0000

mpg	Coefficient	OPG std. err.	z	P>|z|	[95% conf. interval]	
weight	−.0065675	.0014926	−4.40	0.000	−.0094929	−.0036421
displacement	.0052906	.0132628	0.40	0.690	−.0207041	.0312852
_cons	40.0837	2.291817	17.49	0.000	35.59182	44.57558
/sigma	3.384624	.2180343	15.52	0.000	2.957285	3.811964

The BHHH algorithm produces the OPG variance estimate by default. For other algorithms, the default is to use the inverse of the negative Hessian. If you use the BHHH algorithm to fit your model and want the Hessian-based variance estimate, specify technique(bhhh) and vce(oim) together.

```
. sysuse auto
(1978 automobile data)

. ml model lf mynormal1_lf (mpg = weight displacement) /sigma, tech(bhhh)
> vce(oim)

. ml maximize
Initial:       Log likelihood =      -<inf>  (could not be evaluated)
Feasible:      Log likelihood = -10383.274
Rescale:       Log likelihood = -292.89564
Rescale eq:    Log likelihood = -238.45986
Iteration 0:   Log likelihood = -238.45986
(output omitted )
Iteration 22: Log likelihood =  -195.2398
Iteration 23: Log likelihood =  -195.2398
                                                Number of obs =      74
                                                Wald chi2(2)  = 139.24
Log likelihood = -195.2398                      Prob > chi2   = 0.0000
```

mpg	Coefficient	Std. err.	z	P>|z|	[95% conf. interval]	
weight	-.0065675	.0011421	-5.75	0.000	-.008806	-.0043289
displacement	.0052906	.0096656	0.55	0.584	-.0136536	.0242348
_cons	40.0837	1.978354	20.26	0.000	36.2062	43.96121
/sigma	3.384624	.2781334	12.17	0.000	2.839493	3.929756

The technique() option allows you to specify multiple algorithms, and in that case, ml will cycle between the specified algorithms. For example, specifying technique(bhhh nr) will cause ml to use BHHH for five iterations, Newton–Raphson for the next five, BHHH for the five after that, and so on.

```
. sysuse auto
(1978 automobile data)

. ml model lf mynormal1_lf (mpg = weight displacement) /sigma, tech(bhhh nr)

. ml maximize
Initial:      Log likelihood =      -<inf>  (could not be evaluated)
Feasible:     Log likelihood = -10383.274
Rescale:      Log likelihood = -292.89564
Rescale eq:   Log likelihood = -238.45986
(setting technique to bhhh)
Iteration 0:  Log likelihood = -238.45986
Iteration 1:  Log likelihood = -198.28587
Iteration 2:  Log likelihood = -195.80276
Iteration 3:  Log likelihood = -195.63467
Iteration 4:  Log likelihood = -195.54062
(switching technique to nr)
Iteration 5:  Log likelihood =  -195.3671
Iteration 6:  Log likelihood = -195.23992
Iteration 7:  Log likelihood =  -195.2398
Iteration 8:  Log likelihood =  -195.2398
```

				Number of obs =		74
				Wald chi2(2)	= 139.21	
Log likelihood = -195.2398				Prob > chi2	= 0.0000	

mpg	Coefficient	Std. err.	z	P>|z|	[95% conf. interval]	
weight	-.0065671	.0011424	-5.75	0.000	-.0088061	-.0043281
displacement	.0052808	.0096674	0.55	0.585	-.0136671	.0242286
_cons	40.08452	1.978739	20.26	0.000	36.20627	43.96278
/sigma	3.385282	.2782685	12.17	0.000	2.839886	3.930678

Switching between techniques is an excellent way of finding the maximum of a difficult-to-maximize function. By default, `ml` performs five iterations of each algorithm before switching to the next. You can change the number of iterations performed for each algorithm by specifying your own count after each named algorithm in `technique()`. For example,

```
. sysuse auto
(1978 automobile data)
. ml model lf mynormal1_lf (mpg = weight displace) /sigma, tech(bhhh 2 nr 5)
. ml maximize
Initial:      Log likelihood =      -<inf>  (could not be evaluated)
Feasible:     Log likelihood = -10383.274
Rescale:      Log likelihood = -292.89564
Rescale eq:   Log likelihood = -238.45986
(setting technique to bhhh)
Iteration 0:  Log likelihood = -238.45986
Iteration 1:  Log likelihood = -198.28587
(switching technique to nr)
Iteration 2:  Log likelihood = -195.80276
Iteration 3:  Log likelihood = -195.24154
Iteration 4:  Log likelihood =  -195.2398
Iteration 5:  Log likelihood =  -195.2398
```

					Number of obs	=	74
					Wald chi2(2)	=	139.21
Log likelihood = -195.2398					Prob > chi2	=	0.0000

mpg	Coefficient	Std. err.	z	P>|z|	[95% conf. interval]	
weight	-.0065671	.0011424	-5.75	0.000	-.0088061	-.0043281
displacement	.0052808	.0096674	0.55	0.585	-.0136671	.0242286
_cons	40.08452	1.978738	20.26	0.000	36.20627	43.96278
/sigma	3.385282	.2782684	12.17	0.000	2.839886	3.930678

4.6 Maximizing your own likelihood functions

Here is an outline of the steps you will take when writing and debugging your likelihood evaluator.

1. Derive the log-likelihood function from your probability model.

2. Write a likelihood evaluator, the program that calculates the log-likelihood values. Depending on the method, you may have to include derivative calculations.

   See chapter 5 for the method-lf evaluator, chapter 6 for lf-family evaluators, and chapter 7 for d-family evaluators.

3. Specify the method (such as lf, d0, or lf2), the name of your program (*myprog*), and the model to be fit in the ml model statement.

   . ml model *method myprog* (*equation for* θ_1) (*equation for* θ_2) ...

   At this point, you also specify options that affect how the model is fit. See section 4.5 for commonly specified options.

4. Verify that the program you have written is free of syntax errors and other common problems with

   `. ml check`

   See section 8.1 for details, but the output produced by `ml check` is easy to interpret.

5. Optionally, specify starting values using `ml init`; see section 9.3. We do not recommend this unless you have an analytic approximation that gives good starting values. Otherwise, use `ml search`.

6. Optionally, look for starting values using `ml plot`; see section 9.2. This is not as useful as `ml search`.

7. Optionally—and we recommend this—tell `ml` to search for starting values with

   `. ml search`

   See section 9.1.

8. Obtain maximum likelihood estimates with

   `. ml maximize`

   See chapter 10.

9. To graph the convergence, type

   `. ml graph`

   See section 11.1.

10. If you later wish to redisplay the results from the most recently fit model, type

    `. ml display`

    See section 11.2.

11. If you wish to add a new command to Stata to perform this kind of estimation, see chapter 13.

4.7 Appendix: More about scalar parameters

In our classical normal linear regression model, we have the scalar parameter σ, and in the main text, we showed three alternative ways of informing `ml model` about that parameter:

1. `ml model lf mynormal1_lf (mpg = weight displacement) ()`
2. `ml model lf mynormal1_lf (mpg = weight displacement) (sigma:)`
3. `ml model lf mynormal1_lf (mpg = weight displacement) /sigma`

Alternatives 1 and 2 are essentially the same, except that the latter names the equation `sigma` rather than `eq2`; everything we say about alternative 2 below is true of alternative 1 aside from the different name. More substantive are the differences between alternatives 2 and 3, and here we discuss some of them.

Internally, ml uses the concept of a "free parameter" for scalar parameters such as σ. By definition, a free parameter is one that is not part of a linear combination. In contrast, the ml equation that contains our regressors is known as a "linear-form" equation because it is a linear combination of variables (and a constant term). In alternative 2, we are declaring σ to be a linear-form equation with just a constant term, while in alternative 3, we are declaring σ to be a free parameter. The implementation details of free parameters are irrelevant to users of ml, but they do affect the output and stored results it produces.

Regarding output, when we use alternative 2, the output will show our estimate of σ like the following:

sigma						
_cons	3.385282	.2782684	12.17	0.000	2.839886	3.930678

This more expressive output allows for the fact that the equation sigma may include regressors; see the output on page 62 where we fit a model with heteroskedasticity conditional on price. In contrast, when we use alternative 3, the output will be slightly more compact because sigma is a free parameter that will never depend on regressors:

/sigma	3.385282	.2782684	12.17	0.000	2.839886	3.930678

This discussion illustrates one possible limitation of free parameters you may want to consider when implementing your estimator. If you use alternative 2, it is easy to generalize your model and turn scalar parameters into functions of variables, as we did to model conditional heteroskedasticity. If you use alternative 3, you no longer have that flexibility.

In section 4.2, we discussed the _b syntax for obtaining the values of estimated parameters. If you use alternative 2 to specify your model, then after typing ml maximize, you can obtain the estimate of σ using the syntax [sigma]_b[_cons], just as you obtain the value of the constant term in any other equation. In contrast, when you specify your model using alternative 3, then you must use the syntax _b[/sigma]. In old versions of Stata, you could use either syntax regardless of whether you used alternative 2 or alternative 3, but that is no longer the case: you must use the _b syntax corresponding to the way in which you specify σ.

Underlying the difference in output and the need to use different _b syntaxes is the fact that alternatives 2 and 3 result in different column labels for e(b), the system matrix that stores the estimated parameter vector. With alternative 2, where σ is treated as a constant-only equation, e(b) shows we have two equations:

```
e(b)[1,4]
              eq1:           eq1:           eq1:         sigma:
            weight   displacement         _cons          _cons
   y1    -.00656711       .00528078     40.084522      3.3852817
```

With alternative 3, where σ is treated as a free parameter, `e(b)` is labeled differently:

```
e(b)[1,4]
              eq1:           eq1:           eq1:              /:
            weight   displacement         _cons           sigma
   y1    -.00656711       .00528078     40.084522      3.3852817
```

The special / where an equation name would normally be indicates that the parameter sigma is a free parameter. If your model has multiple free parameters, all of them will be listed with / in place of the equation name. The row and column names of `e(V)`, the estimated variance–covariance matrix, will also follow this naming convention. If you intend to work with the column names of `e(b)`, there are several macro functions related to matrices that retrieve the number and names of linear-form equations and the number of free parameters. See [P] **macro**.

❏ Technical note

In the documentation for ml, you will notice that there is an option **freeparm** available when specifying equations. The syntax

```
. ml model lf mynormal1_lf (mpg = weight displacement) (sigma:, freeparm)
```

is equivalent to

```
. ml model lf mynormal1_lf (mpg = weight displacement) /sigma
```

The **freeparm** option tells ml that this constant-only equation is a free parameter and that it should be treated as such. We recommend using the slash notation because it requires less typing and, at least to us, is a more natural way of declaring a scalar parameter.

❏

5   Method lf

Method `lf` is for use when the likelihood function meets the linear-form restrictions and you do not wish to provide first and second derivatives. Use this method whenever possible. Method `lf` is fast, accurate, and easy to use.

The likelihood function is assumed to be of the form

$$\ln \ell_j = \ln \ell_j(\theta_{1j}, \theta_{2j}, \dots, \theta_{Ej}; y_{1j}, y_{2j}, \dots, y_{Dj})$$

$$\ln L = \sum_{j=1}^{N} \ln \ell_j$$

$$\theta_{1j} = \mathbf{x}_{1j}\boldsymbol{\beta}_1$$

$$\theta_{2j} = \mathbf{x}_{2j}\boldsymbol{\beta}_2$$

$$\vdots$$

$$\theta_{Ej} = \mathbf{x}_{Ej}\boldsymbol{\beta}_E$$

where j indexes observations. The method-`lf` evaluator is required to calculate $\ln \ell_j$, the log likelihood, observation by observation. The outline for a method-`lf` likelihood evaluator is

```
program myprog
        version 18
        args lnfj theta1 theta2 ...
        // if you need to create any intermediate results
        tempvar tmp1 tmp2 ...
        quietly generate double 'tmp1' = ...
        ...
        quietly replace 'lnfj' = ...
end
```

You specify that you are using method `lf` with program *myprog* on the `ml model` statement,

```
. ml model lf myprog ...
```

which you issue after you have defined the likelihood-evaluation program.

1. `ml` places the names of any dependent variables in global macros; the first dependent variable is $ML_y1, the second is $ML_y2, ... Use $ML_y1, $ML_y2, ... just as you would any existing variable in the data.

2. `ml` supplies a variable for each equation specified in the `ml model` statement. Each variable contains the values of the linear combination of independent variables and current regression coefficients for the respective equation. Thus '*theta1*' contains the linear combination for the first equation, '*theta2*' contains the linear combination for the second equation, and so on.

   Do not change the contents of variables '*theta1*', '*theta2*', ... That is, never write

       `. replace '`*theta1*`' = ...`

3. Final results are stored in '`lnfj`', which is just a double-precision variable that `ml` creates for you. When `ml` calls your program, '`lnfj`' contains missing values.

4. Any temporary variables you create for intermediate results must be `doubles`. Type

       `. generate double` *name* `= ...`

   Do not omit the word `double`.

Using method `lf` is simply a matter of writing the likelihood function for an observation, identifying the parameters that will make up `ml`'s equations, and coding it in a program.

5.1   The linear-form restrictions

Method `lf` requires that the physical observations in the dataset correspond to independent pieces of the likelihood function:

$$\ln L = \ln \ell_1 + \ln \ell_2 + \cdots + \ln \ell_N$$

This restriction is commonly (but not always) met in statistical analyses. It corresponds to

$$\Pr(\text{dataset}) = \Pr(\text{obs. } 1) \times \Pr(\text{obs. } 2) \times \cdots \times \Pr(\text{obs. } N)$$

with discrete data. Anytime you have a likelihood function and can write it as

    "The log likelihood for the jth observation is $\ln \ell_j = \ldots$."

and that formula uses data for observation j only, and it is obvious that

    "The overall log likelihood function is $\ln L = \sum_{j=1}^{N} \ln \ell_j$."

you have a likelihood that meets this restriction. In other words, the observations are independent.

The following examples of likelihoods do not meet this criterion: the conditional (fixed-effects) logit model, Cox regression, and likelihoods for panel data. Basically, any likelihood that explicitly models independent groups rather than independent observations does not meet the `lf` criterion.

5.2   Examples

The following models meet the lf criterion that the physical observations in the dataset correspond to independent pieces of the likelihood function.

5.2.1   The probit model

The probit log-likelihood function for the jth observation is

$$\ln \ell_j = \left\{ \begin{array}{ll} \ln \Phi(\mathbf{x}_j \boldsymbol{\beta}) & \text{if } y_j = 1 \\ \ln \Phi(-\mathbf{x}_j \boldsymbol{\beta}) & \text{if } y_j = 0 \end{array} \right. \tag{5.1}$$

where $\Phi(\cdot)$ denotes the cumulative probability distribution function of the standard normal distribution. This is a one-equation model, so we skip using θ and identify the equation directly as $\mathbf{x}_j \boldsymbol{\beta}$. Our method-lf likelihood evaluator is

```
───────────────────────────────────── begin myprobit_lf.ado ─────────
program myprobit_lf
        version 18
        args lnfj xb
        quietly replace 'lnfj' = ln(normal( 'xb')) if $ML_y1 == 1
        quietly replace 'lnfj' = ln(normal(-'xb')) if $ML_y1 == 0
end
──────────────────────────────────────── end myprobit_lf.ado ─────────
```

myprobit_lf requires two arguments, as indicated by the line

```
args lnfj xb
```

The first argument, 'lnfj', is the variable where we fill in the log-likelihood values, and the second, 'xb', is the variable that contains the values of $\mathbf{x}_j \boldsymbol{\beta}$ (the first and only equation). The name of the dependent variable is placed in the global macro $ML_y1 by ml.

If we wanted to fit the probit regression of died on i.drug (a factor variable with three levels) and age, we would type

```
. ml model lf myprobit_lf (died = i.drug age)
. ml maximize
```

died = specifies that the dependent variable is died. i.drug age specifies that

$$\mathbf{x}_j \boldsymbol{\beta} = \beta_1 1.\text{drug}_j + \beta_2 2.\text{drug}_j + \beta_3 3.\text{drug}_j + \beta_4 \text{age} + \beta_5 \tag{5.2}$$

Our program never has to deal with the fact that $\mathbf{x}_j\boldsymbol{\beta}$ is given by (5.2), nor is it ever even aware of it. Let's pretend that at some point in the maximization process, $\beta_1 = -0.0$, $\beta_2 = -0.1$, $\beta_3 = -0.002$, $\beta_4 = 0.2$, and $\beta_5 = 8$. We never need to know any of that because all lf will ask our program is questions such as

"What is the log likelihood if the outcome is 1 and $\mathbf{x}_j\boldsymbol{\beta} = 0.5$?"

and all we have to do is reply "-0.36894642", which is $\ln\Phi(0.5)$, or

"What is the log likelihood if the outcome is 0 and $\mathbf{x}_j\boldsymbol{\beta} = 0.1$?"

to which we reply, "-0.77615459", which is $\ln\Phi(-0.1)$. In fact, lf asks both these questions at once, along with many others, when it asks us to fill in the variable 'lnfj' with the log likelihood for each of the observations, based on the values of the variables in $ML_y1 and 'xb'. Here we never need to know that the model being fit contains i.drug and age or whether the model has an intercept.

Here is the result of using this likelihood evaluator and some cancer data, modeling the incidence of death as a function of age and drug type (the example from chapter 1).

```
. sysuse cancer
(Patient survival in drug trial)
. ml model lf myprobit_lf (died = i.drug age)
. ml maximize
Initial:      Log likelihood = -33.271065
Alternative:  Log likelihood = -31.427839
Rescale:      Log likelihood = -31.424556
Iteration 0:  Log likelihood = -31.424556
Iteration 1:  Log likelihood = -21.883453
Iteration 2:  Log likelihood = -21.710901
Iteration 3:  Log likelihood = -21.710799
Iteration 4:  Log likelihood = -21.710799
```

					Number of obs = 48
					Wald chi2(3) = 13.58
Log likelihood = -21.710799					Prob > chi2 = 0.0035

died	Coefficient	Std. err.	z	P>|z|	[95% conf. interval]
drug					
Other	-1.941275	.597057	-3.25	0.001	-3.111485 -.7710644
NA	-1.792924	.5845829	-3.07	0.002	-2.938686 -.6471626
age	.0666733	.0410666	1.62	0.104	-.0138159 .1471624
_cons	-2.044876	2.284283	-0.90	0.371	-6.521988 2.432235

We used factor-variable notation for the categories of the drug variable. ml fully supports factor variables.

5.2.2   Normal linear regression

The jth observation's log-likelihood function for linear regression assuming normally distributed errors is

$$\ln \ell_j = \ln \phi\{(y_j - \theta_{1j})/\theta_{2j}\} - \ln \theta_{2j}$$
$$\theta_{1j} = \mu_j = \mathbf{x}_j \boldsymbol{\beta}$$
$$\theta_{2j} = \sigma$$

where $\phi(z)$ is the standard normal density function evaluated at z:

$$\phi(z) = \frac{1}{\sqrt{2\pi}}\, e^{\frac{1}{2} z^2}$$

Most people think of linear regression as a single-equation model, $\mu_j = \mathbf{x}_j \boldsymbol{\beta}$, with an ancillary parameter σ that also must be estimated. As we discussed in chapter 4, we can either view σ as an equation with just a constant term or view σ as a free parameter. Here we will treat σ as a free parameter to start. Making use of the fact that the function `normalden()` allows us to specify the mean and standard deviation, our method-`lf` likelihood evaluator is

─────────────────────────────────── begin `mynormal1_lf.ado` ───────────

```
program mynormal1_lf
        version 18
        args lnfj mu sigma
        quietly replace `lnfj' = ln(normalden($ML_y1,`mu',`sigma'))
end
```
─────────────────────────────────── end `mynormal1_lf.ado` ───────────

`mynormal1_lf` requires three arguments, as indicated by the line

```
    args lnfj mu sigma
```

The first argument, `'lnfj'`, is the variable where we fill in the log-likelihood values. The second argument, `'mu'`, is the variable that contains the values of the regression equation. And the third argument, `'sigma'`, is the variable that contains the values of the standard deviation. (Even if we treat σ as a free parameter, it is still passed to our evaluator as a variable, not a scalar.) `ml` places the name of the dependent variable in the global macro $ML_y1.

If we wanted to regress `mpg` on `weight` and `displ` with free parameter σ, we would type

```
    . ml model lf mynormal1_lf (mu: mpg = weight displ) /sigma
    . ml maximize
```

Here `mpg =` specifies that the dependent variable is `mpg`, and `weight displ` specifies that

$$\mu_j = \beta_1 \mathtt{weight}_j + \beta_2 \mathtt{displ}_j + \beta_3$$

We specified σ as free parameter /sigma, so the standard deviation is modeled as a constant:

$$\sigma = \beta_4$$

mynormal1_lf can just as easily fit heteroskedastic models with no changes required, but we will need to use the ml model syntax that specifies σ to be an equation, not a free parameter. Suppose we want to model σ as

$$\sigma_j = \beta_4 \texttt{price}_j + \beta_5 \tag{5.3}$$

To fit this model using auto.dta, we type

```
───────────────────────────────────────────── begin ex_normal3 ─────────
. sysuse auto
(1978 automobile data)
. ml model lf mynormal1_lf (mu: mpg = weight displacement) (sigma: price)
. ml maximize, nolog
Initial:       Log likelihood =     -<inf>  (could not be evaluated)
Feasible:      Log likelihood = -10383.274
Rescale:       Log likelihood = -292.89564
Rescale eq:    Log likelihood = -238.45986
                                              Number of obs =       74
                                              Wald chi2(2)  = 126.71
Log likelihood = -195.14254                   Prob > chi2   = 0.0000
```

mpg	Coefficient	Std. err.	z	P>|z|	[95% conf. interval]	
mu						
weight	-.0065258	.0011369	-5.74	0.000	-.008754	-.0042975
displacement	.0041644	.0099923	0.42	0.677	-.01542	.0237489
_cons	40.1942	1.976217	20.34	0.000	36.32089	44.06752
sigma						
price	.0000438	.0001061	0.41	0.680	-.0001641	.0002517
_cons	3.11326	.6915815	4.50	0.000	1.757785	4.468735

```
────────────────────────────────────────────── end ex_normal3 ─────────
```

Say you never intend for σ to be modeled as a function of covariates. Then an advantage to specifying /sigma rather than (sigma:) in your ml model statement is that the former will result in a slightly more compact coefficient table than the latter. See the partial coefficient tables on 77.

❏ **Technical note**

With the way we have parameterized this model, the log likelihood is defined only for $\theta_{2j} = \sigma > 0$. ml is designed to deal with restricted ranges—there is nothing special you need to do—but you should be impressed and realize that you are asking a lot of ml. There could be convergence problems, and even if not, ml is going to perform more work, so the maximization process may not occur as quickly as it otherwise could. Note that ml may not be able to find feasible initial values for restricted ranges such as $(-1, 1)$ or $(0, 1)$.

It would be better if the model were parameterized so that $(-\infty, \infty)$ is a valid range for each element of $\boldsymbol{\theta}_i$. An alternative parameterization of the linear regression likelihood function would be

$$\ln \ell_j = \ln \phi\{(y_j - \theta_{1j})/\exp(\theta_{2j})\} - \theta_{2j}$$
$$\theta_{1j} = \mu_j = \mathbf{x}_j \boldsymbol{\beta}$$
$$\theta_{2j} = \ln \sigma$$

and that parameterization would lead to the following likelihood evaluator:

─────────────── begin `mynormal2_lf.ado` ───────────

```
program mynormal2_lf
        version 18
        args lnfj mu lnsigma
        quietly replace 'lnfj' = ln(normalden($ML_y1,'mu',exp('lnsigma')))
end
```

─────────────── end `mynormal2_lf.ado` ───────────

❑

For more information about restricted parameters, see section 2.4.

5.2.3   The Weibull model

The Weibull log-likelihood function for the jth observation, where t_j is the time of failure or censoring and $d_j = 1$ if failure and 0 if censored, is

$$\ln \ell_j = -(t_j/e^{\theta_{1j}})^{e^{\theta_{2j}}} + d_j\{\theta_{2j} - \theta_{1j} + (e^{\theta_{2j}} - 1)(\ln t_j - \theta_{1j})\}$$
$$\theta_{1j} = \ln \eta_j = \mathbf{x}_{1j}\boldsymbol{\beta}_1$$
$$\theta_{2j} = \ln \gamma_j = \mathbf{x}_{2j}\boldsymbol{\beta}_2$$

For details on how this log likelihood is derived, see section 17.4. For now, it suffices to say that $\ln \eta_j$ is a scale parameter and $\ln \gamma_j$ is a shape parameter, and we wish to model the logs of both as linear combinations of covariates.

This is a two-equation model: the first equation models $\ln \eta_j$ and the second equation models $\ln \gamma_j$. The corresponding method-`1f` evaluator for this likelihood function is

─────────────── begin `myweibull_lf.ado` ───────────

```
program myweibull_lf
        version 18
        args lnfj leta lgam
        tempvar p M R
        quietly {
                generate double 'p' = exp('lgam')
                generate double 'M' = ($ML_y1*exp(-'leta'))^'p'
                generate double 'R' = ln($ML_y1)-'leta'
                replace 'lnfj' = -'M' + $ML_y2*('lgam'-'leta'+('p'-1)*'R')
        }
end
```

─────────────── end `myweibull_lf.ado` ───────────

`myweibull_lf` requires three arguments, as indicated by the line

```
args lnfj leta lgam
```

The first argument, '`lnfj`', is the variable where we fill in the log-likelihood values. The second argument, '`leta`', is the name of the variable that contains the values of $\ln \eta_j$. And the third argument, '`lgam`', is the name of the variable that contains the values of $\ln \gamma_j$.

We use temporary variables to evaluate some of the terms in the log likelihood. This is not necessary, but it makes the program more readable and keeps us from having to evaluate `exp('lgam')` more than once. It is important that these temporary variables were generated as `doubles`.

This model has two dependent variables: t_j and d_j. In `myweibull_lf`, we assume that t_j is the first dependent variable specified (so `$ML_y1` is coded for t_j) and d_j is the second (so `$ML_d2` is coded for d_j). We will arrange this by specifying the dependent variables in this order on the `ml model` statement.

Here is an example of our likelihood evaluator in action using `cancer.dta`, modeling time to death (with $t = $ `studytime` and $d = $ `died`) as a function of age and drug type:

```
. sysuse cancer
(Patient survival in drug trial)
. ml model lf myweibull_lf (lneta: studytime died = i.drug age) /lngamma
. ml maximize
Initial:       Log likelihood =        -744
Alternative:   Log likelihood = -356.14276
Rescale:       Log likelihood = -200.80201
Rescale eq:    Log likelihood = -136.69232
Iteration 0:   Log likelihood = -136.69232  (not concave)
Iteration 1:   Log likelihood = -124.11726
Iteration 2:   Log likelihood =  -113.9591
Iteration 3:   Log likelihood = -110.30683
Iteration 4:   Log likelihood = -110.26748
Iteration 5:   Log likelihood = -110.26736
Iteration 6:   Log likelihood = -110.26736
```

	Number of obs =	48
	Wald chi2(3) =	35.25
Log likelihood = -110.26736	Prob > chi2 =	0.0000

	Coefficient	Std. err.	z	P>|z|	[95% conf. interval]	
drug						
Other	1.012966	.2903917	3.49	0.000	.4438086	1.582123
NA	1.45917	.2821195	5.17	0.000	.9062261	2.012114
age	-.0671728	.0205688	-3.27	0.001	-.1074868	-.0268587
_cons	6.060723	1.152845	5.26	0.000	3.801188	8.320259
/lngamma	.5573333	.1402154	3.97	0.000	.2825162	.8321504

5.3 The importance of generating temporary variables as doubles

In `myweibull_lf`, we used three temporary variables.

```
generate double 'p' = exp('lgam')
generate double 'M' = ($ML_y1*exp(-'leta'))^'p'
generate double 'R' = ln($ML_y1)-'leta'
```

We did not have to approach this problem in this way, but the temporary variables made the equation for the log likelihood,

```
replace 'lnfj' = -'M' + $ML_y2*('lgam'-'leta'+('p'-1)*'R')
```

easier and more readable and so reduced the chances we made an error. When we created these temporary variables, we created them as `doubles`. This is of great importance.

The program works as written. Just to show you what can happen if you do not declare temporary variables to be `doubles`, we made a modified copy of `myweibull_lf` in which the temporary variables are not declared as `double`:

```
 ──────────────────────────────────── begin myweibull_bad_lf.ado ────────────
program myweibull_bad_lf
        version 18
        args lnfj leta lgam
        tempvar p M R
        quietly {
                generate 'p' = exp('lgam')
                generate 'M' = ($ML_y1*exp(-'leta'))^'p'
                generate 'R' = ln($ML_y1)-'leta'
                replace 'lnfj' = -'M' + $ML_y2*('lgam'-'leta'+('p'-1)*'R')
        }
end
 ──────────────────────────────────────── end myweibull_bad_lf.ado ──────────
```

Here is what happened when we used `myweibull_bad_lf`:

```
. sysuse cancer
(Patient survival in drug trial)
. ml model lf myweibull_bad_lf (lneta: studyt died = i.drug age) /lngamma
. ml maximize, iter(100)
Initial:       Log likelihood =       -744
Alternative:   Log likelihood = -356.14277
Rescale:       Log likelihood = -200.80201
Rescale eq:    Log likelihood = -188.26758
Iteration 0:   Log likelihood = -188.26758
Iteration 1:   Log likelihood = -183.36968   (not concave)
Iteration 2:   Log likelihood = -183.36817   (not concave)
   (output omitted, every line of which said "not concave")
Iteration 99: Log likelihood = -113.97176   (not concave)
Iteration 100: Log likelihood = -113.96625   (not concave)
convergence not achieved
```

```
                                                      Number of obs = 48
                                                      Wald chi2(2)  = .
         Log likelihood = -113.96625                  Prob > chi2   = .
```

	Coefficient	Std. err.	z	P>|z|	[95% conf. interval]	
drug						
Other	1.028316	.0971575	10.58	0.000	.8378905	1.218741
NA	1.50256	.1336381	11.24	0.000	1.240634	1.764486
age	-.012124	.	.	.	.	.
_cons	2.985251	.	.	.	.	.
/lngamma	.4347168	.	.	.	.	.

Warning: Convergence not achieved.

In the last iteration, the function was still "not concave". As we warned in chapter 1, when the iterations end in that way, you are in trouble. It is difficult to miss because when the last iteration is still "not concave", one or more of the standard errors will most likely be missing.

If you forget to specify `double`, one of five things will happen:

1. The maximization will have many "not concave" iterations, and importantly, the last iteration will be "not concave". (Intermediate "not concave" iterations should not concern you.)

2. `ml maximize` will get to iteration 1 or 2 and just sit. (In that case, press the *Break* key.)

3. `ml maximize` will iterate forever, never declaring convergence.

4. There is a small chance that `ml maximize` (and `ml check`) will report that it cannot calculate the derivatives. (In most cases when you see this error, however, it is due to a different problem: the likelihood evaluates to the same constant regardless of the value of the parameter vector because of some silly error in the evaluation program; often that constant is a missing value.)

5. There is a small chance that no mechanical problem will arise, but in that case, the reported standard errors are accurate to only a few digits.

If you include intermediate calculations, review the `generate` statements carefully to ensure that you have included the word `double` following `generate`.

❑ Technical note

Suppose that beforehand, we typed

```
. set type double
```

The result would be as if we had specified all the temporary variables as `double`. There are many reasons why Stata's default data type is not `double`; suffice it to say, you cannot assume that the default data type will always be `double`, especially if you share your code with others. Thus, make sure that you code `double` following every `generate`.

❑

5.4   Problems you can safely ignore

Method `lf` handles a variety of issues for you:

- You are not responsible for restricting your calculations to the estimation sub-sample, although your program will run slightly faster if you do. The variable `$ML_samp` contains 1 if the observation is to be used and 0 if it is to be ignored. It is necessary for you to fill in '`lnfj`' only in observations for which `$ML_samp==1`. Thus in our first example, the probit likelihood-evaluation program could read

```
                                         begin myprobit2_lf.ado
program myprobit2_lf
        version 18
        args lnfj xb
        quietly replace `lnfj' = ln(normal( `xb')) if $ML_y1==1 & $ML_samp==1
        quietly replace `lnfj' = ln(normal(-`xb')) if $ML_y1==0 & $ML_samp==1
end
                                         end myprobit2_lf.ado
```

- This program would execute a little more quickly in some cases, but that is all. Even if you do not do this, results reported by `ml` method `lf` will be correct.

- Whether you specify `ml model`'s `nopreserve` option (discussed in appendix B) makes no difference. This is a technical issue; do not bother to see the appendix over this. As we just said, restricting calculations to the estimation subsample is handled by `ml` method `lf` itself.

- You can perform weighted estimation without doing anything special in your program. When you read about `$ML_w` in other parts of this text, ignore it if you are using `ml` method `lf`. Weights are automatically handled by `ml` method `lf`.

- You can specify the `vce(opg)`, `vce(robust)`, `vce(cluster` *varname*`)`, and `svy` options on the `ml model` statement without doing anything special in your program to handle this.

5.5   Nonlinear specifications

The term *linear-form restrictions* suggests that the parameterization of the likelihood must be linear. That is not true. The linear-form restrictions require only that the likelihood function be calculable observation by observation and that the individual equations each be linear.

For instance, suppose you wished to fit

$$y_j = \beta_1 x_{1j} + \beta_2 x_{2j} + \beta_3 x_{3j}^{\beta_4} + \beta_5 + \epsilon_j$$

where ϵ_j is distributed $N(0, \sigma^2)$. Note that the term involving x_3 is nonlinear. Here is how you could write this model:

$$\ln \ell_j = \ln \phi\{(y_j - \theta_{1j} - \theta_{2j} x_{3j}^{\theta_{3j}})/\theta_{4j}\} - \ln \theta_{4j}$$
$$\theta_{1j} = \beta_1 x_{1j} + \beta_2 x_{2j} + \beta_5$$
$$\theta_{2j} = \beta_3$$
$$\theta_{3j} = \beta_4$$
$$\theta_{4j} = \sigma$$

This could be considered a four-equation model in `ml` jargon even though conceptually it has just one equation. That is because β_3, β_4, and σ can be specified as constant-only equations. Alternatively, if you want to be able to parameterize σ as a function of covariates to control for heteroskedasticity, then this would qualify as a two-equation model with two free parameters (β_3 and β_4). Only if you treat β_3, β_4, and σ as free parameters would `ml` jargon align with the mathematical view that there is only one equation in this regression model. Nonlinear terms will add equations or free parameters, but adding them does not cause a model that previously met the linear-form restriction to no longer meet it.[1]

Our method-`lf` evaluator for this likelihood function is

```
                                              begin mynonlin_lf.ado
program mynonlin_lf
        version 18
        args lnfj theta1 theta2 theta3 sigma
        quietly replace `lnfj' = ///
                ln(normalden($ML_y1,`theta1'+`theta2'*X3^`theta3',`sigma'))
end
                                              end mynonlin_lf.ado
```

---

1. In theory, you could allow β_3 or β_4 to be a function of covariates, too, but you would almost surely encounter difficulties obtaining convergence. If the model is too flexible, some of the parameters might not be identified given the variation in the data.

The variable name corresponding to x_{3j} is hardcoded in the evaluation program. We used the name X3, but you would substitute the actual variable name. For instance, if you wished to fit $\mathtt{bp} = \beta_1 \mathtt{age} + \beta_2 \mathtt{sex} + \beta_3 \mathtt{bp0}^{\beta_4} + \epsilon$, you could change the program to read

```
program mynonlin_lf
        version 18
        args lnfj theta1 theta2 theta3 sigma
        quietly replace 'lnfj' = ///
                ln(normalden($ML_y1,'theta1'+'theta2'*bp0^'theta3','sigma'))
end
```

and then fit the model by typing

```
. ml model lf mynonlin_lf (bp=age sex) /beta3 /beta4 /sigma
. ml maximize
```

or

```
. ml model lf mynonlin_lf (bp=age sex) /beta3 /beta4 (sigma:)
. ml maximize
```

Alternatively, you could create a global macro holding the name of the variable corresponding to x_{3j} and refer to that global macro in your evaluator.

5.6   The advantages of lf in terms of execution speed

Because method lf works with the model parameters (the θ's) instead of the coefficients (the β's), it provides a considerable speed advantage over d-family evaluators and approaches the performance of method d2, which requires you to specify first- and second-order derivatives. For similar reasons, the lf-family evaluators we discuss in the next chapter will also execute more quickly than d-family evaluators.

For the purpose of this discussion, and without loss of generality, assume that our model contains only one equation,

$$\theta_j = x_{1j}\beta_1 + x_{2j}\beta_2 + \cdots + x_{kj}\beta_k$$

In addition to the log-likelihood function itself, the ml optimizer needs the first and second derivatives with respect to the coefficients:

$$\mathbf{g} = \left(\frac{\partial \ln L}{\partial \beta_1} \quad \frac{\partial \ln L}{\partial \beta_2} \quad \cdots \quad \frac{\partial \ln L}{\partial \beta_k} \right)$$

$$\mathbf{H} = \begin{pmatrix} \frac{\partial^2 \ln L}{\partial \beta_1^2} & \frac{\partial^2 \ln L}{\partial \beta_1 \partial \beta_2} & \cdots & \frac{\partial^2 \ln L}{\partial \beta_1 \partial \beta_k} \\ \frac{\partial^2 \ln L}{\partial \beta_2 \partial \beta_1} & \frac{\partial^2 \ln L}{\partial \beta_2^2} & \cdots & \frac{\partial^2 \ln L}{\partial \beta_2 \partial \beta_k} \\ \vdots & \vdots & \ddots & \vdots \\ \frac{\partial^2 \ln L}{\partial \beta_k \partial \beta_1} & \frac{\partial^2 \ln L}{\partial \beta_k \partial \beta_2} & \cdots & \frac{\partial^2 \ln L}{\partial \beta_k^2} \end{pmatrix}$$

We write **g** as a row vector for this discussion because that is what `ml` expects of likelihood evaluators that are required to return the gradient.

If we do not have analytic formulas for **g** and **H**, `ml` must calculate each of these $k + k(k+1)/2$ derivatives numerically. For instance, `ml` could calculate the first element of **g** as

$$\frac{\partial \ln L}{\partial \beta_1} = \frac{\ln L(\beta_1 + h/2, \beta_2, \ldots, \beta_k; \mathbf{Z}) - \ln L(\beta_1 - h/2, \beta_2, \ldots, \beta_k; \mathbf{Z})}{h}$$

This involves evaluating the log-likelihood function $\ln L(\cdot)$ twice. The second element of **g** could be obtained as

$$\frac{\partial \ln L}{\partial \beta_2} = \frac{\ln L(\beta_1, \beta_2 + h/2, \ldots, \beta_k; \mathbf{Z}) - \ln L(\beta_1, \beta_2 - h/2, \ldots, \beta_k; \mathbf{Z})}{h}$$

which are two more evaluations, and so on. Using centered derivatives, we would need to evaluate the log-likelihood function $2k$ times, assuming we know the optimum value of h. If we have to search for an optimum h in each calculation, we will need even more evaluations of $\ln L(\cdot)$.

Next we will have to evaluate the $k(k+1)/2$ second derivatives, and as we showed in section 1.3.5, that takes at least three function evaluations each, although one of the evaluations is $\ln L(\beta_1, \beta_2, \ldots, \beta_k; \mathbf{Z})$ and is common to all the terms. Furthermore, $2k$ of the required function evaluations are in common with the function evaluations for the first derivative. Thus we need "only" $k(k-1)$ more function evaluations. The grand total for first and second derivatives is thus $2k + k(k-1)$.

In the case of `lf`, however, $\ln \ell_j = \ln \ell(\theta_j)$ where $\theta_j = \mathbf{x}_j \boldsymbol{\beta}$, and this provides a shortcut for the calculation of the derivatives. To simplify notation, let's write $f(\theta_j)$ for $\ln \ell(\theta_j)$, $f'(\theta_j)$ for its first derivative with respect to θ_j, and $f''(\theta_j)$ for its second. Using the chain rule,

$$\mathbf{g}_j = \frac{\partial f(\theta_j)}{\partial \boldsymbol{\beta}} = \frac{\partial f(\theta_j)}{\partial \theta_j} \frac{\partial \theta_j}{\partial \boldsymbol{\beta}} = f'(\theta_j) \mathbf{x}_j$$

Thus

$$\mathbf{g} = \sum_{j=1}^{N} \mathbf{g}_j = \sum_{j=1}^{N} f'(\theta_j) \mathbf{x}_j$$

The k derivatives of the gradient vector can be obtained by calculating just one derivative numerically: $f'(\theta_j)$. In fact, we must calculate N numeric derivatives because we need $f'(\theta_j)$ separately evaluated for each observation, but that is not as bad as it sounds. The major cost of evaluating numeric derivatives is the time it takes to evaluate the likelihood function. We can obtain all $f'(\theta_j)$, $j = 1, \ldots, N$, with just two calls to the user-written likelihood-evaluation program, at least if we are willing to use the same h for each observation, and if we assume we already know the optimal h to use.

Similarly, the matrix of second derivatives can be written as

$$\mathbf{H} = \sum_{j=1}^{N} f''(\theta_j) \mathbf{x}'_j \mathbf{x}_j$$

and these $k(k+1)/2$ derivatives can similarly be obtained by evaluating one second derivative numerically (requiring two more calls).

This approach results in substantial computer time savings, especially for large values of k.

6 Methods lf0, lf1, and lf2

In the previous chapter, we used method `lf`, which for many applications is the only method you will need. One shortcoming of method `lf` is that there is no facility through which you can specify analytic first and second derivatives. In this chapter, we introduce methods `lf0`, `lf1`, and `lf2`, which we refer to collectively as the `lf`-family evaluators. In exchange for a more complicated syntax for evaluator programs, this family of evaluators allows you to specify derivatives. Method `lf0` is a stepping stone used while developing method-`lf1` and method-`lf2` evaluators. If you do not intend to specify analytic derivatives, you should use the easier method `lf` instead; in this case, method `lf0` provides no added functionality. Method `lf1` requires you to program analytic first derivatives, and method `lf2` requires you to program analytic first and second derivatives.

6.1 Comparing these methods

Methods `lf0`, `lf1`, and `lf2` differ from method `lf` in the arguments that `ml` passes to the likelihood evaluator and what the likelihood evaluator returns:

- In method `lf`, you are passed $\theta_{ij} = \mathbf{x}_{ij}\boldsymbol{\beta}_i$, the linear combinations that have already been evaluated.

  Methods `lf0`, `lf1`, and `lf2` pass your likelihood evaluator the coefficients themselves combined into a single vector $\boldsymbol{\beta} = (\boldsymbol{\beta}_1, \boldsymbol{\beta}_2, \ldots)$. It is your responsibility to obtain the θ_{ij} values from $\boldsymbol{\beta}$. The utility command `mleval`, described below, helps do this.

- Methods `lf0`, `lf1`, and `lf2` pass your program an additional argument 'todo' indicating whether your program is to evaluate the likelihood, the likelihood and first derivatives, or the likelihood, first, and second derivatives.

- Methods `lf1` and `lf2` pass your program additional arguments into which you place the observation-level first derivatives (scores) and the Hessian matrix when requested.

To fit a model using an evaluator of method `lf0`, `lf1`, or `lf2`, you type the command

> . ml model lf0 *myprog* ...

or

> . ml model lf1 *myprog* ...

or

> . ml model lf2 *myprog* ...

Write *myprog*—your likelihood evaluator—according to the `lf0`, `lf1`, or `lf2` specifications.

As long as you do not specify the `nopreserve` option to `ml model`, `ml` automatically restricts computations to the relevant estimation sample when used with `lf`-family evaluators. Therefore, you do not need to handle `if` or `in` conditions or exclude observations with missing values. Weights are also handled automatically. If you do specify the `nopreserve` option, you will need to restrict your calculations to the estimation sample. Also, if you calculate the Hessian matrix for an `lf2` evaluator and do not use the `mlmatsum` command (discussed later), you will need to account for weights and restrict your computations for the Hessian to the relevant subsample.

6.2   Outline of evaluators of methods lf0, lf1, and lf2

Methods `lf0`, `lf1`, and `lf2` can be used with likelihood functions that meet the linear-form restrictions

$$\ln \ell_j = \ln \ell(\theta_{1j}, \theta_{2j}, \ldots, \theta_{Ej}; y_{1j}, y_{2j}, \ldots, y_{Dj})$$

$$\ln L = \sum_{j=1}^{N} \ln \ell_j$$

$$\theta_{ij} = \mathbf{x}_{ij} \boldsymbol{\beta}_i, \qquad i = 1, \ldots, E \text{ and } j = 1, \ldots, N$$

where i is the equation index and j is the observation index.

The outline of an evaluator of method `lf0`, `lf1`, or `lf2` is

```
program myprog
        version 18
        args todo b lnfj g1 g2 ... H
        // if you need to create any intermediate results:
        tempvar tmp1 tmp2
        quietly generate double 'tmp1' = ...
        ...
        quietly replace 'lnfj' = ...
        if ('todo'==0 | 'lnfj' >= .) exit          // method-lf0 evaluators end here
        // fill in score variables
        quietly replace 'g1' = ∂ln ℓ_j/∂θ_{1j}
        quietly replace 'g2' = ∂ln ℓ_j/∂θ_{2j}
        ...
        if ('todo'==1) exit                        // method-lf1 evaluators end here
        // compute Hessian matrix
        tempname d11 d12 d22 ...
        mlmatsum 'lnfj' 'd11' = ∂²ln L/∂θ₁²
        mlmatsum 'lnfj' 'd12' = ∂²ln L/∂θ₁∂θ₂
        mlmatsum 'lnfj' 'd22' = ∂²ln L/∂θ₂²
        matrix 'H' = ('d11', 'd12', ...  'd12'', 'd22', ...)
end
```

6.2.1   The todo argument

The first argument an evaluator of method `lf0`, `lf1`, or `lf2` receives is `'todo'`.

- If `'todo'` = 0, the evaluator must fill in $\ln \ell_j$, the observation-level contributions to the log likelihood.

- If `'todo'` = 1, the evaluator must fill in $\ln \ell_j$ and the observation-level scores, stored as a set of E variables.

- If `'todo'` = 2, the evaluator must fill in $\ln \ell_j$, the observation-level scores, and the Hessian matrix.

Method-`lf1` evaluators receive `'todo'` = 0 or `'todo'` = 1. When called with `'todo'` = 0, the evaluator may calculate the gradient, but it will be ignored; doing so needlessly increases execution time.

Method-`lf2` evaluators receive `'todo'` = 0, `'todo'` = 1, or `'todo'` = 2. Regardless of the value received, the evaluator may calculate more than is requested, but the extra will be ignored; doing so needlessly increases execution time.

6.2.2   The b argument

The second argument an evaluator of method `lf0`, `lf1`, or `lf2` receives is `'b'`, the current guess for $\boldsymbol{\beta}$ (the vector of coefficients) as a row vector. If you type

```
. ml model lf1 myprog (foreign = mpg weight)
```

when specifying your model, then the vector 'b' that *myprog* receives would contain three elements: the coefficient on mpg, the coefficient on weight, and the intercept. The interpretation would be

$$\theta_j = \beta_1 \text{mpg}_j + \beta_2 \text{weight}_j + \beta_3$$

If you type

```
. ml model lf1 myprog (foreign = mpg weight) (displ = weight)
```

then 'b' would contain five elements representing two equations:

$$\theta_{1j} = \beta_{1,1} \text{mpg}_j + \beta_{1,2} \text{weight}_j + \beta_{1,3}$$
$$\theta_{2j} = \beta_{2,1} \text{weight}_j + \beta_{2,2}$$

If you type

```
. ml model lf1 myprog (foreign = mpg weight) (displ = weight) /sigma
```

'b' would contain six elements representing two equations and one free parameter:

$$\theta_{1j} = \beta_{1,1} \text{mpg}_j + \beta_{1,2} \text{weight}_j + \beta_{1,3}$$
$$\theta_{2j} = \beta_{2,1} \text{weight}_j + \beta_{2,2}$$
$$\theta_{3j} = \sigma$$

The equation order and the order of coefficients within the equation are exactly as you specify them in the ml model statement. The intercept coefficient is always the last coefficient of a given equation. ml sets the column names of 'b' with equation and variable names for elements associated with linear-form equations. For free parameters, ml uses the special equation name /, and column names are set equal to the parameter names. This will become clearer with a few examples.

If you type

```
. ml model lf1 myprog (foreign = mpg weight)
```

and *myprog* is called with 'b' = $(1, 2, 3)$, then your program will receive the following vector:

```
. matrix list 'b', noheader
         eq1:    eq1:    eq1:
         mpg   weight   _cons
    r1     1       2       3
```

If you type

```
. ml model lf1 myprog (foreign = mpg weight) (displ = weight) (sigma:)
```

and 'b' = $(1, 2, 3, 4, 5, 6)$, then your program would receive

```
. matrix list 'b', noheader
        eq1:    eq1:    eq1:    eq2:    eq2:   sigma:
        mpg   weight   _cons  weight   _cons   _cons
   r1     1       2       3       4       5       6
```

If you type

```
. ml model lf1 myprog (foreign = mpg weight) (displ = weight) /sigma
```

and 'b' $= (1, 2, 3, 4, 5, 6)$, then your program would receive

```
. matrix list 'b', noheader
        eq1:    eq1:    eq1:    eq2:    eq2:      /:
        mpg   weight   _cons  weight   _cons   sigma
   r1     1       2       3       4       5       6
```

If you type

```
. ml model lf1 myprog (foreign = mpg weight) /sigma_1 (displ = weight) /sigma_2
```

and 'b' $= (1, 2, 3, 4, 5, 6, 7)$, then your program would receive

```
. matrix list 'b', noheader
        eq1:    eq1:    eq1:      /:     eq3:    eq3:      /:
        mpg   weight   _cons  sigma_1  weight   _cons  sigma_2
   r1     1       2       3       4       5       6       7
```

In this example, we specified a free parameter in between our two linear-form equations. Specifying a free parameter increases the equation count, so the second linear-form equation is labeled eq3:.

Using mleval to obtain equation and free-parameter values

The mleval utility calculates linear combinations using a coefficient vector, and it extracts scalar (free) parameters from such a vector. The syntax of mleval to obtain linear combinations is

mleval *newvarname* = *vecname* $\left[\text{, eq}(\#) \right]$

newvarname, stored as a double, is created containing $\theta_{ij} = \mathbf{x}_{ij}\boldsymbol{\beta}_i$, the linear combination of independent variables corresponding to the ith equation.

vecname refers to the coefficient vector passed by ml (that is, 'b').

eq() specifies i, the equation to be evaluated; it defaults to eq(1).

The syntax of mleval to obtain a free parameter or constant-only equation is

mleval *scalarname* = *vecname* , scalar $\left[\text{eq}(\#) \right]$

scalarname, stored as a Stata scalar, is created containing $\theta_{ij} = \beta_i$.

`mleval` is typically used as follows:

```
program myprog
        version 18
        args todo b lnfj g1 g2 ... H
        tempvar theta1 theta2 ...
        mleval 'theta1' = 'b', eq(1)
        mleval 'theta2' = 'b', eq(2)
        ...
end
```

The rest of the program can then be written in terms of 'theta1', 'theta2', ...

If we type

```
. ml model lf2 myprog (foreign = mpg weight)
```

then our evaluator can be written as

```
program myprog
        version 18
        args todo b lnfj g1 H
        tempvar xb
        mleval 'xb' = 'b'
        ...
end
```

because `mleval` defaults to the first equation.

Regarding the `equation()` option, remember that an equation in `ml` jargon refers to either a linear-form equation or a free parameter, not just a linear-form equation. Consider another model:

```
. ml model lf2 myprog (foreign = mpg weight) (displ = weight) /sigma
```

We might write

```
program myprog
        version 18
        args todo b lnfj g1 g2 g3 H
        tempvar theta1 theta2 theta3
        mleval 'theta1' = 'b', eq(1)
        mleval 'theta2' = 'b', eq(2)
        mleval 'theta3' = 'b', eq(3)
        ...
end
```

In this example, $\theta_3 = \sigma$ is a constant. Thus there would be an efficiency gain to making 'theta3' a scalar rather than a variable in the dataset. To do that, we use the syntax of `mleval` for scalar parameters:

```
program myprog3
        version 18
        args todo b lnfj g1 g2 g3 H
        tempvar theta1 theta2
        tempname theta3
        mleval 'theta1' = 'b', eq(1)
        mleval 'theta2' = 'b', eq(2)
        mleval 'theta3' = 'b', eq(3) scalar
        ...
end
```

If you tried to use the `scalar` option on an inappropriate equation (such as `eq(1)` in the example above), `mleval` would issue the following error:

```
. mleval 'theta1' = 'b', eq(1) scalar
mleval, scalar: eq(1) not constant
r(198);
```

If your `ml model` statement is

```
. ml model lf1 myprog4 (foreign = mpg weight) /sigma_1 (displ = weight) /sigma_2
```

then the start of your program should be

```
program myprog4
        version 18
        args todo b lnfj g1 g2 g3 H
        tempvar  theta1 theta3
        tempname theta2 theta4
        mleval 'theta1' = 'b', eq(1)
        mleval 'theta2' = 'b', eq(2) scalar
        mleval 'theta3' = 'b', eq(3)
        mleval 'theta4' = 'b', eq(4) scalar
        ...
end
```

When evaluating equations, use `mleval` without the `scalar` option for equations that produce values that vary across observations or that might vary across observations. To produce more efficient code, specify the `scalar` option for equations that produce values that are constant.

6.2.3   The lnfj argument

The third argument an evaluator of method `lf0`, `lf1`, or `lf2` receives is 'lnfj', the name of a variable whose values you are to replace with the observation-level log-likelihood values evaluated at 'b'. If the log likelihood cannot be calculated for an observation, that observation's value of 'lnfj' is to contain a missing value, ".". If `ml` encounters any missing values of 'lnfj' in the estimation sample, it considers the current value of 'b' to be inadmissible and searches for a new value.

To illustrate how to use the 'b' and 'lnfj' arguments, here is how we would write a method-`lf0` evaluator for the probit model:

```
─────────────────────────────────────────────── begin myprobit_lf0.ado ───────────
program myprobit_lf0
        version 18
        args todo b lnfj
        tempvar xb
        mleval 'xb' = 'b'
        quietly replace 'lnfj' = ln(normal( 'xb')) if $ML_y1 == 1
        quietly replace 'lnfj' = ln(normal(-'xb')) if $ML_y1 == 0
end
─────────────────────────────────────────────────── end myprobit_lf0.ado ───────────
```

Because the probit model includes just one equation, we did not bother to specify the eq(1) option with `mleval` because `eq(1)` is the default.

❑ **Technical note**

The probability of observing a positive outcome in the probit model is $\Phi(\mathbf{x}_{ij}\boldsymbol{\beta}_i)$, so the probability of observing a negative outcome is $1 - \Phi(\mathbf{x}_{ij}\boldsymbol{\beta}_i)$. You might therefore be tempted to write the penultimate line of `myprobit_lf0.ado` as

```
quietly replace 'lnfj' = ln(1-normal('xb')) if $ML_y1 == 0
```

However, calculating the log likelihood that way has poor numerical properties when the outcome is zero. `ln(1-normal(9))` evaluates to missing, whereas `ln(normal(-9))` evaluates to the correct answer of -43.63. The difficulty arises because `normal(9)` is close to one, and evaluating the natural logarithm of one minus a number close to one is inherently unstable.

❑

6.2.4 Arguments for scores

A method-`lf1` evaluator receives $3 + E$ arguments, and a method-`lf2` evaluator receives $4 + E$ arguments, where E is the number of equations in the model. We have already discussed the first three arguments: `todo`, `b`, and `lnfj`. The next E arguments these two evaluators receive are 'g1', 'g2', ..., variables into which you are to place the observation-level scores for the E equations.

The general outline for a method-`lf1` evaluator is

```
program myprog
        version 18
        args todo b lnfj g1 g2 ...
        // if you need to create any intermediate results:
        tempvar tmp1 tmp2
        quietly generate double 'tmp1' = ...
        ...
        quietly replace 'lnfj' = ...
        if ('todo'==0) exit                    // method-lf0 evaluators end here
        // fill in score variables
        quietly replace 'g1' = ∂ln ℓ_j/∂θ_{1j}
        quietly replace 'g2' = ∂ln ℓ_j/∂θ_{2j}
        ...
end
```

The argument g1 is the name of a temporary variable whose values we are to replace with the scores for the first equation, $\partial \ln \ell_j / \partial \theta_{1j}$, evaluated at the current values of the coefficient vector b. Similarly, if our model had two or more equations, g2's values would need to be replaced with $\partial \ln \ell_j / \partial \theta_{2j}$.

In some cases, you might be able to evaluate the likelihood function at the current value of 'b' but cannot evaluate one or more of the scores, perhaps because of a division-by-zero error. In these cases, you can place a missing value in any observation of any of the score variables, or else exit the evaluator program as soon as you realize a problem because the score variables contain missing values when passed to your program. If ml encounters any missing values in any of the score variables in the estimation sample, it considers the current value of 'b' to be inadmissible and searches for a new value.

▷ **Example**

Using the probit likelihood function in (5.1), the score equation is

$$\frac{d \ln \ell_j}{d \theta_j} = \begin{cases} \phi(\theta_j)/\Phi(\theta_j) & \text{if } y_j = 1 \\ -\phi(\theta_j)/\Phi(-\theta_j) & \text{if } y_j = 0 \end{cases} \tag{6.1}$$

In our likelihood-evaluator program, we would therefore write

```
replace 'g1' =  normalden('xb')/normal( 'xb') if $ML_y1 == 1
replace 'g1' = -normalden('xb')/normal(-'xb') if $ML_y1 == 0
```

◁

Later in the chapter, when we demonstrate complete examples, we will show how to fill in the score variables for multiple-equation models.

6.2.5   The H argument

The final argument a method-lf2 evaluator receives is 'H', the name of a matrix you are to create containing the Hessian (the matrix of second derivatives). 'H' is to be filled in when 'todo' is 2.

The second derivatives are to be stored as a matrix that is conformable with the 'b' vector you were passed. This means that if 'b' is $1 \times k$, 'H' is to be $k \times k$, and it is to have the derivatives in the same order as the coefficients in 'b'.

'H' is to contain the second partial derivatives of $\ln L$ with respect to each of the coefficients:

$$\mathbf{H} = \begin{pmatrix} \frac{\partial^2 \ln L}{\partial \beta_1^2} & \frac{\partial^2 \ln L}{\partial \beta_1 \partial \beta_2} & \cdots & \frac{\partial^2 \ln L}{\partial \beta_1 \partial \beta_k} \\ \frac{\partial^2 \ln L}{\partial \beta_2 \partial \beta_1} & \frac{\partial^2 \ln L}{\partial \beta_2^2} & \cdots & \frac{\partial^2 \ln L}{\partial \beta_2 \partial \beta_k} \\ \vdots & \vdots & \ddots & \vdots \\ \frac{\partial^2 \ln L}{\partial \beta_k \partial \beta_1} & \frac{\partial^2 \ln L}{\partial \beta_k \partial \beta_2} & \cdots & \frac{\partial^2 \ln L}{\partial \beta_k^2} \end{pmatrix}$$

The row and column names placed on 'H' do not matter because ml will set them.

Unlike scores, which are observation-specific, 'H' represents the second derivative of the overall log-likelihood function. However, likelihoods that meet the linear-form restrictions, as both lf1 and lf2 evaluators must, afford a major simplification in computing the first and second derivatives of the overall log likelihood with respect to β. We have

$$\ln L = \sum_{j=1}^{N} \ln \ell(\theta_{1j}, \theta_{2j}, \dots, \theta_{Ej})$$

$$\theta_{ij} = \mathbf{x}_{ij}\boldsymbol{\beta}_i$$

and $\boldsymbol{\beta} = (\boldsymbol{\beta}_1, \boldsymbol{\beta}_2, \dots, \boldsymbol{\beta}_E)$. Thus the gradient vector can be written as a sum

$$\frac{\partial \ln L}{\partial \boldsymbol{\beta}} = \sum_{j=1}^{N} \frac{\partial \ln \ell_j}{\partial \boldsymbol{\beta}}$$

where

$$\frac{\partial \ln \ell_j}{\partial \boldsymbol{\beta}} = \left(\frac{\partial \ln \ell_j}{\partial \theta_{1j}} \frac{\partial \theta_{1j}}{\partial \boldsymbol{\beta}_1}, \quad \frac{\partial \ln \ell_j}{\partial \theta_{2j}} \frac{\partial \theta_{2j}}{\partial \boldsymbol{\beta}_2}, \quad \dots, \quad \frac{\partial \ln \ell_j}{\partial \theta_{Ej}} \frac{\partial \theta_{Ej}}{\partial \boldsymbol{\beta}_E} \right)$$

$$= \left(\frac{\partial \ln \ell_j}{\partial \theta_{1j}} \mathbf{x}_{1j}, \quad \frac{\partial \ln \ell_j}{\partial \theta_{2j}} \mathbf{x}_{2j}, \quad \dots, \quad \frac{\partial \ln \ell_j}{\partial \theta_{Ej}} \mathbf{x}_{Ej} \right)$$

The last equation shows why we can define our scores with respect to $\theta_{1j}, \theta_{2j}, \dots, \theta_{Ej}$ and have our likelihood evaluator work regardless of the number of regressors we specify for each equation. For notational simplicity, we focus on the second derivative of $\ln L$ first with respect to β_i and then with respect to β_k. We have

$$\frac{\partial^2 \ln L}{\partial \beta_k \partial \beta_i} = \frac{\partial}{\partial \beta_k} \left(\frac{\partial \ln L}{\partial \beta_i} \right) = \sum_{j=1}^{N} \frac{\partial^2 \ln \ell_j}{\partial \beta_k \partial \beta_i}$$

where

$$\frac{\partial^2 \ln \ell_j}{\partial \beta_k \partial \beta_i} = \frac{\partial}{\partial \beta_k} \left(\frac{\partial \ln \ell_j}{\partial \theta_{ij}} \mathbf{x}_{ij} \right)$$

$$= \frac{\partial^2 \ln \ell_j}{\partial \theta_{kj} \partial \theta_{ij}} \mathbf{x}'_{ij} \mathbf{x}_{kj}$$

This result provides us with a considerable reduction in complexity, both mathematically and with respect to our likelihood-evaluator program. As with scores, we need only obtain second derivatives with respect to the θs instead of the individual βs; and because the chain rule applies with linear-form models, we can change the independent variables in our equations at will without having to touch our likelihood-evaluator program.

Like with scores, in some cases you might be able to evaluate the log likelihood at the given value of 'b' but not the second derivative for one or more observations. In those cases, you can either set one or more observations on 'lnfj' to missing, or you can set one or more elements of 'H' to missing. ml then considers the current value of 'b' to be infeasible and searches for a new value.

The second derivative of the overall log-likelihood function is a sum over the observations in the dataset. Were we to compute this sum ourselves, we would need to limit our sum to only those observations within the estimation sample, we would need to take into account any weights if specified, and we would need to know the variables represented by the $\mathbf{x}_i$'s. Fortunately, ml provides the mlmatsum command, which makes computing $\mathbf{H}$ easy once we obtain the second derivatives with respect to the θs. mlmatsum obviates the three details we would need to consider if we were to compute the sum ourselves.

Using mlmatsum to define H

mlmatsum assists in calculating the Hessian matrix of the overall log-likelihood function when the linear-form restrictions are met. As we just saw, in this case we can write the Hessian as

$$\mathbf{H} = \sum_{j=1}^{N} \frac{\partial^2 \ln \ell_j}{\partial \beta \partial \beta'} \tag{6.2}$$

$$\frac{\partial^2 \ln \ell_j}{\partial \beta \partial \beta'} = \begin{pmatrix} \frac{\partial^2 \ln \ell_j}{\partial \theta_{1j}^2}\mathbf{x}'_{1j}\mathbf{x}_{1j} & \frac{\partial^2 \ln \ell_j}{\partial \theta_{1j}\partial \theta_{2j}}\mathbf{x}'_{1j}\mathbf{x}_{2j} & \cdots & \frac{\partial^2 \ln \ell_j}{\partial \theta_{1j}\partial \theta_{kj}}\mathbf{x}'_{1j}\mathbf{x}_{kj} \\ \frac{\partial^2 \ln \ell_j}{\partial \theta_{2j}\partial \theta_{1j}}\mathbf{x}'_{2j}\mathbf{x}_{1j} & \frac{\partial^2 \ln \ell_j}{\partial \theta_{2j}^2}\mathbf{x}'_{2j}\mathbf{x}_{2j} & \cdots & \frac{\partial^2 \ln \ell_j}{\partial \theta_{2j}\partial \theta_{kj}}\mathbf{x}'_{2j}\mathbf{x}_{kj} \\ \vdots & \vdots & \ddots & \vdots \\ \frac{\partial^2 \ln \ell_j}{\partial \theta_{kj}\partial \theta_{1j}}\mathbf{x}'_{kj}\mathbf{x}_{1j} & \frac{\partial^2 \ln \ell_j}{\partial \theta_{kj}\partial \theta_{2j}}\mathbf{x}'_{kj}\mathbf{x}_{2j} & \cdots & \frac{\partial^2 \ln \ell_j}{\partial \theta_{kj}^2}\mathbf{x}'_{kj}\mathbf{x}_{kj} \end{pmatrix} \tag{6.3}$$

Now $\partial^2 \ln \ell_j/\partial \theta_{ij}\partial \theta_{kj}$ is just a scalar value for each observation, so we can store the values for all the observations in a variable. Moreover, in general, log-likelihood functions satisfy Young's Theorem so that $\partial^2 \ln \ell_j/\partial \theta_{ij}\partial \theta_{kj} = \partial^2 \ln \ell_j/\partial \theta_{kj}\partial \theta_{ij}$. We have E equations, so we have a total of $E(E+1)/2$ distinct submatrices in $\mathbf{H}$.

The syntax for mlmatsum is

mlmatsum *varname*lnfj *matrixname* = *exp* [*if*] [, eq(#[, #])]

Obtaining $\partial^2 \ln L/\partial \beta \partial \beta'$ is simply a matter of issuing one mlmatsum per equation pair and then joining the results into a single matrix.

In a single-equation system, we have

```
mlmatsum 'lnfj' 'H' = formula for ∂² ln ℓ_j/∂θ²_1j
```
mlmatsum `'lnfj'` `'H'` = *formula for* $\partial^2 \ln \ell_j/\partial\theta^2_{1j}$

In a two-equation system, we have

```
tempname d11 d12 d22
mlmatsum 'lnfj' 'd11' = formula for ∂² ln ℓ_j/∂θ²_1j     , eq(1)
mlmatsum 'lnfj' 'd12' = formula for ∂² ln ℓ_j/∂θ_1j∂θ_2j, eq(1,2)
mlmatsum 'lnfj' 'd22' = formula for ∂² ln ℓ_j/∂θ²_2j     , eq(2)
matrix 'H' = ('d11','d12' \ ('d12')','d22')
```

In the last line, we wrote (`'d12'`)' instead of `'d12'`' to emphasize that we are taking the transpose of `'d12'`.

Note that `mlmatsum` has two left-hand-side arguments:

```
. mlmatsum 'lnfj' 'whatever' = exp
```

`mlmatsum` checks that every observation in the estimation subsample has a nonmissing value for *exp* before computing the sum. If there are any missing values for *exp*, `mlmatsum` replaces the first observation on `'lnfj'` to missing to indicate to `ml` that the Hessian cannot be computed at the current value of `'b'`; `ml` will then search for a different value of `'b'`. Further advantages of using `mlmatsum` are that it restricts its calculations to the estimation sample and automatically applies the user's weights if specified.

▷ **Example**

Using the log likelihood from the probit model (5.1) and its first derivative (6.1), the Hessian is

$$\frac{d^2 \ln \ell_j}{d\theta_j^2} = \begin{cases} -R(\theta_j)\{R(\theta_j) + \theta_j\} & \text{if } y_j = 1 \\ -S(\theta_j)\{S(\theta_j) - \theta_j\} & \text{if } y_j = 0 \end{cases}$$

where

$$R(\theta_j) = \phi(\theta_j)/\Phi(\theta_j)$$
$$S(\theta_j) = \phi(\theta_j)/\Phi(-\theta_j)$$

`mlmatsum` can be used to obtain the Hessian by programming

```
tempvar R S h
generate double 'R' = normalden('xb')/normal('xb')
generate double 'S' = normalden('xb')/normal(-'xb')
generate double 'h' = -'R'*('R'+'xb') if $ML_y1 == 1
replace     'h' = -'S'*('S'-'xb') if $ML_y1 == 0
mlmatsum 'lnfj' 'H' = 'h'
```

Because the right-hand side of `mlmatsum` can be an expression, the above could be compressed into fewer lines using the `cond()` function, if preferred.

◁

6.2.6   Aside: Stata's scalars

To use methods lf1 and lf2 effectively—and other ml methods—you must use Stata's scalars if you make intermediate scalar calculations. Instead of programming

```
local eta = ...
generate double ... = ... `eta' ...
```

you should program

```
tempname eta
scalar `eta' = ...
generate double ... = ... `eta' ...
```

You do this for the same reason that you specify that the intermediate variables you generate are doubles: scalars are more accurate than macros. Macros provide about 12 digits of accuracy, sometimes more but never fewer. Scalars are full double-precision binary numbers, meaning that they provide about 16.5 digits of accuracy in all cases.

A Stata scalar is just that—it contains one number (which could be a missing value). The scalar command defines scalars, and you use scalars in expressions just as you would any variable:

```
. scalar x = 3
. display x
3
. scalar y = 1/sqrt(2)
. display y
.70710678
. scalar z = 1/ln(x-3)
. display z
.
```

We have been sloppy in the above example, although there is no reason you would know that. The example would have better read

```
. scalar x = 3
. display scalar(x)              // we type scalar(x), not x
3
. scalar y = 1/sqrt(2)
. display scalar(y)              // we type scalar(y), not y
.70710678
. scalar z = 1/ln(scalar(x)-3)  // we type scalar(x), not x
. display scalar(z)             // we type scalar(z), not z
.
```

The scalar() function says "use the value of the named scalar here", and as you have seen, we can omit it. We should not omit it, however, unless we are certain there is no variable in the dataset by the same name. Pretend that we had a variable named x and a scalar named x:

```
. clear all
. input x                          // create the variable name x

            x
  1. 1
  2. 2
  3. 3
  4. end
. scalar x = 57                    // create the scalar name x
. generate y = x+1                 // what is in y?
```

Notice that `generate y = x+1` did not result in an error, nor did defining a scalar named x when a variable named x already existed; neither would it have been an error to do things the other way around and create a variable named x when a scalar named x already existed. This, however, leads to an interesting question: to which are we referring when we type x in `generate y = x+1`? Is it the data variable, in which case y would contain 2, 3, and 4? Or is it the scalar, in which case y would contain 58, 58, and 58?

When Stata has a variable and a scalar of the same name, it chooses the variable over the scalar:

```
. list
```

	x	y
1.	1	2
2.	2	3
3.	3	4

We previously typed `generate y = x+1`. There is a scalar named x that contains 57; nevertheless, we obtained y = variable x + 1. If we really want Stata to use the scalar called x, we must tell it:

```
. generate z = scalar(x) + 1
. list
```

	x	y	z
1.	1	2	58
2.	2	3	58
3.	3	4	58

The same logic applies to all Stata commands, including `scalar` itself:

```
. scalar new = x
. display new
1
```

Why 1 and not 57? Because `scalar new = x` was interpreted as

```
. scalar new = /*VARIABLE*/ x
```

and that in turn was interpreted as

```
. scalar new = /*VARIABLE*/ x /*IN THE FIRST OBSERVATION*/
```

Had there instead been no variable called x, `scalar new = x` would have been interpreted as

```
. scalar new = /*SCALAR*/ x
```

An error would result if there were no variable and no scalar named x:

```
. scalar new = x
x not found
r(111);
```

Stata's preference for variables over scalars is even stronger than we have indicated. Here is another example:

```
. clear all
. input mpg weight
          mpg       weight
  1. 22 2100
  2. 23 1900
  3. 12 3800
  4. end
. scalar m = 9.5
. scalar z = m
. display z
22
```

Scalar z contains 22 and not 9.5 because m was taken as an abbreviation for variable mpg! Thus it appears that about the only safe way we can use scalars is to reference them inside the `scalar()` function:

```
. scalar z = scalar(m)
. display scalar(z)
9.5
```

In fact, we seldom do that. Instead, we write our programs using the `tempname` statement:

```
tempname x y
scalar 'x' = (some complicated function)
scalar 'y' = ... 'x' ...
```

The last line could have read

```
scalar 'y' = ... scalar('x') ...
```

but the `scalar()` is not necessary. `tempname` will create a unique name for the scalar, and we can be certain that there is no variable by the same name, abbreviated or not.

6.3   Summary of methods lf0, lf1, and lf2

6.3.1   Method lf0

The method-`lf0` evaluator calculates the observation-level log likelihood but does not compute any derivatives. Method `lf0` is a stepping stone toward writing an `lf1` or `lf2` evaluator. If you have no intention of ever programming the derivatives, you should use method `lf` instead, because it is easier to use.

The outline of a method-`lf0` evaluator is

```
program myprog
        version 18
        args todo b lnfj
        // if you need to create any intermediate results:
        tempvar tmp1 tmp2
        quietly generate double 'tmp1' = ...
        ...
        quietly replace 'lnfj' = ...
end
```

In section 6.2, we showed that for methods `lf0`, `lf1`, and `lf2`, the `args` statement has the form

```
args todo b lnfj g1 g2 ... H
```

Method-`lf0` evaluators are always called with 'todo' $= 0$ only and are never requested to compute scores or the Hessian matrix, so the shorter version

```
args todo b lnfj
```

will work just as well, though you must remember to modify that line when you convert your `lf0` evaluator to an `lf1` or `lf2` evaluator.

You specify that you are using method `lf0` with program *myprog* on the `ml model` statement:

```
. ml model lf0 myprog ...
```

Unlike method `lf`, which passes to your evaluator the parameters θ_i, method-`lf0` evaluators receive the coefficient vector 'b'. Because the likelihood function is generally written as a function of the parameters θ_i from the probability model, the first step is to obtain their values from the coefficients vector 'b':

```
program myprog
        version 18
        args todo b lnf g
        tempvar theta1 theta2 ...
        mleval 'theta1' = 'b'
        mleval 'theta2' = 'b', eq(2)            // if there is a θ₂
        ...
end
```

The following points enumerate what `ml` supplies and expects of a method-`lf0` likelihood evaluator.

- You are passed the vector of coefficients. Use `mleval` to obtain the values of each of the linear predictors of each equation from it; see section 6.2.2.

- The names of the dependent variables (if any) are placed in global macros by `ml`; the first dependent variable is `$ML_y1`, the second is `$ML_y2`, ... Use `$ML_y1`, `$ML_y2`, ... just as you would any existing variable in the data.

- You must create any temporary variables for intermediate results as `doubles`. Type

    generate double *name* = ...

  Do not omit the word `double`.

- The estimation subsample is identified by `$ML_samp==1`. You may safely ignore this if you do not specify `ml model`'s `nopreserve` option. When your program is called, only relevant data will be in memory. Even if you do specify `nopreserve`, `mleval` automatically restricts itself to the estimation subsample; it is not necessary to include `if $ML_samp==1` with this command. However, you will need to restrict other calculation commands to the `$ML_samp==1` subsample if you specify `nopreserve`.

- The observation-level log-likelihood value is to be stored in the variable '`lnfj`'.

6.3.2   Method lf1

The method-`lf1` evaluator calculates the observation-level log likelihood and scores. The outline for a method-`lf1` likelihood evaluator is

```
program myprog
        version 18
        args todo b lnfj g1 g2 ...
        // if you need to create any intermediate results:
        tempvar tmp1 tmp2
        quietly generate double 'tmp1' = ...
        ...
        quietly replace 'lnfj' = ...
        if ('todo'==0 | 'lnfj' >= .) exit        // method-lf0 evaluators end here
        // fill in score variables
        quietly replace 'g1' = ∂ ln ℓ_j/∂θ_{1j}
        quietly replace 'g2' = ∂ ln ℓ_j/∂θ_{2j}
end
```

The final argument that `lf`-family evaluators receive is H, used to store the Hessian matrix. Method-`lf1` evaluators are never required to compute the Hessian, so we can omit H from the `args` statement. You specify that you are using method `lf1` with program *myprog* on the `ml model` statement, which you issue after you have defined the likelihood-evaluation program.

```
. ml model lf1 myprog ...
```

As with method lf0, the likelihood evaluator receives the coefficient vector 'b', so typically, the first part of your program will need to extract the θ parameters using mleval.

Method lf1 also requires you to calculate the scores $\partial \ln \ell_j / \partial \theta_{1j}$, $\partial \ln \ell_j / \partial \theta_{2j}$, ... In many cases, the log-likelihood function and scores contain the same terms [such as $\phi(\mathbf{x}\boldsymbol{\beta})$], so to increase readability of your program, you should first create temporary variables (as type double) to hold those terms.

```
program myprog
        version 18
        args todo b lnfj g1 g2 ...
        tempvar theta1 theta2 ...
        mleval 'theta1' = 'b'
        mleval 'theta2' = 'b', eq(2)              // if there is a θ2
        ...
        tempvar tmp1 tmp2
        quietly generate double 'tmp1' = ...
        quietly generate double 'tmp2' = ...
        ...
        quietly replace 'lnfj' = expression in terms of 'tmp1' and 'tmp2'
        if ('todo'==0 | 'lnfj'>=.) exit

        replace 'g1' = formula for ∂ln ℓⱼ/∂θ₁ⱼ in terms of 'tmp1' and 'tmp2'
        replace 'g2' = formula for ∂ln ℓⱼ/∂θ₂ⱼ in terms of 'tmp1' and 'tmp2'
        ...
end
```

The following points enumerate what ml supplies and expects of a method-lf1 likelihood evaluator.

- You are passed the vector of coefficients. Use mleval to obtain the values of each of the linear predictors of each equation from it; see section 6.2.2.

- The names of the dependent variables (if any) are placed in global macros by ml; the first dependent variable is \$ML_y1, the second is \$ML_y2, ... Use \$ML_y1, \$ML_y2, ... just as you would any existing variable in the data.

- You must create any temporary variables for intermediate results as doubles. Type

    ```
    generate double name = ...
    ```

  Do not omit the word double.

- The estimation subsample is identified by \$ML_samp==1. You may safely ignore this if you do not specify ml model's nopreserve option. When your program is called, only relevant data will be in memory. Even if you do specify nopreserve, mleval automatically restricts itself to the estimation subsample; it is not necessary to include if \$ML_samp==1 with this command. However, you will need to restrict other calculation commands to the \$ML_samp==1 subsample if you specify nopreserve.

- The observation-level log-likelihood value is to be stored in the variable 'lnfj'.

- Observation-level scores for each equation are to be stored in variables 'g1', 'g2', and so on.

6.3.3   Method lf2

The method-lf2 evaluator calculates the observation-level log likelihood and scores and the Hessian matrix of the overall log likelihood. The outline for the program is

```
program myprog
        version 18
        args todo b lnfj g1 g2 ... H
        // if you need to create any intermediate results:
        tempvar tmp1 tmp2
        quietly generate double 'tmp1' = ...
        ...
        quietly replace 'lnfj' = ...
        if ('todo'==0 | 'lnfj' >= .) exit          // method-lf0 evaluators end here
        // fill in score variables
        quietly replace 'g1' = ∂ ln ℓ_j/∂θ_{1j}
        quietly replace 'g2' = ∂ ln ℓ_j/∂θ_{2j}
        ...
        if ('todo'==1) exit                        // method-lf1 evaluators end here
        // compute Hessian matrix
        tempname d11 d12 d22 ...
        mlmatsum 'lnfj' 'd11' = ∂² ln L/∂θ₁²
        mlmatsum 'lnfj' 'd12' = ∂² ln L/∂θ₁∂θ₂
        mlmatsum 'lnfj' 'd22' = ∂² ln L/∂θ₂²
        matrix 'H' = ('d11', 'd12', ...  'd12'', 'd22', ...)
end
```

You specify that you are using method lf2 on the ml model statement, which you issue after you have defined the likelihood-evaluation program.

```
. ml model lf2 myprog ...
```

The method-lf2 evaluator continues from where the method-lf1 evaluator left off. Taking out the common code, here is how the evaluators differ:

```
program myprog
        version 18
        args todo b lnfj g1 g2 ... H
        ...
        if ('todo'==1 | 'lnfj'==.) exit            // from here down is new
        tempname d11 d12 d22 ...
        mlmatsum 'lnfj' 'd11' = formula for ∂² ln ℓ_j/∂θ_{1j}², eq(1)
        mlmatsum 'lnfj' 'd12' = formula for ∂² ln ℓ_j/∂θ_{1j}∂θ_{2j}, eq(1,2)
        mlmatsum 'lnfj' 'd22' = formula for ∂² ln ℓ_j/∂θ_{2j}², eq(2)
        ...
        matrix 'H' = ('d11','d12',... \ 'd12'','d22',...)
end
```

The Hessian matrix you are to define is

$$\mathbf{H} = \frac{\partial^2 \ln L}{\partial\boldsymbol{\beta}\partial\boldsymbol{\beta}'} = \begin{pmatrix} \frac{\partial^2 \ln L}{\partial\beta_1^2} & \frac{\partial^2 \ln L}{\partial\beta_1\partial\beta_2} & \cdots & \frac{\partial^2 \ln L}{\partial\beta_1\partial\beta_k} \\ \frac{\partial^2 \ln L}{\partial\beta_2\partial\beta_1} & \frac{\partial^2 \ln L}{\partial\beta_2^2} & \cdots & \frac{\partial^2 \ln L}{\partial\beta_2\partial\beta_k} \\ \vdots & \vdots & \ddots & \vdots \\ \frac{\partial^2 \ln L}{\partial\beta_k\partial\beta_1} & \frac{\partial^2 \ln L}{\partial\beta_k\partial\beta_2} & \cdots & \frac{\partial^2 \ln L}{\partial\beta_k^2} \end{pmatrix}$$

`ml` provides the coefficient vector as a $1 \times k$ row vector and expects the Hessian matrix to be returned as a $k \times k$ matrix. Because method-`lf2` evaluators satisfy the linear-form restrictions, you simply need to compute the second derivatives with respect to the equation-level $\boldsymbol{\theta}$s. Then you can use `mlmatsum` to create the required submatrices of $\mathbf{H}$ without having to worry about the dimensions of β or the variables in the model; see section 6.2.5 for more information.

The following points enumerate what `ml` supplies and expects of a method-`lf2` likelihood evaluator.

- You are passed the vector of coefficients. Use `mleval` to obtain the values of each of the linear predictors of each equation from it; see section 6.2.2.

- The names of the dependent variables (if any) are placed in global macros by `ml`; the first dependent variable is `$ML_y1`, the second is `$ML_y2`, ... Use `$ML_y1`, `$ML_y2`, ... just as you would any existing variable in the data.

- You must create any temporary variables for intermediate results as `doubles`. Type

      generate double *name* = ...

  Do not omit the word `double`.

- The estimation subsample is identified by `$ML_samp==1`. You may safely ignore this if you do not specify `ml model`'s `nopreserve` option. When your program is called, only relevant data will be in memory. Even if you do specify `nopreserve`, `mleval` automatically restricts itself to the estimation subsample; it is not necessary to include `if $ML_samp==1` with this command. However, you will need to restrict other calculation commands to the `$ML_samp==1` subsample if you specify `nopreserve`.

- The observation-level log-likelihood value is to be stored in the variable '`lnfj`'.

- Observation-level scores for each equation are to be stored in variables '`g1`', '`g2`', and so on.

- The Hessian of the overall log likelihood is to be stored in matrix '`H`'. Use the `mlmatsum` command to produce the various submatrices; it automatically handles weights and properly restricts the summations to the estimation sample.

6.4   Examples

We will now illustrate the implementation of method-lf1 and method-lf2 likelihood evaluators for the examples of chapter 5: probit, linear regression, and Weibull models. All of these models meet the linear-form restrictions. Because there is no point in implementing these estimators with method lf0 evaluator functions other than as a first step toward implementing lf1 and lf2 evaluators, we do not consider lf0 evaluators except for debugging purposes.

6.4.1   The probit model

The probit log-likelihood function for the jth observation is

$$\ln \ell_j = \begin{cases} \ln \Phi(\mathbf{x}_j\boldsymbol{\beta}) & \text{if } y_j = 1 \\ \ln \Phi(-\mathbf{x}_j\boldsymbol{\beta}) & \text{if } y_j = 0 \end{cases}$$

Taking derivatives with respect to $\mathbf{x}_j\boldsymbol{\beta}$, the jth term for the gradient is

$$g_j = \frac{\partial \ln \ell_j}{\partial \mathbf{x}_j\boldsymbol{\beta}} = \begin{cases} \phi(\mathbf{x}_j\boldsymbol{\beta})/\Phi(\mathbf{x}_j\boldsymbol{\beta}) & \text{if } y_j = 1 \\ -\phi(\mathbf{x}_j\boldsymbol{\beta})/\Phi(-\mathbf{x}_j\boldsymbol{\beta}) & \text{if } y_j = 0 \end{cases}$$

where $\phi(\cdot)$ is the probability density function for the standard normal distribution. Thus our method-lf1 likelihood evaluator is

————————————————————— begin myprobit_lf1.ado —————————

```
program myprobit_lf1
        version 18
        args todo b lnfj g1
        tempvar xb lj
        mleval `xb' = `b'
        quietly {
                // Create temporary variable used in both likelihood
                // and scores
                generate double `lj' = normal( `xb') if $ML_y1 == 1
                replace      `lj' = normal(-`xb') if $ML_y1 == 0
                replace      `lnfj' = log(`lj')
                if (`todo'==0) exit

                replace `g1' =  normalden(`xb')/`lj' if $ML_y1 == 1
                replace `g1' = -normalden(`xb')/`lj' if $ML_y1 == 0
        }
end
```

————————————————————————— end myprobit_lf1.ado —————————

Taking the second derivative with respect to $\mathbf{x}_j\boldsymbol{\beta}$, the jth term of the Hessian is

$$H_j = \frac{\partial^2 \ln \ell_j}{\partial^2 \mathbf{x}_j\boldsymbol{\beta}} = -g_j(g_j + \mathbf{x}_j\boldsymbol{\beta})$$

and our method-lf2 likelihood evaluator is

```
                                         begin myprobit_lf2.ado
program myprobit_lf2
        version 18
        args todo b lnfj g1 H
        tempvar xb lj
        mleval 'xb' = 'b'
        quietly {
                // Create temporary variable used in both likelihood
                // and scores
                generate double 'lj' = normal( 'xb')  if $ML_y1 == 1
                replace    'lj' = normal(-'xb')  if $ML_y1 == 0
                replace    'lnfj' = log('lj')
                if ('todo'==0) exit

                replace 'g1' =  normalden('xb')/'lj'  if $ML_y1 == 1
                replace 'g1' = -normalden('xb')/'lj'  if $ML_y1 == 0
                if ('todo'==1) exit

                mlmatsum 'lnfj' 'H' = -'g1'*('g1'+'xb'), eq(1,1)
        }
end
                                         end myprobit_lf2.ado
```

▷ **Example**

Using cancer.dta and myprobit_lf2, we fit a probit model of the incidence of death on drug type and age:

```
. sysuse cancer, clear
(Patient survival in drug trial)
. ml model lf2 myprobit_lf2 (died = i.drug age)
. ml maximize
Initial:      Log likelihood = -33.271065
Alternative:  Log likelihood = -31.427839
Rescale:      Log likelihood = -31.424556
Iteration 0:  Log likelihood = -31.424556
Iteration 1:  Log likelihood = -21.884411
Iteration 2:  Log likelihood = -21.710897
Iteration 3:  Log likelihood = -21.710799
Iteration 4:  Log likelihood = -21.710799
```

		Number of obs =	48
		Wald chi2(3) =	13.58
Log likelihood = -21.710799		Prob > chi2 =	0.0035

died	Coefficient	Std. err.	z	P>|z|	[95% conf. interval]	
drug						
Other	-1.941275	.597057	-3.25	0.001	-3.111485	-.7710644
NA	-1.792924	.5845829	-3.07	0.002	-2.938686	-.6471626
age	.0666733	.0410666	1.62	0.104	-.0138159	.1471624
_cons	-2.044876	2.284283	-0.90	0.371	-6.521988	2.432235

◁

The first time we tried fitting this model, here is what we obtained:

```
. ml model lf2 myprobit_lf2 (died = i.drug age)
. ml maximize
Initial:       Log likelihood = -33.271065
Alternative:   Log likelihood = -31.427839
Rescale:       Log likelihood = -31.424556
varlist required
r(100);
```

How did we go about finding our mistake? In chapter 8, we will learn about tools available to check our derivative calculations, but in this case, we did not need them. `myprobit_lf2.ado` is an `lf2` evaluator, but it also satisfies the requirements of `lf0` and `lf1` evaluators. If we type

```
. ml model lf1 myprobit_lf2 (died = i.died age)
. ml maximize
```

and do not receive any errors, we can deduce that the problem with our likelihood evaluator must lie in the portion of the program that computes the Hessian. If we do receive an error, our next step would be to try

```
. ml model lf0 myprobit_lf2 (died = i.died age)
. ml maximize
```

If that works, we at least know we are computing the log likelihood correctly.

6.4.2 Normal linear regression

The classical normal linear regression log-likelihood function for the jth observation is

$$\ln \ell_j = \ln \left\{ \frac{1}{\sigma_j} \phi \left(\frac{y_j - \mu_j}{\sigma_j} \right) \right\}$$

For this example, we will model the standard deviation in the log space to ensure it is positive. Moreover, we have made the standard deviation observation specific (notice the j subscript) so that we can incorporate conditional heteroskedasticity in our model. Using the notation from section 4.2, our second equation is

$$\theta_{2j} = \ln \sigma_j = \mathbf{x}_{2j} \boldsymbol{\beta}_2$$

We can use `normalden(`y, μ, σ`)` to quickly compute

$$\frac{1}{\sigma} \phi \left(\frac{y - \mu}{\sigma} \right)$$

without having to do the subtraction and two divisions ourselves. The derivative of $\phi(z)$ with respect to z is

$$\phi'(z) = -z\phi(z)$$

Taking the derivatives of $\ln \ell_j$ with respect to μ_j and $\ln \sigma_j$ yields the scores for the jth observation:

$$g_{1j} = \frac{\partial \ln \ell_j}{\partial \mu_j} = \frac{y_j - \mu_j}{\sigma_j^2}$$

$$g_{2j} = \frac{\partial \ln \ell_j}{\partial \ln \sigma_j} = \left(\frac{y_j - \mu_j}{\sigma_j}\right)^2 - 1$$

Our method-`lf1` likelihood evaluator is

———————————————————————————— begin `mynormal_lf1.ado` ————

```
program mynormal_lf1
        version 18
        args todo b lnfj g1 g2
        tempvar mu lnsigma sigma
        mleval `mu' = `b', eq(1)
        mleval `lnsigma' = `b', eq(2)
        quietly {
                generate double `sigma' = exp(`lnsigma')
                replace `lnfj' = ln( normalden($ML_y1,`mu',`sigma') )
                if (`todo'==0) exit

                tempvar z
                generate double `z' = ($ML_y1-`mu')/`sigma'
                replace `g1' = `z'/`sigma'
                replace `g2' = `z'*`z'-1
        }
end
```

———————————————————————————— end `mynormal_lf1.ado` ————

Taking the second derivatives of $\ln \ell_j$ yields the jth element's observations to the Hessian matrix:

$$H_{11j} = \frac{\partial^2 \ell_j}{\partial^2 \mu_j} = -\frac{1}{\sigma_j^2}$$

$$H_{12j} = \frac{\partial^2 \ell_j}{\partial \mu_j \partial \ln \sigma_j} = -2\frac{y_j - \mu_j}{\sigma_j^2}$$

$$H_{22j} = \frac{\partial^2 \ell_j}{\partial^2 \ln \sigma_j} = -2\left(\frac{y_j - \mu_j}{\sigma_j}\right)^2$$

Our method-lf2 likelihood evaluator is

——————————————————————————— begin `mynormal_lf2.ado` ———————

```
program mynormal_lf2
        version 18
        args todo b lnfj g1 g2 H
        tempvar mu lnsigma sigma
        mleval 'mu' = 'b', eq(1)
        mleval 'lnsigma' = 'b', eq(2)
        quietly {
                generate double 'sigma' = exp('lnsigma')
                replace 'lnfj' = ln( normalden($ML_y1,'mu','sigma') )
                if ('todo'==0) exit

                tempvar z
                generate double 'z' = ($ML_y1-'mu')/'sigma'
                replace 'g1' = 'z'/'sigma'
                replace 'g2' = 'z'*'z'-1
                if ('todo'==1) exit

                tempname d11 d12 d22
                mlmatsum 'lnfj' 'd11' = -1/'sigma'^2      , eq(1)
                mlmatsum 'lnfj' 'd12' = -2*'z'/'sigma'    , eq(1,2)
                mlmatsum 'lnfj' 'd22' = -2*'z'*'z'        , eq(2)
                matrix 'H' = ('d11', 'd12' \ 'd12'', 'd22')
        }
end
```

——————————————————————————— end `mynormal_lf2.ado` ———————

▷ **Example**

Using `auto.dta` and `mynormal_lf1`, we fit a linear regression with the `vce(robust)` option.

```
─────────────────────────────── begin ex_normal_lf1 ───────────
. sysuse auto
(1978 automobile data)

. ml model lf1 mynormal_lf1 (mu: mpg = weight displacement) /lnsigma,
> vce(robust)

. ml maximize
Initial:      Log pseudolikelihood =      -<inf>  (could not be evaluated)
Feasible:     Log pseudolikelihood = -6441.8168
Rescale:      Log pseudolikelihood = -368.12496
Rescale eq:   Log pseudolikelihood =  -257.3948
Iteration 0:  Log pseudolikelihood =  -257.3948  (not concave)
Iteration 1:  Log pseudolikelihood = -241.78314  (not concave)
Iteration 2:  Log pseudolikelihood =  -231.8097  (not concave)
Iteration 3:  Log pseudolikelihood = -222.78552  (not concave)
Iteration 4:  Log pseudolikelihood = -212.75192
Iteration 5:  Log pseudolikelihood = -197.06449
Iteration 6:  Log pseudolikelihood = -195.24283
Iteration 7:  Log pseudolikelihood =  -195.2398
Iteration 8:  Log pseudolikelihood =  -195.2398
```

Number of obs =	74
Wald chi2(2) =	111.39
Prob > chi2 =	0.0000

Log pseudolikelihood = -195.2398

mpg	Coefficient	Robust std. err.	z	P>|z|	[95% conf. interval]
weight	-.0065671	.0009195	-7.14	0.000	-.0083692 -.004765
displacement	.0052808	.0073381	0.72	0.472	-.0091017 .0196633
_cons	40.08452	2.045516	19.60	0.000	36.07538 44.09366
/lnsigma	1.219437	.1342357	9.08	0.000	.9563401 1.482534

```
─────────────────────────────── end ex_normal_lf1 ───────────
```

◁

▷ **Example**

When developing this estimator, we made σ, the error variance, observation specific. That is, in the previous example, we had as our second equation

$$\theta_{2j} = \ln \sigma_j = \mathbf{x}_{2j}\boldsymbol{\beta}_2$$

where $\mathbf{x}_{2j} = 1$ for all j. By examining evaluators `mynormal_lf1` and `mynormal_lf2`, you will notice that we have followed all the ordinary prescriptions, meaning that a linear equation can be estimated for $\ln \sigma_j$.

In the example below, we fit a model in which we assume that the variance of the disturbance is given by

$$\ln \sigma_j = \beta_{2,1}\texttt{price}_j + \beta_{2,2}$$

```
. sysuse auto
(1978 automobile data)

. ml model lf1 mynormal_lf1 (mu: mpg = weight displacement) (lnsigma: price),
> vce(robust)

. ml maximize
Initial:      Log pseudolikelihood =      -<inf>  (could not be evaluated)
Feasible:     Log pseudolikelihood = -6441.8168
Rescale:      Log pseudolikelihood = -368.12496
Rescale eq:   Log pseudolikelihood = -257.3948
Iteration 0:  Log pseudolikelihood = -257.3948  (not concave)
Iteration 1:  Log pseudolikelihood = -241.3443  (not concave)
Iteration 2:  Log pseudolikelihood = -230.49029 (not concave)
Iteration 3:  Log pseudolikelihood = -222.66544 (not concave)
Iteration 4:  Log pseudolikelihood = -218.79323
Iteration 5:  Log pseudolikelihood = -202.46255
Iteration 6:  Log pseudolikelihood = -195.15622
Iteration 7:  Log pseudolikelihood = -195.13889
Iteration 8:  Log pseudolikelihood = -195.13887
Iteration 9:  Log pseudolikelihood = -195.13887

                                              Number of obs =      74
                                              Wald chi2(2)  = 115.98
Log pseudolikelihood = -195.13887             Prob > chi2   = 0.0000
```

mpg	Coefficient	Robust std. err.	z	P>|z|	[95% conf. interval]	
mu						
weight	-.0065244	.0009002	-7.25	0.000	-.0082887	-.0047601
displacement	.0041361	.0071429	0.58	0.563	-.0098638	.018136
_cons	40.19516	2.032263	19.78	0.000	36.212	44.17832
lnsigma						
price	.0000129	.0000267	0.48	0.630	-.0000396	.0000653
_cons	1.138677	.2280395	4.99	0.000	.6917278	1.585626

This is an exciting example: we have just fit a model with heteroskedasticity without changing our code one bit. The way `ml` is designed, you get the flexibility of allowing a parameter to depend on covariates with no extra work on your part. Arguably, you have to think harder to make terms like $\ln \sigma$ be free parameters rather than equations.

In the case of `mynormal_lf2`, the sixth line reads

```
mleval 'lnsigma' = 'b', eq(2)
```

If we wanted to treat $\ln \sigma$ as a free parameter, we would have had to code

```
mleval 'lnsigma' = 'b', eq(2) scalar
```

That is the only change we would need to make, though the technical note below points out several additional changes we should make. The advantage to making $\ln \sigma$ a free parameter is that our program would run a little faster. The disadvantage is that we would lose the ability to fit heteroskedastic models. We like our program as it is.

◁

❑ Technical note

In the example above, we said that we need to make only one change to our program if we want to force $\ln \sigma$ to be estimated as a free parameter, modifying the line

```
mleval 'lnsigma' = 'b', eq(2)
```

to be

```
mleval 'lnsigma' = 'b', eq(2) scalar
```

There is a second change we should make. Earlier in the program, where it says

```
tempvar mu lnsigma sigma
```

we should change it to read

```
tempvar mu sigma
tempname lnsigma
```

We should make that change because 'lnsigma' will no longer be a temporary variable in our dataset but instead will be a temporary scalar. That was, after all, the point of adding the scalar option to the mleval command. It does not matter whether you make that extra change. The way Stata works, you may use tempvars for tempnames and vice versa. What you must not do is use the same name twice for two different things, and we are not doing that.

If tempvar and tempname are interchangeable, why does Stata have both? The answer concerns programming style: your program will be more readable if it clearly indicates how the name 'lnsigma' will be used. On that basis alone, you ought to make the change.

There is yet another change you ought to make, too. This one has to do with efficiency. Later in the code, after

```
mleval 'lnsigma' = 'b', eq(2) scalar
```

is a line that reads

```
generate double 'sigma' = exp('lnsigma')
```

That line ought to be change to read

```
scalar 'sigma' = exp('lnsigma')
```

Making that change will make your program run faster. The idea here is that if we are going to force the parameter 'lnsigma' to be a scalar, then we should look throughout the rest of our code and exploit the speed advantages associated with scalars wherever we can.

Here is a version of `mynormal_lf2` incorporating those changes:

─────────────────────────────────── begin `mynormal_nohet_lf2.ado` ───────────

```
program mynormal_nohet_lf2
        version 18
        args todo b lnfj g1 g2 H
        tempvar mu
        tempname lnsigma sigma
        mleval 'mu' = 'b', eq(1)
        mleval 'lnsigma' = 'b', eq(2) scalar
        scalar 'sigma' = exp('lnsigma')
        quietly {
                replace 'lnfj' = ln( normalden($ML_y1,'mu','sigma') )
                if ('todo'==0) exit

                tempvar z
                generate double 'z' = ($ML_y1-'mu')/'sigma'
                replace 'g1' = 'z'/'sigma'
                replace 'g2' = 'z'*'z'-1
                if ('todo'==1) exit

                tempname d11 d12 d22
                mlmatsum 'lnfj' 'd11' = -1/'sigma'^2      , eq(1)
                mlmatsum 'lnfj' 'd12' = -2*'z'/'sigma'    , eq(1,2)
                mlmatsum 'lnfj' 'd22' = -2*'z'*'z'        , eq(2)
                matrix 'H' = ('d11', 'd12' \ 'd12'', 'd22')
        }
end
```

─────────────────────────────── end `mynormal_nohet_lf2.ado` ───────────

The program `mynormal_nohet_lf2` is not strictly better than the original `mynormal_lf2`. Indeed, one could argue that it is not as good because it does not let us estimate heteroskedastic models. `mynormal_nohet_lf2` is better only in that it is faster, and in some situations, speed may be more important than flexibility.

❏

In the previous example, we modeled the heteroskedasticity as

$$\ln \sigma_j = \beta_{2,1}\texttt{price}_j + \beta_{2,2}$$

whereas in (5.3), we modeled the heteroskedasticity as

$$\sigma_j = \beta_4\texttt{price}_j + \beta_5$$

Equation (5.3) does not guarantee that σ_j will be positive for all observations as a standard deviation must be, though after we showed that example, we marveled at the fact that `ml` was smart enough to recognize invalid parameter values and work around them. In contrast, the version we used in the previous example does ensure that $\sigma_j > 0$ for all values of $\beta_{2,1}$ and $\beta_{2,2}$.

More substantive is the fact that the example in this chapter is of *multiplicative* heteroskedasticity, meaning that the variance of the error term in a given observation is proportional to the regressor. This is the version of heteroskedastic linear regression introduced by Harvey (1976) and taught in most textbooks (for example, Goldberger [1991]). In practice, we would use the version implemented by `mynormal_lf2` rather than the version implemented in (5.3).

6.4.3   The Weibull model

A Weibull log-likelihood function for the jth observation, where t_j is the time of failure or censoring and $d_j = 1$ if the observation represents a failure and 0 if the observation is censored, is

$$\ln \ell_j = -(t_j/e^{\theta_{1j}})^{e^{\theta_{2j}}} + d_j\{\theta_{2j} - \theta_{1j} + (e^{\theta_{2j}} - 1)(\ln t_j - \theta_{1j})\}$$
$$\theta_{1j} = \ln \eta_j = \mathbf{x}_{1j}\boldsymbol{\beta}_2$$
$$\theta_{2j} = \ln \gamma_j = \mathbf{x}_{2j}\boldsymbol{\beta}_1$$

Before we take the derivatives of the log likelihood, we rewrite this function using slightly different notation. Define the variables

$$p_j = e^{\theta_{2j}}$$
$$M_j = (t_j e^{-\theta_{1j}})^{e^{\theta_{2j}}}$$
$$R_j = \ln t_j - \theta_{1j}$$

so that we can rewrite our log likelihood as

$$\ln \ell_j = -M_j + d_j\{\theta_{2j} - \theta_{1j} + (p_j - 1)R_j\}$$

Taking the derivatives of p_j, R_j, and M_j with respect to θ_{1j} and θ_{2j} yields

$$\frac{\partial p_j}{\partial \theta_{1j}} = 0 \qquad\qquad\qquad \frac{\partial p_j}{\partial \theta_{2j}} = p_j$$
$$\frac{\partial M_j}{\partial \theta_{1j}} = -p_j M_j \qquad\qquad \frac{\partial M_j}{\partial \theta_{2j}} = R_j p_j M_j$$
$$\frac{\partial R_j}{\partial \theta_{1j}} = -1 \qquad\qquad\qquad \frac{\partial R_j}{\partial \theta_{2j}} = 0$$

Then by using the chain rule, the jth observation's scores are

$$g_{1j} = \frac{\partial \ln \ell_j}{\partial \theta_{1j}} = p_j(M_j - d_j)$$
$$g_{2j} = \frac{\partial \ln \ell_j}{\partial \theta_{2j}} = d_j - R_j p_j(M_j - d_j)$$

and our method-`lf1` likelihood evaluator is

-- begin `myweibull_lf1.ado` ----------

```
program myweibull_lf1
        version 18
        args todo b lnf g1 g2
        tempvar leta lgam p M R
        mleval 'leta' = 'b', eq(1)
        mleval 'lgam' = 'b', eq(2)
        local t "$ML_y1"
        local d "$ML_y2"
        quietly {
                generate double 'p' = exp('lgam')
                generate double 'M' = ('t'*exp(-'leta'))^'p'
                generate double 'R' = ln('t')-'leta'
                replace 'lnf' = -'M' + 'd'*('lgam'-'leta' + ('p'-1)*'R')
                if ('todo'==0) exit
                replace 'g1' = 'p'*('M'-'d')
                replace 'g2' = 'd' - 'R'*'p'*('M'-'d')
        }
end
```

------------------------------------- end `myweibull_lf1.ado` ----------

Taking the second derivatives yields the jth observation's contribution to the Hessian:

$$H_{11j} = \frac{\partial^2 \ln \ell_j}{\partial^2 \theta_{1j}} = -p_j^2 M_j$$

$$H_{22j} = \frac{\partial^2 \ln \ell_j}{\partial^2 \theta_{2j}} = -p_j R_j (R_j p_j M_j + M_j - d_j)$$

$$H_{12j} = \frac{\partial^2 \ln \ell_j}{\partial \theta_{1j} \partial \theta_{2j}} = p_j (M_j - d_j + R_j p_j M_j)$$

Our method-`lf2` likelihood evaluator is

─────────────────────────────────── begin myweibull_lf2.ado ───────

```
program myweibull_lf2
        version 18
        args todo b lnf g1 g2 H
        tempvar leta lgam p M R
        mleval `leta' = `b', eq(1)
        mleval `lgam' = `b', eq(2)
        local t "$ML_y1"
        local d "$ML_y2"
        quietly {
                generate double `p' = exp(`lgam')
                generate double `M' = (`t'*exp(-`leta'))^`p'
                generate double `R' = ln(`t')-`leta'
                replace `lnf' = -`M' + `d'*(`lgam'-`leta' + (`p'-1)*`R')
                if (`todo'==0) exit

                replace `g1' = `p'*(`M'-`d')
                replace `g2' = `d' - `R'*`p'*(`M'-`d')
                if (`todo'==1) exit

                tempname d11 d12 d22
                mlmatsum `lnf' `d11' = -`p'^2 * `M'                    , eq(1)
                mlmatsum `lnf' `d12' = `p'*(`M'-`d'+`R'*`p'*`M')     , eq(1,2)
                mlmatsum `lnf' `d22' = -`p'*`R'*(`R'*`p'*`M'+`M'-`d') , eq(2)
                matrix `H' = (`d11',`d12' \ `d12'',`d22')
        }
end
```

───────────────────────────────────── end myweibull_lf2.ado ───────

▷ **Example**

Using `cancer.dta` and `myweibull_lf2`, we fit the Weibull model of time of death with censoring on drug type and age.

```
. sysuse cancer
(Patient survival in drug trial)
. ml model lf2 myweibull_lf2 (lneta: studytime died = i.drug age) /lngamma
. ml maximize, nolog
```

				Number of obs	=	48
				Wald chi2(3)	=	35.25
Log likelihood = -110.26736				Prob > chi2	=	0.0000

	Coefficient	Std. err.	z	P>|z|	[95% conf. interval]	
drug						
Other	1.012966	.2903917	3.49	0.000	.4438086	1.582123
NA	1.45917	.2821195	5.17	0.000	.9062261	2.012114
age	-.0671728	.0205688	-3.27	0.001	-.1074868	-.0268587
_cons	6.060723	1.152845	5.26	0.000	3.801188	8.320259
/lngamma	.5573333	.1402154	3.97	0.000	.2825163	.8321504

◁

▷ **Example**

Just as with the second example of section 6.4.2, we can parameterize the second equation of our model ($\ln \gamma_j$) with a linear equation simply by modifying what we type on the ml model statement:

```
. ml model lf2 myweibull_lf2
>      (lneta: studytime died = i.drug age)
>      (lngamma: female)
```

In the above command, we would be specifying

$$\ln \gamma_j = \beta_{2,1} \mathtt{female}_j + \beta_{2,2}$$
$$= \left\{ \begin{array}{ll} \beta_{2,2} & \text{if male} \\ \beta_{2,1} + \beta_{2,2} & \text{if female} \end{array} \right.$$

Linear equations can be specified for any of the estimated equations. As in the linear regression example in section 6.4.2, if we wanted to limit our program so that $\ln \gamma_j$ were a free parameter, we would change

```
mleval 'lgam' = 'b', eq(2)
```

to

```
mleval 'lgam' = 'b', eq(2) scalar
```

and we would declare lgam to be a tempname rather than a tempvar. After those changes, our program would run slightly faster.

◁

7 Methods d0, d1, and d2

Methods d0, d1, and d2 are used when the likelihood function does not meet the linear-form restrictions. Method d0 requires you to evaluate the overall log likelihood, and it calculates the gradient and Hessian numerically. Method d1 requires you to evaluate the overall log likelihood and analytic gradient, and it calculates the Hessian numerically based on your analytic gradient. Method d2 requires you to do all the work yourself, evaluating the overall log likelihood and the analytic gradient and Hessian.

These d-family evaluators can be used to fit models that satisfy the linear-form requirements, but you are better off using method lf or one of the lf-family evaluators. The d-family evaluators do not use observation-level scores, so you cannot obtain robust or cluster–robust standard errors, work with survey data and sampling weights, or use the BHHH maximization algorithm. Also, d-family evaluators are slightly more difficult to write. Use the d-family evaluators only when the other alternatives are insufficient.

We assume that you have already read chapter 6.

7.1 Comparing these methods

Methods d0, d1, and d2 differ from lf-family evaluators in several respects.

- In lf-family evaluators, you fill in the variable 'lnfj' with the log likelihoods $\ln \ell_j$ for each observation. Those methods sum the components to form the overall log likelihood $\ln L = \ln \ell_1 + \ln \ell_2 + \cdots + \ln \ell_N$.

  Methods d0, d1, and d2 require you to fill in the *scalar* 'lnf' with $\ln L$, the overall log likelihood. The utility command mlsum, described below, helps you do this.

- Methods lf1 and lf2 receive arguments g1, g2, etc., into which you store observation-level scores $\partial \ln \ell_j / \partial \theta_{ij}$. In contrast, methods d1 and d2 receive one argument, g, into which you place the $1 \times k$ row vector $(\partial \ln L/\partial \beta_1, \partial \ln L/\partial \beta_2, \ldots, \partial \ln L/\partial \beta_k)$ where k is the number of coefficients in the model. The utility command mlvecsum, described below, helps you do this.

- lf-family evaluators gracefully handle many problems for you, such as when the likelihood or scores cannot be calculated for a given set of parameter values. Utility routines mleval, mlsum, mlvecsum, mlmatsum, and mlmatbysum can assist you in handling these problems when you use d-family evaluators. These utilities are documented in this chapter.

129

Methods `d0`, `d1`, and `d2` differ from each other only in the way they calculate derivatives.

- Method `d0` requires that you fill in only the overall log likelihood. No analytical derivatives are required.
- Method `d1` requires that you fill in the overall log likelihood and the overall gradient vector (`mlvecsum` helps do this).
- Method `d2` requires that you fill in the overall log likelihood, the overall gradient vector, and the overall negative Hessian (`mlmatsum` and `mlmatbysum` help do this).

There are two reasons why you would use method `d0` instead of methods `lf`, `lf0`, `lf1`, or `lf2`:

1. Method `d0` does not assume that the overall log likelihood $\ln L$ is the sum of the log likelihoods from each observation. Some models—such as the Cox proportional hazards model, conditional logistic regression, and panel-data models—produce log likelihoods for groups of observations and so do not lend themselves to methods that assume the likelihood function satisfies the linear-form restrictions.

2. Method `d0` is the first step along the path to writing a method-`d1` or method-`d2` evaluator.

You would choose methods `d1` and `d2` for the same reasons. Method `d1` is faster and more accurate than method `d0`—but it is more work to program—and method `d2` is faster and more accurate than that—and even more work to program.

To specify method `d0`, `d1`, or `d2`, you type d0, d1, or d2 on the `ml model` line:

```
. ml model d0 myprog ...
```

or

```
. ml model d1 myprog ...
```

or

```
. ml model d2 myprog ...
```

Write *myprog*—your likelihood evaluator—according to the d0, d1, or d2 specifications.

7.2   Outline of method d0, d1, and d2 evaluators

Methods `d0`, `d1`, and `d2` can be used with any likelihood function,

$$\ln L = \ln L\{(\theta_{1j}, \theta_{2j}, \ldots, \theta_{Ej}; y_{1j}, y_{2j}, \ldots, y_{Dj}) : j = 1, \ldots, N\}$$
$$\theta_{ij} = \mathbf{x}_{ij}\boldsymbol{\beta}_i, \qquad i = 1, \ldots, E \text{ and } j = 1, \ldots, N$$

where i is the equation index and j is the observation index. We wrote the log likelihood in terms of θs because most models include regressors, but we emphasize that by letting $x_{ik} = 1$, a constant term, our notation allows θ_{ik} to represent a scalar parameter in the log likelihood function.

The outline of the evaluator of method d0, d1, or d2 is

```
program myprog
        version 18
        args todo b lnf g H
        Form scalar 'lnf' = ln L
        if ('todo'==0 | 'lnf' >= .) exit        // method-d0 evaluators end here
        Form row vector 'g' = ∂ ln L/∂b
        if ('todo'==0 | 'lnf' >= .) exit        // method-d1 evaluators end here
        Form matrix 'H' = ∂² ln L/∂b∂b'
end
```

7.2.1   The todo argument

The first argument an evaluator of method d0, d1, or d2 receives is 'todo'.

- If 'todo' = 0, the evaluator must fill in the log likelihood.

- If 'todo' = 1, the evaluator must fill in the log likelihood and the gradient vector (stored as a row vector).

- If 'todo' = 2, the evaluator must fill in the log likelihood, the gradient vector, and the Hessian matrix.

Method-d0 evaluators receive 'todo' = 0 only.

Method-d1 evaluators receive 'todo' = 0 or 'todo' = 1. When called with 'todo' = 0, the evaluator may calculate the gradient vector, but it will be ignored; calculating the gradient vector needlessly increases execution time.

Method-d2 evaluators receive 'todo' = 0, 'todo' = 1, or 'todo' = 2. Regardless of the value received, the evaluator may calculate more than is requested, but the extra will be ignored; calculating more than requested needlessly increases execution time.

7.2.2   The b argument

The second argument an evaluator of method d0, d1, or d2 receives is 'b', the current guess for β (the vector of coefficients) as a row vector. The 'b' argument for these evaluators works the same way it does for lf-family evaluators, so we do not dwell on it here. You will again typically use the mleval utility command to calculate linear combinations and obtain free parameters based on the coefficient vector 'b'. See section 6.2.2.

7.2.3   The lnf argument

The third argument an evaluator of method d0, d1, or d2 receives is 'lnf', the name of a scalar you will fill in with the overall log-likelihood value evaluated at 'b'. If the log likelihood cannot be calculated, 'lnf' is to contain a missing value, ".".

To aid understanding, here is how you might write the method-d0 likelihood evaluator for the probit model:

```
                                        begin myprobit_bad_d0.ado
program myprobit_bad_d0
        version 18
        args todo b lnf
        tempvar xb lnfj
        mleval 'xb' = 'b'
        quietly {
                generate double 'lnfj' = ln( normal('xb'))  if $ML_y1 == 1
                replace 'lnfj' = ln(1-normal('xb'))  if $ML_y1 == 0
                replace 'lnfj' = sum('lnfj')
                scalar 'lnf' = 'lnfj'[_N]
        }
end
                                        end myprobit_bad_d0.ado
```

Do not take this example too seriously; it has several problems:

- When you use methods d0, d1, or d2, it is your responsibility to restrict your calculations to the estimation subsample. Note how we sum over all the observations, including, possibly, observations that should be excluded. Perhaps the user typed

  `. ml model d0 myprobit (foreign=mpg weight) if mpg>20`

  Do not make too much of this problem because it is your responsibility to restrict calculations to the estimation subsample only when you specify ml model's nopreserve option. Otherwise, ml automatically saves your data and then drops the irrelevant observations, so you can be sloppy. (ml does not do this with method lf and lf-family methods because they carefully sum the observations.) Nevertheless, preserving and restoring the data takes time, and you might be tempted to specify ml model's nopreserve option. With the nopreserve option, the program above would then produce incorrect results.

- We did nothing to account for weights, should they be specified. That will not be a problem as long as no weights are specified on the ml model statement, but a better draft of this program would replace

  `replace 'lnfj' = sum('lnfj')`

  with

  `replace 'lnfj' = sum($ML_w*'lnfj')`

  If you use method d0, d1, or d2, weights are your responsibility.

- We assumed that the log likelihood can be calculated for every observation. Let's pretend, however, that the values we received for 'b' are odd and result in the calculation of the log likelihood being missing in the third observation. The sum() function used at the last step will treat the missing value as contributing 0 and so calculate an overall log-likelihood value that will be incorrect. Worse, treating $\ln \ell_3 = 0$ is tantamount to $\ell_3 = 1$. In this case, because likelihoods correspond to probabilities, the impossible value is treated as producing a probability 1 result.

  You might think this unlikely, but it is not. We direct your attention to line 8 of our program:

  ```
  replace 'lnfj' = ln(1-normal('xb')) if $ML_y1 == 0
  ```

  As we discussed in the technical note in section 6.2.3, ln(1-normal('xb')) turns out to be a poor way of calculating $\ln \ell$ when the observed outcome is 0. In fact, ln(1-normal(9)) evaluates to missing though the correct value is -43.63. Method lf and the lf-family methods protected you by watching for when things went badly; methods d0, d1, and d2 do not protect you because they cannot— they never see the results for each observation, and indeed, for applications that require d-family evaluators, there often is nothing to watch.

This third problem is potentially serious, and therefore we strongly urge you to avoid Stata's sum() function and instead use the mlsum command, described below, to fill in the log likelihood. Using mlsum, an adequate version of our probit evaluator would read

```
──────────────────────────────── begin myprobit_d0.ado ────────
program myprobit_d0
        version 18
        args todo b lnf
        tempvar xb lj
        mleval 'xb' = 'b'
        quietly {
                generate double 'lj'  = normal( 'xb')  if $ML_y1 == 1
                replace 'lj'  = normal(-'xb')  if $ML_y1 == 0
                mlsum 'lnf' = ln('lj')
        }
end
──────────────────────────────── end myprobit_d0.ado ────────
```

mlsum addresses all three of the shortcomings we have mentioned.

Using lnf to indicate that the likelihood cannot be calculated

Your evaluator might be called with values of 'b' for which the log likelihood cannot be evaluated for substantive as well as numerical reasons; for instance, perhaps one of the parameters to be estimated is a variance, and the maximizer attempts to evaluate the function with this parameter set to a negative value.

If the likelihood function cannot be evaluated at the coefficient values 'b' supplied by ml, your evaluator is to return a missing value in 'lnf'. At this point, it is best to exit and skip any remaining calculations.

This issue—not being able to calculate the log likelihood—never arose with methods designed for models that satisfy the linear-form restrictions because `ml` could detect the problem by examining what your evaluator returned. With those methods, you calculated log-likelihood values for each observation. If some of the returned values turned out to be missing, whether for numerical or substantive reasons, `ml` could spot the missing values and take the appropriate action.

Methods `d0`, `d1`, and `d2` cannot do this because they never see the values from each observation. The evaluator you write returns the overall log likelihood, so it is your responsibility to watch for impossible coefficient values and let the maximizer know. If you encounter impossible values, set '`lnf`' to contain a missing value and exit.

Using mlsum to define lnf

Most likelihood-evaluation programs generate a log likelihood for each observation or group of observations and then sum the contributions to obtain the overall log-likelihood value. `mlsum` will perform the summation. The syntax of `mlsum` is

mlsum *scalarname*$_{lnf}$ = *exp* $\left[\, if\, \right]$

There are other ways you could make this sum in Stata, but there are three reasons why you should use `mlsum`:

1. `mlsum` automatically restricts the summation to the estimation subsample (observations for which `$ML_samp==1`) and thus makes it more likely that results are correct even if you specify `ml model`'s `nopreserve` option.

2. `mlsum` automatically applies weights, if they were specified, in forming the sum.

3. `mlsum` verifies that what is being summed contains no missing values or sets *scalarname*$_{lnf}$ to missing if it does.

`mlsum` is typically used as follows:

```
program myprog
        version 18
        args todo b lnf g H
        ...
        mlsum 'lnf' = ...
        if ('todo'==0 | 'lnf'>=.) exit
        ...
end
```

The right side of `mlsum` is filled in with something that, when summed across the observations, yields the overall log-likelihood value. The above outline is relevant when log-likelihood values are calculated for each observation in the estimation subsample. `mlsum` itself restricts the summation to the relevant observations (the observations for which `$ML_samp==1`).

When log-likelihood values instead exist for only groups of observations, as is typically true for models that require d-family evaluators, the outline is

```
program myprog
        version 18
        args todo b lnf g H
        ...
        mlsum 'lnf' = ... if ...
        if ('todo'==0 | 'lnf'>=.) exit
        ...
end
```

The if *exp* must be filled in to be true for observations where you expect log-likelihood values to exist. mlsum does not skip over missing values in the estimation subsample. It takes missing as an indication that the log-likelihood value could not be calculated, so it sets the 'lnf' also to contain missing.

▷ **Example**

Using the θ notation, the log likelihood for the probit model is

$$\ln \ell_j = \begin{cases} \ln \Phi(\theta_j) & \text{if } y_j = 1 \\ \ln \Phi(-\theta_j) & \text{if } y_j = 0 \end{cases} \tag{7.1}$$
$$\theta_j = \mathbf{x}_j \boldsymbol{\beta}$$

The dots in mlsum must be filled in with the formula for $\ln \ell_j$:

```
tempvar xb lnfj
mleval 'xb' = 'b'
generate double 'lnfj' = ln(normal( 'xb')) if $ML_y1 == 1
replace    'lnfj' = ln(normal(-'xb')) if $ML_y1 == 0
mlsum 'lnf' = 'lnfj'
```

Note that we can specify an expression following mlsum, so the above lines could more compactly be written as

```
tempvar xb
mleval 'xb' = 'b'
mlsum 'lnf' = ln(normal(cond($ML_y1==1,'xb',-'xb')))
```

◁

▷ **Example**

As mentioned above, when log-likelihood values exist not observation by observation but instead for groups of observations, you must include the if restriction with the mlsum command:

```
mlsum 'lnf' = ... if ...
```

This step is necessary, for example, when coding likelihoods for panel-data models. We will show examples of this in sections 7.4 and 7.5.

◁

7.2.4   The g argument

The fourth argument an evaluator of method d0, d1, or d2 receives is 'g', the name of
a row vector you are to fill in with the gradient of the overall log likelihood. You need
only create 'g' if 'todo' is 1 or 2 (which would happen, of course, only if you are using
methods d1 or d2).

Method-d0 evaluators never need to fill in 'g', and most people would write a
method-d0 evaluator as

```
program myprog
        version 18
        args todo b lnf
        Form scalar 'lnf' = ln L
end
```

In fact, even method-d0 evaluators receive the fourth argument 'g' and fifth argument 'H'; however, these evaluators are called only with 'todo' = 0, so they can ignore
the fourth and fifth arguments. This allows a method-d1 or method-d2 evaluator to be
used with method d0. Like we saw with lf-family evaluators, this is a useful feature
for debugging and checking results.

A method-d1 or method-d2 evaluator will be called upon to fill in 'g'. The first
derivatives (gradient) are to be stored as a row vector that is conformable with the 'b'
vector passed by ml. This means that if 'b' is $1 \times k$, 'g' will be $1 \times k$ and will have
the derivatives in the same order as the coefficients in 'b'.

'g' will contain the partial derivatives of $\ln L$ with respect to each of the coefficients:

$$\mathbf{g} = \left(\frac{\partial \ln L}{\partial \beta_1}, \frac{\partial \ln L}{\partial \beta_2}, \ldots, \frac{\partial \ln L}{\partial \beta_k} \right)$$

The row and column names placed on 'g' do not matter because they will be ignored
by ml. If 'g' cannot be calculated at the current values of the coefficient vector 'b',
then 'lnf' is to be set to contain a missing value, and the contents of 'g', or even
whether 'g' is defined, will not matter.

Using mlvecsum to define g

mlvecsum assists in calculating the gradient vector when the linear-form restrictions are
met. In that case,

$$\ln L = \sum_{j=1}^{N} \ln \ell(\theta_{1j}, \theta_{2j}, \ldots, \theta_{Ej})$$

$$\theta_{ij} = \mathbf{x}_{ij}\boldsymbol{\beta}_i$$

and $\boldsymbol{\beta} = (\boldsymbol{\beta}_1, \boldsymbol{\beta}_2, \ldots, \boldsymbol{\beta}_E)$. Thus the gradient vector can be written as a sum:

$$\frac{\partial \ln L}{\partial \boldsymbol{\beta}} = \sum_{i=1}^{N} \frac{\partial \ln \ell_j}{\partial \boldsymbol{\beta}}$$

$$\frac{\partial \ln \ell_j}{\partial \boldsymbol{\beta}} = \left(\frac{\partial \ln \ell_j}{\partial \theta_{1j}} \frac{\partial \theta_{1j}}{\partial \boldsymbol{\beta}_1}, \quad \frac{\partial \ln \ell_j}{\partial \theta_{2j}} \frac{\partial \theta_{2j}}{\partial \boldsymbol{\beta}_2}, \quad \ldots, \quad \frac{\partial \ln \ell_j}{\partial \theta_{Ej}} \frac{\partial \theta_{Ej}}{\partial \boldsymbol{\beta}_E} \right)$$

$$= \left(\frac{\partial \ln \ell_j}{\partial \theta_{1j}} \mathbf{x}_{1j}, \quad \frac{\partial \ln \ell_j}{\partial \theta_{2j}} \mathbf{x}_{2j}, \quad \ldots, \quad \frac{\partial \ln \ell_j}{\partial \theta_{Ej}} \mathbf{x}_{Ej} \right)$$

$\partial \ln \ell_j / \partial \theta_{ij}$ is simply one value per observation. You supply $\partial \ln \ell_j / \partial \theta_{ij}$, and `mlvecsum` returns

$$\partial \ln L / \partial \boldsymbol{\beta}_i = \sum_{j=1}^{N} \frac{\partial \ln \ell_j}{\partial \theta_{ij}} \mathbf{x}_{ij}$$

which is one component of the vector above.

`mlvecsum`'s syntax is

`mlvecsum` *scalarname*_{lnf} *rowvecname* = *exp* [*if*] [, eq(#)]

where *exp* evaluates to $\partial \ln \ell_j / \partial \theta_{ij}$ for equation $i = \#$. Thus obtaining $\partial \ln L / \partial \boldsymbol{\beta}$ is simply a matter of issuing one `mlvecsum` per equation and then joining the results into a single row vector.

In a single-equation system,

```
mlvecsum 'lnf' 'g' = formula for ∂ ln ℓj/∂θ1j
```

In a two-equation system,

```
tempname d1 d2
mlvecsum 'lnf' 'd1' = formula for ∂ ln ℓj/∂θ1j, eq(1)
mlvecsum 'lnf' 'd2' = formula for ∂ ln ℓj/∂θ2j, eq(2)
matrix 'g' = ('d1','d2')
```

In a three-equation system,

```
tempname d1 d2 d3
mlvecsum 'lnf' 'd1' = formula for ∂ ln ℓj/∂θ1j, eq(1)
mlvecsum 'lnf' 'd2' = formula for ∂ ln ℓj/∂θ2j, eq(2)
mlvecsum 'lnf' 'd3' = formula for ∂ ln ℓj/∂θ3j, eq(3)
matrix 'g' = ('d1','d2','d3')
```

Note that `mlvecsum` has two left-hand-side arguments:

```
. mlvecsum 'lnf' 'whatever' = exp
```

That is so `mlvecsum` can reset `'lnf'` to be a missing value if *exp* cannot be calculated for some relevant observation. Evaluators of methods d0, d1, and d2 are to set `'lnf'` to missing when called with impossible values of the coefficient vector. It is possible that 'lnf' can be calculated but the gradient cannot. `mlvecsum` appropriately handles such cases, and it applies weights if they were specified.

▷ **Example**

Using the probit likelihood from (7.1), the derivative is

$$\frac{d \ln \ell_j}{d\theta_j} = \begin{cases} \phi(\theta_j)/\Phi(\theta_j) & \text{if } y_j = 1 \\ -\phi(\theta_j)/\Phi(-\theta_j) & \text{if } y_j = 0 \end{cases} \tag{7.2}$$

Thus `mlvecsum` could be used to obtain the entire gradient vector by coding

```
tempvar gj
generate double 'gj' =  normalden('xb')/normal( 'xb') if $ML_y1 == 1
replace     'gj' = -normalden('xb')/normal(-'xb') if $ML_y1 == 0
mlvecsum 'lnf' 'g' = 'gj'
```

Because the right-hand side of `mlvecsum` can be an expression, the above could be reduced to one line using the `cond()` function if preferred.

◁

Like with `mlsum`, the real power of `mlvecsum` lies in its acceptance of an `if` clause. As we will see by way of example later in the chapter, this acceptance simplifies the computation of gradients of some models that do not satisfy the linear-form restrictions.

7.2.5   The H argument

The fifth argument an evaluator of method d0, d1, or d2 receives is 'H'. You fill in this argument just as you do with lf-family evaluators; see section 6.2.5. The Hessian is to be stored as a matrix that is conformable with the 'b' vector you were passed. If 'b' is $1 \times k$, then 'H' is to be $k \times k$, and it is to have the derivatives in the same order as the coefficients in 'b'. If 'H' cannot be calculated at the current value of coefficient vector 'b', then 'lnf' is to be set to contain a missing value, and you need not fill in 'H'.

'H' is to contain the second partial derivatives of $\ln L$ with respect to each of the coefficients:

$$\mathbf{H} = \begin{pmatrix} \frac{\partial^2 \ln L}{\partial \beta_1^2} & \frac{\partial^2 \ln L}{\partial \beta_1 \partial \beta_2} & \cdots & \frac{\partial^2 \ln L}{\partial \beta_1 \partial \beta_k} \\ \frac{\partial^2 \ln L}{\partial \beta_2 \partial \beta_1} & \frac{\partial^2 \ln L}{\partial \beta_2^2} & \cdots & \frac{\partial^2 \ln L}{\partial \beta_2 \partial \beta_k} \\ \vdots & \vdots & \ddots & \vdots \\ \frac{\partial^2 \ln L}{\partial \beta_k \partial \beta_1} & \frac{\partial^2 \ln L}{\partial \beta_k \partial \beta_2} & \cdots & \frac{\partial^2 \ln L}{\partial \beta_k^2} \end{pmatrix}$$

Unlike scores, which are observation-specific, 'H' represents the second derivative of the overall log-likelihood function. If the likelihood function satisfies the linear-form restrictions, then `mlmatsum` can be used to simplify the construction of 'H'. For likelihood functions that are defined at the group level, an additional command, `mlmatbysum`, exists to aide in computing the Hessian matrix. `mlmatbysum` is most easily understood in the context of a specific example, so we will wait until section 7.4.3 to discuss it.

▷ Example

Using the log likelihood from the probit model (7.1) and its first derivative (7.2), the second derivative is

$$\frac{d^2 \ln \ell_j}{d\theta_j^2} = \begin{cases} -R(\theta_j)\{R(\theta_j) + \theta_j\} & \text{if } y_j = 1 \\ -S(\theta_j)\{S(\theta_j) - \theta_j\} & \text{if } y_j = 0 \end{cases}$$

where

$$R(\theta_j) = \phi(\theta_j)/\Phi(\theta_j)$$
$$S(\theta_j) = \phi(\theta_j)/\Phi(-\theta_j)$$

Our likelihood function is written in terms of θ, so we can use `mlmatsum` to obtain the Hessian:

```
tempvar R S h
generate double `R' = normalden(`xb')/normal(`xb')
generate double `S' = normalden(`xb')/normal(-`xb')
generate double `h' = -`R'*(`R'+`xb') if $ML_y1 == 1
replace       `h' = -`S'*(`S'-`xb') if $ML_y1 == 0
mlmatsum `lnf' `H' = `h'
```

Because the right-hand side of `mlmatsum` can be an expression, we could compress it into fewer lines by using the `cond()` function.

◁

7.3   Summary of methods d0, d1, and d2

7.3.1   Method d0

The method-d0 evaluator calculates the overall log likelihood. The outline for a method-d0 likelihood evaluator is

```
program myprog
        version 18
        args todo b lnf                          // 'todo' is subsequently ignored

        tempvar theta1 theta2 ...
        mleval 'theta1' = 'b'
        mleval 'theta2' = 'b', eq(2)             // if there is a θ₂
        ...
        // if you need to create any intermediate results
        tempvar tmp1 tmp2 ...
        generate double 'tmp1' = ...
        generate double 'tmp2' = ...
        ...
        mlsum 'lnf' = ...
end
```

You specify that you are using method d0 with program *myprog* on the `ml model` statement,

```
. ml model d0 myprog ...
```

which you issue after you have defined the likelihood-evaluation program.

Using method d0 is a matter of writing down the likelihood function and coding it. Unlike methods designed for models that satisfy the linear-form restrictions, however, you must sum the likelihood to produce the overall log-likelihood value.

Method-d0 evaluators will be called with 'todo'$= 0$ only, so method-d0 evaluators generally ignore 'todo' and also ignore the fact that they receive arguments 'g' and 'H'. That is, formally, a method-d0 evaluator could be written

```
program myprog
        version 18
        args todo b lnf g H
        ...
end
```

but it is perfectly acceptable to program

```
program myprog
        version 18
        args todo b lnf
        ...
end
```

Because the likelihood function is typically written as a function of the parameters θ_i from the probability model, the first step is to obtain their values from the coefficients vector 'b':

```
program myprog
        version 18
        args todo b lnf
        tempvar theta1 theta2 ...
        mleval 'theta1' = 'b'
        mleval 'theta2' = 'b', eq(2)          // if there is a θ₂
        ...
end
```

The final step is to calculate the overall log-likelihood value and store it in the scalar 'lnf'. d-family evaluators are often used with models that produce group-level log-likelihood values, the leading examples being models for panel data. You can use mlsum in those cases to produce 'lnf':

```
program myprog
        version 18
        args todo b lnf
        tempvar theta1 theta2 ...
        mleval 'theta1' = 'b'
        mleval 'theta2' = 'b', eq(2)          // if there is a θ₂
        ...
        mlsum 'lnf' = ... if ...
end
```

where the first set of dots following mlsum is filled in with an expression producing the log-likelihood value for a group in terms of 'theta1', 'theta2', ..., and $ML_y1, $ML_y2, ... The dots following the keyword if restrict the sample so that mlsum uses just one observation from each group.

In more complicated cases, you will need to generate temporary variables containing partial calculations before invoking mlsum to sum the contributions:

```
program myprog
        version 18
        args todo b lnf
        tempvar theta1 theta2 ...
        mleval 'theta1' = 'b'
        mleval 'theta2' = 'b', eq(2)          // if there is a θ₂
        ...
        tempvar tmp1 tmp2
        quietly generate double 'tmp1' = ...
        quietly generate double 'tmp2' = ...
        mlsum 'lnf' = expression in terms of 'tmp1' and 'tmp2' if ...
end
```

The following points enumerate what ml supplies and expects of a method-d0 likelihood evaluator.

- You are passed the vector of coefficients. Use mleval to obtain the values of each of the linear predictors of each equation from it; see section 6.2.2.

- The names of the dependent variables (if any) are placed in global macros by ml; the first dependent variable is $ML_y1, the second is $ML_y2, ... Use $ML_y1, $ML_y2, ... just as you would any existing variable in the data.

- You must create any temporary variables for intermediate results as `doubles`. Type

      generate double *name* = ...

  Do not omit the word `double`.

- The estimation subsample is identified by `$ML_samp==1`. You may safely ignore this if you do not specify `ml model`'s `nopreserve` option. When your program is called, only relevant data will be in memory. If you do specify `nopreserve`, `mleval` and `mlsum` automatically restrict themselves to the estimation subsample; it is not necessary to include `if $ML_samp==1` with these commands. However, you will need to restrict other calculation commands to the `$ML_samp==1` subsample.

- `ml` stores the name of the weight variable in `$ML_w`, which contains 1 in every observation if no weights are specified. If you use `mlsum` to produce the log likelihood, you may ignore this because `mlsum` handles that itself.

- The resulting log-likelihood value is to be stored in scalar '`lnf`'. Use `mlsum` to produce this value if possible; see section 7.2.3.

- Though evaluators of methods `d0`, `d1`, and `d2` can be used with `fweights`, `aweights`, and `iweights`, `pweights` cannot be specified. Moreover, you cannot specify `ml model`'s `vce(opg)`, `vce(robust)`, `vce(cluster` *clustvar*`)`, or `svy` options; nor can you use the BHHH optimization technique.

7.3.2   Method d1

The method-`d1` evaluator calculates the overall log likelihood and its first derivatives. The outline for the program is

```
program myprog
        version 18
        args todo b lnf g
        tempvar theta1 theta2 ...
        mleval 'theta1' = 'b'
        mleval 'theta2' = 'b', eq(2)              // if there is a θ₂
        ...
        // if you need to create any intermediate results:
        tempvar tmp1 tmp2 ...
        generate double 'tmp1' = ...
        generate double 'tmp2' = ...
        ...
        mlsum 'lnf' = ...
        if ('todo'==0 | 'lnf'>=.) exit

        tempname d1 d2 ...
        mlvecsum 'lnf' 'd1' = formula for ∂ln ℓⱼ/∂θ₁ⱼ if ..., eq(1)
        mlvecsum 'lnf' 'd2' = formula for ∂ln ℓⱼ/∂θ₂ⱼ if ..., eq(2)
        ...
        matrix 'g' = ('d1','d2',...)
    end
```

You specify that you are using method **d1** with program *myprog* on the **ml model** statement,

```
. ml model d1 myprog ...
```

which you issue after you have defined the likelihood-evaluation program.

The method-**d1** evaluator simply picks up where the method-**d0** evaluator left off. Taking out the code in common, here is how they differ:

```
program myprog
        version 18
        args todo b lnf g                       // g is added
        ...
        if ('todo'==0 | 'lnf'>=.) exit          // from here down is new

        tempname d1 d2 ...
        mlvecsum 'lnf' 'd1' = formula for ∂ln ℓ_j/∂θ_{1j} if ..., eq(1)
        mlvecsum 'lnf' 'd2' = formula for ∂ln ℓ_j/∂θ_{2j} if ..., eq(2)
        ...
        matrix 'g' = ('d1','d2',...)
end
```

The gradient vector you define is

$$\mathbf{g} = \frac{\partial \ln L}{\partial \boldsymbol{\beta}} = \left(\frac{\partial \ln L}{\partial \boldsymbol{\beta}_1}, \frac{\partial \ln L}{\partial \boldsymbol{\beta}_2}, \dots, \frac{\partial \ln L}{\partial \boldsymbol{\beta}_k} \right)$$

ml provides the coefficient vector as a $1 \times k$ row vector and expects the gradient vector to be returned as a $1 \times k$ row vector.

Many likelihood functions produce group-level log-likelihood values, and in those cases, you can use **mlvecsum** to sum over the groups when computing the derivatives.

The following points enumerate what **ml** supplies and expects of a method-**d1** likelihood evaluator.

- You are passed the vector of coefficients. Use **mleval** to obtain the values of each of the linear predictors of each equation from it; see section 6.2.2.

- The names of the dependent variables (if any) are placed in global macros by **ml**; the first dependent variable is $ML_y1, the second is $ML_y2, ... Use $ML_y1, $ML_y2, ... just as you would any existing variable in the data.

- You must create any temporary variables for intermediate results as **doubles**. Type

    ```
    generate double name = ...
    ```

  Do not omit the word **double**.

- The estimation subsample is identified by $ML_samp==1. You may safely ignore this if you do not specify **ml model**'s **nopreserve** option. When your program is called, only relevant data will be in memory. If you do specify **nopreserve**, **mleval** and **mlsum** automatically restrict themselves to the estimation subsample;

it is not necessary to include `if $ML_samp==1` with these commands. However, you will need to restrict other calculation commands to the `$ML_samp==1` subsample.

- `ml` stores the name of the weight variable in `$ML_w`, which contains 1 in every observation if no weights are specified. If you use `mlsum` to produce the log likelihood, you may ignore this because `mlsum` handles that itself.

- The resulting log-likelihood value is to be stored in scalar 'lnf'. Use `mlsum` to produce this value if possible; see section 7.2.3.

- The resulting gradient vector is to be stored in vector 'g'. You can use `mlvecsum` to produce it when the likelihood function is produced at the group level. See sections 7.4 and 7.5 for examples.

7.3.3   Method d2

The method-d2 evaluator calculates the overall log likelihood along with its first and second derivatives. The outline for the program is

```
program myprog
        version 18
        args todo b lnf g H

        tempvar theta1 theta2 ...
        mleval 'theta1' = 'b', eq(1)
        mleval 'theta2' = 'b', eq(2) // if there is a θ_2
        ...

        // if you need to create any intermediate results:
        tempvar tmp1 tmp2 ...
        generate double 'tmp1' = ...
        ...

        mlsum 'lnf' = ...
        if ('todo'==0 | 'lnf'>=.) exit

        tempname d1 d2 ...
        mlvecsum 'lnf' 'd1' = formula for ∂ln ℓ_j/∂θ_1j if ..., eq(1)
        mlvecsum 'lnf' 'd2' = formula for ∂ln ℓ_j/∂θ_2j if ..., eq(2)
        ...
        matrix 'g' = ('d1','d2',...)
        if ('todo'==1 | 'lnf'>=.) exit

        tempname d11 d12 d22 ...
        mlmatsum 'lnf' 'd11' = formula for ∂²ln ℓ_j/∂θ²_1j if ..., eq(1)
        mlmatsum 'lnf' 'd12' = formula for ∂²ln ℓ_j/∂θ_1j∂θ_2j if ..., eq(1,2)
        mlmatsum 'lnf' 'd22' = formula for ∂²ln ℓ_j/∂θ²_2j if ..., eq(2)
        ...
        matrix 'H' = ('d11','d12',... \ 'd12'',
                'd22',...)
end
```

You specify that you are using method d2 on the `ml model` statement,

```
. ml model d2 myprog ...
```

which you issue after you have defined the likelihood-evaluation program.

The method-**d2** evaluator continues from where the method-**d1** evaluator left off. Taking out the common code, here is how the evaluators differ:

```
program myprog
        version 18
        args todo b lnf g H                    // H is added
        ...
        if ('todo'==1 | 'lnf'==.) exit         // from here down is new
        tempname d11 d12 d22 ...
        mlmatsum 'lnf' 'd11' = formula for ∂² ln ℓⱼ/∂θ²₁ⱼ if ..., eq(1)
        mlmatsum 'lnf' 'd12' = formula for ∂² ln ℓⱼ/∂θ₁ⱼ∂θ₂ⱼ if ..., eq(1,2)
        mlmatsum 'lnf' 'd22' = formula for ∂² ln ℓⱼ/∂θ²₂ⱼ if ..., eq(2)
        ...
        matrix 'H' = ('d11','d12',... \ 'd12'','d22',...)
end
```

The Hessian matrix you are to define is

$$\mathbf{H} = \frac{\partial^2 \ln L}{\partial \boldsymbol{\beta} \partial \boldsymbol{\beta}'} = \begin{pmatrix} \frac{\partial^2 \ln L}{\partial \beta_1^2} & \frac{\partial^2 \ln L}{\partial \beta_1 \partial \beta_2} & \cdots & \frac{\partial^2 \ln L}{\partial \beta_1 \partial \beta_k} \\ \frac{\partial^2 \ln L}{\partial \beta_2 \partial \beta_1} & \frac{\partial^2 \ln L}{\partial \beta_2^2} & \cdots & \frac{\partial^2 \ln L}{\partial \beta_2 \partial \beta_k} \\ \vdots & \vdots & \ddots & \vdots \\ \frac{\partial^2 \ln L}{\partial \beta_k \partial \beta_1} & \frac{\partial^2 \ln L}{\partial \beta_k \partial \beta_2} & \cdots & \frac{\partial^2 \ln L}{\partial \beta_k^2} \end{pmatrix}$$

`ml` provides the coefficient vector as a $1 \times k$ row vector and expects the negative Hessian matrix to be returned as a $k \times k$ matrix.

The following points enumerate what `ml` supplies and expects of a method-**d2** likelihood evaluator.

- You are passed the vector of coefficients. Use `mleval` to obtain the values of each of the linear predictors of each equation from it; see section 6.2.2.

- The names of the dependent variables (if any) are placed in global macros by `ml`; the first dependent variable is `$ML_y1`, the second is `$ML_y2`, ... Use `$ML_y1`, `$ML_y2`, ... just as you would any existing variable in the data.

- You must create any temporary variables for intermediate results as `doubles`. Type

    ```
    generate double name = ...
    ```

  Do not omit the word `double`.

- The estimation subsample is identified by `$ML_samp==1`. You may safely ignore this if you do not specify `ml model`'s `nopreserve` option. When your program is called, only relevant data will be in memory. If you do specify `nopreserve`, `mleval` and `mlsum` automatically restrict themselves to the estimation subsample; it is not necessary to include `if $ML_samp==1` with these commands. However, you will need to restrict other calculation commands to the `$ML_samp==1` subsample.

- `ml` stores the name of the weight variable in `$ML_w`, which contains 1 in every observation if no weights are specified. If you use `mlsum` to produce the log likelihood, you may ignore this because `mlsum` handles that itself.

- The resulting log-likelihood value is to be stored in scalar 'lnf'. Use `mlsum` to produce this value if possible; see section 7.2.3.

- The resulting gradient vector is to be stored in vector 'g'. You can use `mlvecsum` to produce it when the likelihood function is produced at the group level. See sections 7.4 and 7.5 for examples.

- The Hessian matrix is to be stored in matrix 'H'. The `mlmatsum` command can be used if the log likelihood meets the linear-form restrictions, and the `mlmatbysum` can be used when the log-likelihood function is defined for groups of observations, as is the case with panel-data models.

7.4   Panel-data likelihoods

In section 7.2, we discussed methods `d0`, `d1`, and `d2` in the context of linear-form likelihood functions for simplicity. However, as we have said before, if you have a linear-form likelihood function, you should instead use methods `lf`, `lf0`, `lf1`, or `lf2` because they are easier to use.

Not all likelihood functions meet the linear-form restrictions, and it is in those cases that methods `d0`, `d1`, and `d2` are required. In this section, we will consider a special case when the linear-form restrictions are violated, that of panel-data estimators. In section 7.5, we will consider other likelihood functions.

In panel data, you have repeated observations for each of a collection of individual persons, firms, etc. Each observation in the data represents one of the observations for one of those individuals. For instance, if we had 5 observations for each of 20 individuals, our dataset would contain $5 \times 20 = 100$ observations in total. This is illustrated in table 7.1.

Table 7.1. Layout for panel data. Variable i identifies the panels, and t identifies the time or instant an observation was made or a measurement was taken.

Obs. no.	i	t	y_{it}	$\mathbf{x}_{it}$
1.	1	1	y_{11}	$\mathbf{x}_{11}$
2.	1	2	y_{12}	$\mathbf{x}_{12}$
3.	1	3	y_{13}	$\mathbf{x}_{13}$
4.	1	4	y_{14}	$\mathbf{x}_{14}$
5.	2	1	y_{21}	$\mathbf{x}_{21}$
6.	2	2	y_{22}	$\mathbf{x}_{22}$
7.	2	3	y_{23}	$\mathbf{x}_{23}$
8.	2	4	y_{24}	$\mathbf{x}_{24}$
9.	3	1	y_{31}	$\mathbf{x}_{31}$
10.	3	2	y_{32}	$\mathbf{x}_{32}$
11.	3	3	y_{33}	$\mathbf{x}_{33}$
12.	3	4	y_{34}	$\mathbf{x}_{34}$

For panel-data estimators, the observations are not independent, but groups of the observations are independent. Panel datasets are usually recorded in the long form, as are the data given in table 7.1. In this dataset, we have independent information on three separate groups i (perhaps i indexes persons), and we have four observations on each of the i's.

Likelihood functions for this type of data will violate the linear-form restrictions because the observations are not independent from others in the same group. We still have independence between groups of observations, however, and in writing our evaluator, it is merely a matter of adjusting our program to exploit this more complicated structure.

To fix ideas, we consider the particular case of random-effects linear regression.

$$y_{it} = \mathbf{x}_{it}\boldsymbol{\beta} + u_i + e_{it}, \quad u_i \sim N(0, \sigma_u^2), \quad e_{it} \sim N(0, \sigma_e^2)$$

In this model, there are three parameters we need to estimate: $\boldsymbol{\beta}$ (presumably a vector) and σ_u^2 and σ_e^2 (presumably scalars). The likelihood function for random-effects regression is

$$\ln L = \sum_{i=1}^{N} \ln L_i$$

where i indexes group, and the individual log-likelihood values for each group are themselves based on the calculation of a cluster of observations within group:

$$\ln L_i = \ln \int_{-\infty}^{\infty} f(u_i) \prod_{t=1}^{T_i} f(y_{it}|u_i)\, du_i$$

$$= \ln \int_{-\infty}^{\infty} \frac{1}{\sigma_u \sqrt{2\pi}} \exp\left(-\frac{u_i^2}{2\sigma_u^2}\right) \prod_{t=1}^{T_i} \left[\frac{1}{\sigma_e \sqrt{2\pi}} \exp\left\{-\frac{(y_{it} - u_i - \mathbf{x}_{it}\boldsymbol{\beta})^2}{2\sigma_e^2}\right\}\right] du_i$$

$$= -\frac{1}{2}\left\{\frac{\sum_t z_{it}^2 - a_i(\sum_t z_{it})^2}{\sigma_e^2} + \ln(T_i\sigma_u^2/\sigma_e^2 + 1) + T_i\ln(2\pi\sigma_e^2)\right\} \tag{7.3}$$

where

$$T_i = \text{number of observations in the } i\text{th group}$$
$$z_{it} = y_{it} - \mathbf{x}_{it}\boldsymbol{\beta} \tag{7.4}$$
$$a_i = \sigma_u^2/(T_i\sigma_u^2 + \sigma_e^2)$$

[We give some hints on how we derived this in (7.10).]

7.4.1   Calculating lnf

The trick to calculating the log-likelihood function in panel datasets is to create a new variable containing the likelihood for each group in the last observation of the group, and then sum only those last observations.

Obs. no.	i	t	y_{it}	$\mathbf{x}_{it}$	New variable	
1.	1	1	y_{11}	$\mathbf{x}_{11}$	.	(missing)
2.	1	2	y_{12}	$\mathbf{x}_{12}$	.	
3.	1	3	y_{13}	$\mathbf{x}_{13}$	.	
4.	1	4	y_{14}	$\mathbf{x}_{14}$	$\ln L_1$	
5.	2	1	y_{21}	$\mathbf{x}_{21}$	.	
6.	2	2	y_{22}	$\mathbf{x}_{22}$	.	
7.	2	3	y_{23}	$\mathbf{x}_{23}$	.	
8.	2	4	y_{24}	$\mathbf{x}_{24}$	$\ln L_2$	
9.	3	1	y_{31}	$\mathbf{x}_{31}$	.	
10.	3	2	y_{32}	$\mathbf{x}_{32}$	.	
11.	3	3	y_{33}	$\mathbf{x}_{33}$	.	
12.	3	4	y_{34}	$\mathbf{x}_{34}$	$\ln L_3$	
					$\ln L$	(sum of nonmissing)

Making the calculation in this way is easy to do. Calculating $\ln L_1$, $\ln L_2$, etc., is easy because most of the terms will be based on sums within i, so we can code

```
     by i: generate double newvar = sum(...)
```

although sometimes multiple sums may appear, such as

```
     by i: generate double newvar = sum(...)*sum(...)/sum(...)
```

To compute the overall value of the log likelihood, we will still use `mlsum`, but we must remember to sum only the last observation in each group because those are the observations that contain the likelihood contribution for their respective group:

```
     tempvar last
     by i:  generate byte 'last' = (_n==_N)
     mlsum 'lnf' = ... if 'last'
```

In the above sketch of code, we did not bother to ensure that missing values were placed in the intermediate observations, but that does not matter because we explicitly excluded them by coding `if 'last'` at the end of our `mlsum` command.

For random-effects linear regression, the likelihood function is (7.3) with the intermediate calculations as defined in (7.4), and our method-d0 evaluator is

--- begin `myrereg_d0.ado` ---------------
```
program myrereg_d0
        version 18
        args todo b lnf
        tempvar xb z T S_z2 Sz_2 a last
        tempname s_u s_e
        mleval 'xb'  = 'b', eq(1)
        mleval 's_u' = 'b', eq(2) scalar
        mleval 's_e' = 'b', eq(3) scalar
        scalar 's_u' = exp('s_u')
        scalar 's_e' = exp('s_e')
        // MY_panel contains the panel ID
        local by $MY_panel
        sort 'by'
        local y $ML_y1
        quietly {
                generate double 'z' = 'y' - 'xb'
                by 'by': generate 'last' = _n==_N
                by 'by': generate 'T' = _N
                by 'by': generate double 'S_z2' = sum('z'^2)
                by 'by': generate double 'Sz_2' = sum('z')^2
                generate double 'a' = 's_u'^2 / ('T'*'s_u'^2 + 's_e'^2)
                mlsum 'lnf'= -.5 *                                      ///
                        (                                              ///
                                ('S_z2'-'a'*'Sz_2')/'s_e'^2 +          ///
                                ln('T'*'s_u'^2/'s_e'^2 + 1) +          ///
                                'T'*ln(2*c(pi)*'s_e'^2)                ///
                        )                                              ///
                        if 'last' == 1
        }
end
```
--- end `myrereg_d0.ado` ---------------

There is a lot of code here, so let's go over it in blocks. We start by retrieving the values from each equation:

$$\theta_{1it} = \mathbf{x}_{it}\boldsymbol{\beta}$$

$$\theta_{2it} = \ln \sigma_u$$

$$\theta_{3it} = \ln \sigma_e$$

We are modeling the variances in the log space, so we must exponentiate the values from the second and third equations:

```
version 18
args todo b lnf
tempvar xb z T S_z2 Sz_2 a last
tempname s_u s_e
mleval 'xb'  = 'b', eq(1)
mleval 's_u' = 'b', eq(2) scalar
mleval 's_e' = 'b', eq(3) scalar
scalar 's_u' = exp('s_u')
scalar 's_u' = exp('s_e')
```

Next we set up some local macros for our own convenience and sort the data by panel. (The technical notes from the Cox model, pages 169–170, regarding the **sort** command apply equally here.) myrereg_d0 assumes that $MY_panel contains the panel ID variable, so we must remember to set this global macro before running **ml maximize**.

```
// MY_panel contains the panel ID
local by $MY_panel
sort 'by'
local y $ML_y1
```

Inside the **quietly** block, we perform the calculations. The 'last' variable indicates the last observation of each panel within the estimation subsample. We use this variable to indicate which observations contain contributions to the log likelihood. The 'S_z2' variable contains the sum of the squared deviations, and 'Sz_2' contains the square of the summed deviations.

```
quietly {
        generate double 'z' = 'y' - 'xb'
        by 'by': generate 'last' = _n==_N
        by 'by': generate 'T' = _N
        by 'by': generate double 'S_z2' = sum('z'^2)
        by 'by': generate double 'Sz_2' = sum('z')^2
        generate double 'a' = 's_u'^2 / ('T'*'s_u'^2 + 's_e'^2)
        mlsum 'lnf'= -.5 *                                   ///
                (                                            ///
                        ('S_z2'-'a'*'Sz_2')/'s_e'^2 +        ///
                        ln('T'*'s_u'^2/'s_e'^2 + 1) +        ///
                        'T'*ln(2*c(pi)*'s_e'^2)              ///
                )                                            ///
                if 'last' == 1
}
```

❏ Technical note

As written, `myrereg_d0` does not accept weights—$ML_w does not appear in any of the calculations above, and the `mlsum` command will not work if you specify weights and use an `if` condition to limit its calculations. The solution is to set the log likelihood equal to zero for observations other than the last one in each panel. Then we can omit the `if` condition with `mlsum` and still have it apply weights appropriately. We modify the last few lines of `myrereg_d0` to read

```
        generate double 'a' = 's_u'^2 / ('T'*'s_u'^2 + 's_e'^2)
        tempvar mylnf                               //  <- New
        generate double 'mylnf' = 0                 //  <- New
        replace 'mylnf' = -.5 *                     /// Modified
                (                                   ///
                        ('S_z2'-'a'*'Sz_2')/'s_e'^2 +   ///
                        ln('T'*'s_u'^2/'s_e'^2 + 1) +   ///
                        'T'*ln(2*c(pi)*'s_e'^2)     ///
                )                                   ///
                if 'last' == 1
        mlsum 'lnf' = 'mylnf'                        //  <- New
end
```

There is one more issue to consider when working with weights and panel data. Panel-data estimators typically assume that weights are constant within each panel. You are responsible for verifying that this is true of your dataset; `ml` cannot tell when it is optimizing a panel likelihood. Our method of handling weights with panel data only considers the weight of the last observation in each panel.

❏

▷ **Example**

To verify that `myrereg_d0` works, we use data from Greene (2018):

```
. use http://www.stata-press.com/data/ml5/tablef7-1
. global MY_panel i        // the panel variable
. ml model d0 myrereg_d0 (xb: lnC = lnQ lnPF lf) /lns_u /lns_e
. ml max, nolog
```

```
                                           Number of obs  =         90
                                           Wald chi2(3)   = 11468.30
Log likelihood = 114.72904                 Prob > chi2    =    0.0000
```

lnC	Coefficient	Std. err.	z	P>|z|	[95% conf. interval]	
lnQ	.90531	.0253759	35.68	0.000	.8555742	.9550459
lnPF	.4233757	.013888	30.48	0.000	.3961557	.4505957
lf	-1.064456	.1962308	-5.42	0.000	-1.449061	-.6798508
_cons	9.618648	.2066219	46.55	0.000	9.213677	10.02362
/lns_u	-2.170817	.3026643	-7.17	0.000	-2.764028	-1.577606
/lns_e	-2.828404	.077319	-36.58	0.000	-2.979947	-2.676862

◁

7.4.2   Calculating g

If we wish to convert our `d0` evaluator into a `d1` evaluator and thus gain speed and precision, we must program the gradient vector calculation. The code to do that is nearly identical to the code for the linear-form case. You use `mlvecsum` to create the sums.

You will discover that gradient calculations in panel-data models come in two flavors: one that varies observation by observation within each group and another that is calculated over the entire group, just as the likelihood value itself.

For the first, you will program the calculation just as you would program a linear-form likelihood:

Obs. no.	i	t	y_{it}	$\mathbf{x}_{it}$	New variable
1.	1	1	y_{11}	$\mathbf{x}_{11}$	g_{11}
2.	1	2	y_{12}	$\mathbf{x}_{12}$	g_{12}
3.	1	3	y_{13}	$\mathbf{x}_{13}$	g_{13}
4.	1	4	y_{14}	$\mathbf{x}_{14}$	g_{14}
5.	2	1	y_{21}	$\mathbf{x}_{21}$	g_{21}
6.	2	2	y_{22}	$\mathbf{x}_{22}$	g_{22}
7.	2	3	y_{23}	$\mathbf{x}_{23}$	g_{23}
8.	2	4	y_{24}	$\mathbf{x}_{24}$	g_{24}
9.	3	1	y_{31}	$\mathbf{x}_{31}$	g_{31}
10.	3	2	y_{32}	$\mathbf{x}_{32}$	g_{32}
11.	3	3	y_{33}	$\mathbf{x}_{33}$	g_{33}
12.	3	4	y_{34}	$\mathbf{x}_{34}$	g_{34}

You obtain the sum using `mlvecsum`:

```
mlvecsum 'lnf' 'g' = expression based on new variable
```

For any term that applies only to the entire group, you will use the same scheme that you used for calculating the likelihood value:

Obs. no.	i	t	y_{it}	$\mathbf{x}_{it}$	New variable	
1.	1	1	y_{11}	$\mathbf{x}_{11}$	.	(missing)
2.	1	2	y_{12}	$\mathbf{x}_{12}$	.	
3.	1	3	y_{13}	$\mathbf{x}_{13}$	.	
4.	1	4	y_{14}	$\mathbf{x}_{14}$	g_1	
5.	2	1	y_{21}	$\mathbf{x}_{21}$	.	
6.	2	2	y_{22}	$\mathbf{x}_{22}$	.	
7.	2	3	y_{23}	$\mathbf{x}_{23}$	.	
8.	2	4	y_{24}	$\mathbf{x}_{24}$	g_2	
9.	3	1	y_{31}	$\mathbf{x}_{31}$	.	
10.	3	2	y_{32}	$\mathbf{x}_{32}$	.	
11.	3	3	y_{33}	$\mathbf{x}_{33}$	.	
12.	3	4	y_{34}	$\mathbf{x}_{34}$	g_3	

You obtain the sum using `mlvecsum`, but this time you must be careful to sum only the last observation for each group:

```
mlvecsum 'lnf' 'g' = expression based on new variable if 'last'
```

For instance, the derivatives of our likelihood function for random-effects regression are

$$\frac{\partial \ln L_i}{\partial \mathbf{x}_{it}\boldsymbol{\beta}} = \frac{1}{\sigma_e^2}\left(z_{it} - a_i \sum_{s=1}^{T_i} z_{is}\right) \tag{7.5}$$

$$\frac{\partial \ln L_i}{\partial \ln \sigma_u} = \frac{a_i^2}{\sigma_u^2}\left(\sum_{t=1}^{T_i} z_{it}\right)^2 - T_i a_i \tag{7.6}$$

$$\frac{\partial \ln L_i}{\partial \ln \sigma_e} = \frac{1}{\sigma_e^2}\left\{\sum_{t=1}^{T_i} z_{it}^2 - a_i\left(\sum_{t=1}^{T_i} z_{it}\right)^2\right\} - (T_i - 1) - \frac{a_i^2}{\sigma_u^2}\left(\sum_{t=1}^{T_i} z_{it}\right)^2 - \frac{\sigma_e^2}{\sigma_u^2} a_i \tag{7.7}$$

Consider writing the code to calculate $\partial \ln L/\partial\boldsymbol{\beta}$:

```
by 'i': generate double 'S_z' = sum('z')
by 'i': replace 'S_z' = 'S_z'[_N]
generate double 'dxb' = ('z' - 'a'*'S_z') / 's_e'^2
mlvecsum 'lnf' 'g1' = 'dxb', eq(1)
```

There are other ways we could have coded this; perhaps you would prefer

```
by 'i': generate double 'S_z' = sum('z')
by 'i': generate double 'dxb' = ('z' - 'a'*'S_z'[_N]) / 's_e'^2
mlvecsum 'lnf' 'g1' = 'dxb'
```

which eliminates a line of code and makes use of `mlvecsum`'s defaulting to the first equation. Either way, think carefully about what is in `'dxb'`, and you will realize that each observation in `'dxb'` enters the sum. We can use `'dxb'` with `mlvecsum` to compute $\partial \ln L/\partial\boldsymbol{\beta}$ because

$$\frac{\partial \ln L_i}{\partial \boldsymbol{\beta}} = \sum_{t=1}^{T_i} \frac{\partial \ln L_i}{\partial \mathbf{x}_{it}\boldsymbol{\beta}} \mathbf{x}_{it}$$

Now consider writing the code to calculate $\partial \ln L/\partial \ln \sigma_e$:

```
generate double 'de' = ('S_z2'-'a'*'Sz_2')/'s_e'^2
         - 'T' + 1
         - 'a'^2*'Sz_2'/'s_u'^2 -
         - 'a'*'s_e'^2/'s_u'^2
         if 'last'==1
mlvecsum 'lnf' 'g3' = 'de' if 'last'==1 , eq(3)
```

All the ingredients to compute $\partial \ln L_i/\partial \ln \sigma_e$ for each group are present in the last observation of the group; thus we use `if 'last'==1`.

The lines that differ between `myrereg_d0` and `myrereg_d1` (our method-d1 evaluator) are

```
———————————————————————— fragment myrereg_d1.ado ————————
program myrereg_d1
        version 18
        args todo b lnf g
        ...
        quietly {
                ...
                by 'by': generate double 'S_z2' = sum('z'^2)
                by 'by': replace     'S_z2' = 'S_z2'[_N]           // new line
                by 'by': generate double 'Sz_2' = sum('z')^2
                by 'by': replace     'Sz_2' = 'Sz_2'[_N]           // new line
                generate double 'a' = ...
                mlsum 'lnf'= -...
                // the following lines are new
                if ('todo'==0 | 'lnf'==.) exit

                // compute the gradient
                tempvar S_z
                tempname dxb du de
                by 'by': generate double 'S_z' = sum('z')
                by 'by': replace 'S_z' = 'S_z'[_N]
                mlvecsum 'lnf' 'dxb' = ('z'-'a'*'S_z')/'s_e'^2   , eq(1)
                mlvecsum 'lnf' 'du' = 'a'^2*'Sz_2'/'s_u'^2       ///
                        -'T'*'a'                                 ///
                        if 'last'==1                             , eq(2)
                mlvecsum 'lnf' 'de' = 'S_z2'/'s_e'^2 -           ///
                        'a'*'Sz_2'/'s_e'^2 -                     ///
                        'a'^2*'Sz_2'/'s_u'^2 -                   ///
                        'T'+1-'a'*'s_e'^2/'s_u'^2                ///
                        if 'last'==1                             , eq(3)
                mat 'g' = ('dxb','du','de')
        }
end
———————————————————————————— end myrereg_d1.ado ————————
```

▷ **Example**

Below we verify that the results from `myrereg_d1` agree with those from `myrereg_d0`.

```
. use http://www.stata-press.com/data/ml5/tablef7-1, clear
. global MY_panel i        // the panel variable
. ml model d1 myrereg_d1 (xb: lnC = lnQ lnPF lf) /lns_u /lns_e
. ml max, nolog
```

		Number of obs	=	90
		Wald chi2(3)	=	11468.30
Log likelihood = 114.72904		Prob > chi2	=	0.0000

lnC	Coefficient	Std. err.	z	P>|z|	[95% conf. interval]	
lnQ	.90531	.0253759	35.68	0.000	.8555742	.9550459
lnPF	.4233757	.013888	30.48	0.000	.3961557	.4505957
lf	-1.064456	.1962308	-5.42	0.000	-1.449061	-.6798508
_cons	9.618648	.2066218	46.55	0.000	9.213677	10.02362
/lns_u	-2.170818	.3026641	-7.17	0.000	-2.764028	-1.577607
/lns_e	-2.828404	.077319	-36.58	0.000	-2.979947	-2.676862

◁

7.4.3 Calculating H

If we wish to convert our d1 evaluator into a d2 evaluator, we must program the Hessian matrix calculation. As with the gradient, the Hessian calculations in panel-data models come in two flavors. This is further complicated by the fact that `mlmatsum` is not sufficient to perform all the required calculations; `mlmatbysum` completes the tool set for panel-data models.

❏ **Using mlmatbysum to help define H**

For the following discussion, suppose that we have E equations in our model and that the observations are grouped. Let i be the group index and t the observation index within a group. $\mathbf{x}_{eit}$ denotes a vector of values of independent variables for equation e from observation t of group i.

Many models that do not meet the linear-form restrictions have Hessian matrices such that the within-equation submatrices have the form

$$\frac{\partial L}{\partial \boldsymbol{\beta}_1' \partial \boldsymbol{\beta}_1} = \sum_{i=1}^{N} \left\{ \sum_{t=1}^{T_i} d_{it} \mathbf{x}_{1it}' \mathbf{x}_{1it} + \left(\sum_{t=1}^{T_i} a_{it} \right) \left(\sum_{t=1}^{T_i} b_{it} \mathbf{x}_{1it}' \right) \left(\sum_{t=1}^{T_i} b_{it} \mathbf{x}_{1it} \right) \right\}$$

and the between-equation submatrices have the form

$$\frac{\partial L}{\partial \boldsymbol{\beta}_1' \partial \boldsymbol{\beta}_2} = \sum_{i=1}^{N} \left\{ \sum_{t=1}^{T_i} d_{it} \mathbf{x}_{1it}' \mathbf{x}_{2it} + \left(\sum_{t=1}^{T_i} a_{it} \right) \left(\sum_{t=1}^{T_i} b_{it} \mathbf{x}_{1it}' \right) \left(\sum_{t=1}^{T_i} c_{it} \mathbf{x}_{2it} \right) \right\}$$

where a_{it}, b_{it}, c_{it}, and d_{it} are not necessarily the same between derivative formulas. In this case, the overall Hessian matrix may be put together using both `mlmatsum` and `mlmatbysum`. The observation-level outer product matrices

$$\sum_{i=1}^{N} \sum_{t=1}^{T_i} d_{it} \mathbf{x}_{1it}' \mathbf{x}_{1it} \quad \text{and} \quad \sum_{i=1}^{N} \sum_{t=1}^{T_i} d_{it} \mathbf{x}_{1it}' \mathbf{x}_{2it}$$

can be calculated using `mlmatsum`. The group-level outer product matrices

$$\sum_{i=1}^{N} \left(\sum_{t=1}^{T_i} a_{it} \right) \left(\sum_{t=1}^{T_i} b_{it} \mathbf{x}_{1it}' \right) \left(\sum_{t=1}^{T_i} b_{it} \mathbf{x}_{1it} \right)$$

and

$$\sum_{i=1}^{N} \left(\sum_{t=1}^{T_i} a_{it} \right) \left(\sum_{t=1}^{T_i} b_{it} \mathbf{x}_{1it}' \right) \left(\sum_{t=1}^{T_i} c_{it} \mathbf{x}_{2it} \right)$$

can be calculated using `mlmatbysum`.

The syntax for `mlmatbysum` is

`mlmatbysum` *scalarname*_{lnf} *matrixname varname*_a *varname*_b [*varname*_c] [*if*] ,
   `by`(*varname*) [`eq(`#[`,`#]`)`]

Thus obtaining $\partial^2 \ln L / \partial \boldsymbol{\beta} \partial \boldsymbol{\beta}'$ is simply a matter of issuing `mlmatsum` and `mlmatbysum` for each equation pair and joining the results into a single matrix.

In a single-equation system,

```
tempname xx sx
mlmatsum 'lnf' 'xx' = 'd'
mlmatbysum 'lnf' 'sx' 'a' 'b' , by(group_variable)
matrix 'H' = ('xx'+'sx')
```

In a two-equation system,

```
tempname xx11 xx12 xx22 sx11 sx12 sx22 d11 d12 d22
mlmatsum 'lnf' 'xx11' = 'd11', eq(1)
mlmatsum 'lnf' 'xx12' = 'd12', eq(1,2)
mlmatsum 'lnf' 'xx22' = 'd22', eq(2)
mlmatbysum 'lnf' 'sx11' 'a11' 'b11'         , by(group_variable) eq(1)
mlmatbysum 'lnf' 'sx12' 'a12' 'b12' 'c12'   , by(group_variable) eq(1,2)
mlmatbysum 'lnf' 'sx22' 'a22' 'b22'         , by(group_variable) eq(2)
matrix 'd11' = 'xx11'+'sx11'
matrix 'd12' = 'xx12'+'sx12'
matrix 'd22' = 'xx22'+'sx22'
matrix 'H' = ('d11','d12' \ 'd12'','d22')
```

`mlmatbysum` requires '`lnf`' as an argument.

```
mlmatbysum 'lnf' 'whatever' vara varb , eq(1)
mlmatbysum 'lnf' 'whatever' vara varb varc , eq(1,2)
```

This is so `mlmatbysum` can reset '`lnf`' to contain missing if var_b or var_c contains a missing value for some relevant observation. Method-**d2** evaluators are to set '`lnf`' to missing when called with impossible values of the coefficient vector. It is possible that '`lnf`' can be calculated but the Hessian cannot. `mlmatbysum` appropriately handles such cases.

❑

Returning to our random-effects regression model, the first and second derivatives of $\ln L_i$ with respect to $\boldsymbol{\beta}$ are

$$\frac{\partial \ln L_i}{\partial \boldsymbol{\beta}} = \frac{1}{\sigma_e^2} \sum_{t=1}^{T_i} \left(z_{it} - a_i \sum_{s=1}^{T_i} z_{is} \right) \mathbf{x}_{it}$$

$$\frac{\partial^2 \ln L_i}{\partial \boldsymbol{\beta} \partial \boldsymbol{\beta}'} = -\frac{1}{\sigma_e^2} \left\{ \sum_{t=1}^{T_i} \mathbf{x}_{it}' \mathbf{x}_{it} - a_i \left(\sum_{s=1}^{T_i} \mathbf{x}_{is}' \right) \left(\sum_{t=1}^{T_i} \mathbf{x}_{it} \right) \right\} \tag{7.8}$$

We cannot use `mlmatsum` to calculate $\partial^2 \ln L_i / \partial \boldsymbol{\beta} \partial \boldsymbol{\beta}'$ directly because

$$\frac{\partial^2 \ln L_i}{\partial \boldsymbol{\beta} \partial \boldsymbol{\beta}'} \neq \sum_{t=1}^{T_i} \frac{\partial^2 \ln L_i}{\partial^2 \mathbf{x}_{it} \boldsymbol{\beta}} \mathbf{x}_{it}' \mathbf{x}_{it}$$

but we can use it to calculate

$$\sum_{t=1}^{T_i} \mathbf{x}_{it}' \mathbf{x}_{it}$$

and we can use `mlmatbysum` to calculate

$$a_i \left(\sum_{s=1}^{T_i} \mathbf{x}_{is}' \right) \left(\sum_{t=1}^{T_i} \mathbf{x}_{it} \right)$$

The second derivatives with respect to $\ln \sigma_u$ and $\ln \sigma_e$ are

$$\frac{\partial^2 \ln L_i}{\partial^2 \ln \sigma_u} = -\left\{ \frac{2a_i^2}{\sigma_u^2} \left(\sum_{t=1}^{T_i} z_{it} \right)^2 - \frac{4a_i^3 \sigma_e^2}{\sigma_u^4} \left(\sum_{t=1}^{T_i} z_{it} \right)^2 + \frac{2T_i a_i^2 \sigma_e^2}{\sigma_u^2} \right\}$$

$$\frac{\partial^2 \ln L_i}{\partial^2 \ln \sigma_e} = -\left[\frac{2}{\sigma_e^2} \left\{ \sum_{t=1}^{T_i} z_{it}^2 - a_i \left(\sum_{t=1}^{T_i} z_{it} \right)^2 \right\} \right.$$
$$\left. - \left\{ \frac{2a_i^2}{\sigma_u^2} + \frac{4a_i^3 \sigma_e^2}{\sigma_u^4} \right\} \left(\sum_{t=1}^{T_i} z_{it} \right)^2 + \frac{2a_i \sigma_e^2}{\sigma_u^2} - \frac{2a_i^2 \sigma_e^4}{\sigma_u^4} \right]$$

$$\frac{\partial^2 \ln L_i}{\partial \ln \sigma_e \partial \ln \sigma_u} = -\left\{ \frac{4a_i^3 \sigma_e^2}{\sigma_u^4} \left(\sum_{t=1}^{T_i} z_{it} \right)^2 - \frac{2T_i a_i^2 \sigma_e^2}{\sigma_u^2} \right\}$$

As with the gradient calculations in `myrereg_d1`, the component of the Hessian derived based on these two parameters can be calculated using the `mlmatsum` command (with `if 'last'==1` because these calculations are constant within group).

The second derivatives first with respect to $\mathbf{x}_{it}\boldsymbol{\beta}$ and then to $\ln \sigma_u$ and $\ln \sigma_e$ are

$$\frac{\partial^2 \ln L_i}{\partial \mathbf{x}_{it}\boldsymbol{\beta} \partial \ln \sigma_u} = -\frac{2a_i^2}{\sigma_u^2} \sum_{t=1}^{T_i} z_{it}$$

$$\frac{\partial^2 \ln L_i}{\partial \mathbf{x}_{it}\boldsymbol{\beta} \partial \ln \sigma_e} = -\left[\frac{2}{\sigma_e^2} \left\{ z_{it} - a_i \sum_{t=1}^{T_i} z_{it} \right\} - \frac{2a_i^2}{\sigma_u^2} \sum_{t=1}^{T_i} z_{it} \right]$$

We can use the above formulas in the `mlmatsum` command to compute the respective components of the Hessian matrix because

$$\frac{\partial^2 \ln L_i}{\partial \boldsymbol{\beta} \partial \ln \sigma_u} = \sum_{t=1}^{T_i} \frac{\partial^2 \ln L_i}{\partial \mathbf{x}_{it}\boldsymbol{\beta} \partial \ln \sigma_u} \mathbf{x}_{it}$$

$$\frac{\partial^2 \ln L_i}{\partial \boldsymbol{\beta} \partial \ln \sigma_e} = \sum_{t=1}^{T_i} \frac{\partial^2 \ln L_i}{\partial \mathbf{x}_{it}\boldsymbol{\beta} \partial \ln \sigma_e} \mathbf{x}_{it}$$

The lines that differ between `myrereg_d1` and `myrereg_d2` (our method-`d2` evaluator) are

```
─────────────────────────────────────────── fragment myrereg_d2.ado ───────
program myrereg_d2
        version 18
        args todo b lnf g H
        ...
        quietly {
                ...
                mat 'g' = ('dxb','du','de')
                // the following lines are new
                if ('todo'==1 | 'lnf'==.) exit

                // compute the Hessian
                tempname d2xb1 d2xb2 d2xb d2u d2e dxbdu dxbde dude one
                mlmatsum 'lnf' 'd2u' = -(2*'a'^2*'Sz_2'/'s_u'^2 -        ///
                        4*'s_e'^2*'a'^3*'Sz_2'/'s_u'^4 +                ///
                        2*'T'*'a'^2*'s_e'^2/'s_u'^2)                    ///
                        if 'last'==1                          , eq(2)
                mlmatsum 'lnf' 'd2e' =                                  ///
                        -(2*('S_z2'-'a'*'Sz_2')/'s_e'^2 -               ///
                        2*'a'^2*'Sz_2'/'s_u'^2 -                        ///
                        4*'a'^3*'Sz_2'*'s_e'^2/'s_u'^4 +                ///
                        2*'a'*'s_e'^2/'s_u'^2 -                         ///
                        2*'a'^2*'s_e'^4/'s_u'^4)                        ///
                        if 'last'==1                          , eq(3)
                mlmatsum 'lnf' 'dude' =                                 ///
                        -(4*'a'^3*'Sz_2'*'s_e'^2/'s_u'^4 -              ///
                        2*'T'*'a'^2*'s_e'^2/'s_u'^2)                    ///
                        if 'last'==1                          , eq(2,3)
                mlmatsum 'lnf' 'dxbdu' = -2*'a'^2*'S_z'/'s_u'^2 , eq(1,2)
                mlmatsum 'lnf' 'dxbde' =                                ///
                        -(2*('z'-'a'*'S_z')/'s_e'^2 -                   ///
                        2*'a'^2*'S_z'/'s_u'^2                 , eq(1,3)
                // 'a' is constant within panel; and
                // -mlmatbysum- treats missing as 0 for 'a'
                by 'by': replace 'a' = . if !'last'
                mlmatsum 'lnf' 'd2xb2' = 1                     , eq(1)
                generate double 'one' = 1
                mlmatbysum 'lnf' 'd2xb1' 'a' 'one', by($MY_panel) eq(1)
                mat 'd2xb' = -('d2xb2'-'d2xb1')/'s_e'^2
                mat 'H' = (                                   ///
                        'd2xb',    'dxbdu',  'dxbde'        \ ///
                        'dxbdu'',  'd2u',    'dude'         \ ///
                        'dxbde'',  'dude',   'd2e'            ///
                )
        }
end
──────────────────────────────────────────── end myrereg_d2.ado ───────
```

▷ **Example**

We can now verify that the results from `myrereg_d2` agree with those of `myrereg_d0` and `myrereg_d1`.

```
. use http://www.stata-press.com/data/ml5/tablef7-1
. global MY_panel i       // the panel variable
. ml model d2 myrereg_d2 (xb: lnC = lnQ lnPF lf) /lns_u /lns_e
. ml max, nolog
```

				Number of obs	=	90
				Wald chi2(3)	=	11468.30
Log likelihood = 114.72904				Prob > chi2	=	0.0000

lnC	Coefficient	Std. err.	z	P>|z|	[95% conf. interval]	
lnQ	.90531	.0253759	35.68	0.000	.8555742	.9550459
lnPF	.4233757	.013888	30.48	0.000	.3961557	.4505957
lf	-1.064456	.1962308	-5.42	0.000	-1.449061	-.6798508
_cons	9.618648	.2066218	46.55	0.000	9.213677	10.02362
/lns_u	-2.170818	.3026641	-7.17	0.000	-2.764028	-1.577607
/lns_e	-2.828404	.077319	-36.58	0.000	-2.979947	-2.676862

◁

❑ **Technical note**

Suppose that we wanted to model heteroskedasticity. That is, instead of

$$\theta_{1it} = \mathbf{x}_{it}\boldsymbol{\beta}$$
$$\theta_{2it} = \ln \sigma_u$$
$$\theta_{3it} = \ln \sigma_e$$

we decided to write a likelihood evaluator in which $\ln \sigma_e$ was a linear equation:

$$\theta_{1it} = \mathbf{x}_{1it}\boldsymbol{\beta}_1$$
$$\theta_{2it} = \ln \sigma_u \qquad\qquad (7.9)$$
$$\theta_{3it} = \ln \sigma_{eit} = \mathbf{x}_{3it}\boldsymbol{\beta}_3$$

We must admit that we were tempted to start with the likelihood as it is given in (7.3); after all, we went through a lot of algebra to complete the squares and integrate out the random effects u_i. The problem is that (7.9) has a substantial effect on the form of the likelihood function. To see why, let's look at the steps we took in deriving (7.3).

$$\ln L_i = \ln \int_{-\infty}^{\infty} f(u_i) \prod_{t=1}^{T_i} f(y_{it}|u_i) \, du_i$$

$$= \ln \int_{-\infty}^{\infty} \frac{1}{\sigma_u \sqrt{2\pi}} \exp\left(-\frac{u_i^2}{2\sigma_u^2}\right) \prod_{t=1}^{T_i} \left[\frac{1}{\sigma_e \sqrt{2\pi}} \exp\left\{-\frac{(y_{it} - u_i - \mathbf{x}_{it}\boldsymbol{\beta})^2}{2\sigma_e^2}\right\}\right] du_i$$

$$= \ln \int_{-\infty}^{\infty} \frac{1}{\sigma_u \sqrt{2\pi}} \exp\left(-\frac{u_i^2}{2\sigma_u^2}\right) \prod_{t=1}^{T_i} \left[\frac{1}{\sigma_e \sqrt{2\pi}} \exp\left\{-\frac{(z_{it} - u_i)^2}{2\sigma_e^2}\right\}\right] du_i$$

$$= \ln \mathcal{K}_1 + \ln \int_{-\infty}^{\infty} \exp\left\{-\frac{u_i^2}{2\sigma_u^2} - \sum_{t=1}^{T_i} \frac{(z_{it} - u_i)^2}{2\sigma_e^2}\right\} du_i \tag{7.10}$$

where

$$\mathcal{K}_1 = \frac{1}{\sigma_u \sqrt{2\pi}} \prod_{t=1}^{T_i} \frac{1}{\sigma_e \sqrt{2\pi}}$$

We then complete the square in (7.10) and integrate with respect to u_i to get (7.3).

For the heteroskedastic model, we have σ_{eit} instead of σ_e, thus

$$\ln L_i = \ln \mathcal{K}_2 + \ln \int_{-\infty}^{\infty} \exp\left\{-\frac{u_i^2}{2\sigma_u^2} - \sum_{t=1}^{T_i} \frac{(z_{it} - u_i)^2}{2\sigma_{eit}^2}\right\} du_i \tag{7.11}$$

where

$$\mathcal{K}_2 = \frac{1}{\sigma_u \sqrt{2\pi}} \prod_{t=1}^{T_i} \frac{1}{\sigma_{eit} \sqrt{2\pi}}$$

After we complete the square in (7.11) and integrate with respect to u_i, the log-likelihood function looks like

$$\ln L_i = -\frac{1}{2} \left\{ \sum_{t=1}^{T_i} \frac{z_{it}^2}{\sigma_{eit}^2} - a_i \left(\sum_{t=1}^{T_i} \frac{z_{it}}{\sigma_{eit}^2}\right)^2 - \ln a_i + \sum_{t=1}^{T_i} \ln(2\pi\sigma_{eit}^2) + \ln \sigma_u^2 \right\}$$

where

$$\frac{1}{a_i} = \frac{1}{\sigma_u^2} + \sum_{t=1}^{T_i} \frac{1}{\sigma_{eit}^2}$$

and our method-d0 evaluator is

```
                                              begin myrereghet_d0.ado
program myrereghet_d0
        version 18
        args todo b lnf
        tempvar xb z T S_z2 Sz_2 a last s_e S_se2
        tempname s_u
        mleval 'xb'  = 'b', eq(1)
        mleval 's_u' = 'b', eq(2) scalar
        mleval 's_e' = 'b', eq(3)
        scalar 's_u' = exp('s_u')
        // MY_panel contains the panel ID
        local by $MY_panel
        sort 'by'
        local y $ML_y1
        quietly {
                replace 's_e' = exp('s_e')
                by 'by': generate double 'S_se2' = sum(ln(2*c(pi)*'s_e'^2))
                generate double 'z' = 'y' - 'xb'
                by 'by': generate double 'a' = 1 / (1/'s_u'^2 + sum(1/'s_e'^2))
                by 'by': generate 'last' = _n==_N
                by 'by': generate 'T' = _N
                by 'by': generate double 'S_z2' = sum('z'^2/'s_e'^2)
                by 'by': generate double 'Sz_2' = sum('z'/'s_e'^2)^2
                mlsum 'lnf'= -.5 *                              ///
                        (                                       ///
                                ('S_z2'-'a'*'Sz_2') -           ///
                                ln('a') +                       ///
                                'S_se2' +                       ///
                                2*ln('s_u')                     ///
                        )                                       ///
                        if 'last' == 1
        }
end
                                              end myrereghet_d0.ado
```

Using the example at the end of section 7.4.1, we can verify that myrereghet_d0 will reproduce the results of the homoskedastic model:

```
. use http://www.stata-press.com/data/ml5/tablef7-1
. global MY_panel i        // the panel variable
. ml model d0 myrereghet_d0 (xb: lnC = lnQ lnPF lf) /lns_u /lns_e
. ml max, nolog
```

```
                                                 Number of obs =          90
                                                 Wald chi2(3)  = 11468.30
          Log likelihood = 114.72904             Prob > chi2   =    0.0000
```

lnC	Coefficient	Std. err.	z	P>|z|	[95% conf. interval]	
lnQ	.90531	.0253759	35.68	0.000	.8555742	.9550459
lnPF	.4233757	.013888	30.48	0.000	.3961557	.4505957
lf	-1.064456	.1962308	-5.42	0.000	-1.449061	-.6798507
_cons	9.618648	.2066219	46.55	0.000	9.213677	10.02362
/lns_u	-2.170817	.3026643	-7.17	0.000	-2.764028	-1.577605
/lns_e	-2.828404	.077319	-36.58	0.000	-2.979947	-2.676862

When developing the gradient calculations for a method-d1 evaluator, you will find that

$$\frac{\partial L}{\partial \boldsymbol{\beta}_3}$$

can be calculated using `mlvecsum`.

For the method-d2 evaluator, you will find that

$$\frac{\partial^2 L}{\partial \boldsymbol{\beta}_3 \partial \boldsymbol{\beta}_3'}$$

can be calculated using `mlmatsum` and `mlmatbysum`, similarly to (7.8) for $\boldsymbol{\beta}_1$. You will also find that

$$\frac{\partial^2 L}{\partial \boldsymbol{\beta}_1 \partial \boldsymbol{\beta}_3'}$$

can be calculated using `mlmatsum` and `mlmatbysum`.

❏

7.5   Other models that do not meet the linear-form restrictions

Here we discuss other models that do not meet the linear-form restrictions. It is hard to talk about specific challenges you will face when writing an evaluator for these models' likelihood functions, so we will use the Cox proportional hazards model as an example.

Let $\mathbf{x}_j$ be a (fixed) row vector of covariates for a subject. The hazard of failure is assumed to be

$$h_j(t) = h_0(t) \exp(\mathbf{x}_j \boldsymbol{\beta})$$

The Cox proportional hazards model provides estimates of $\boldsymbol{\beta}$ without estimating $h_0(t)$. Given a dataset of N individuals in which each individual is observed from time 0 to t_j, at which time the individual either is observed to fail or is censored, the log-likelihood function is

$$\ln L = \sum_{i=1}^{N_d} \left[\sum_{k \in D_i} \mathbf{x}_k \boldsymbol{\beta} - d_i \ln \left\{ \sum_{j \in R_i} \exp(\mathbf{x}_j \boldsymbol{\beta}) \right\} \right] \qquad (7.12)$$

where i indexes the ordered failure times $t_{(i)}$ $(i = 1, \ldots, N_d)$, N_d is the number of unique failure times, d_i is the number that fail at $t_{(i)}$, D_i is the set of observations that fail at $t_{(i)}$, and R_i is the set of observations that are at risk at time $t_{(i)}$. R_i is called the risk pool (or risk set) at time $t_{(i)}$.

Actually, (7.12) is just one of several possible log likelihoods used to fit a Cox model. This particular form is attributable to Peto (1972) and Breslow (1974). This likelihood has the following properties:

1. There are likelihood values only when failures occur.

2. The likelihood is a function of all the subjects who have not failed.

As such, it violates the linear-form assumptions. Moreover, it is not an easy likelihood function to calculate in any language.

Let's rewrite the likelihood function as

$$\ln L = \sum_{i=1}^{N_d} \left\{ A_i - d_i \ln(B_i) \right\}$$

where

$$A_i = \sum_{k \in D_i} \mathbf{x}_k \boldsymbol{\beta} \qquad \text{and} \qquad B_i = \sum_{j \in R_i} \exp(\mathbf{x}_j \boldsymbol{\beta})$$

As before, we tend to break a problem into manageable pieces. Specifically, we remove a complicated term from the likelihood function, give it a letter, substitute the letter in the original formula, and write the definition of the letter on the side. Once we have done that, we can think about how to program each of the ingredients. As things stand right now, if we knew A_i, d_i, and B_i, calculating $\ln L$ would be easy. So now we think about how to obtain A_i, d_i, and B_i.

To make things a little more concrete, let's pretend that our data are as given in table 7.2. Here t is time of death or censoring, and d is 1 if death or 0 if censored. A person is in the risk pool until he or she dies or is censored.

Table 7.2. Layout for survival data.

i	t	d	$\mathbf{x}_i$
1	1	1	$\mathbf{x}_1$
2	3	0	$\mathbf{x}_2$
3	5	1	$\mathbf{x}_3$
4	5	1	$\mathbf{x}_4$
5	5	0	$\mathbf{x}_5$
6	6	1	$\mathbf{x}_6$

For calculating B_i, let's think about the pool of still-living persons:

i	t	d	$\mathbf{x}_i$	
1	1	1	$\mathbf{x}_1$	first death; every person was alive
2	3	0	$\mathbf{x}_2$	
3	5	1	$\mathbf{x}_3$	second and third deaths;
4	5	1	$\mathbf{x}_4$	$i = 1, 2$ are out of the pool
5	5	0	$\mathbf{x}_5$	
6	6	1	$\mathbf{x}_6$	fourth death; everyone above is out

If you read the above table from bottom to top, the calculation of who is in the pool becomes easy: at the bottom, no one is in the pool, then person 6 is added to the pool, then person 5, and so on. So we could reverse the data and calculate as follows:

i	t	d	$\mathbf{x}_i$	empty pool
6	6	1	$\mathbf{x}_6$	add to pool, $B_6 = \exp(\mathbf{x}_6\boldsymbol{\beta})$
5	5	0	$\mathbf{x}_5$	add to pool, $B_5 = B_6 + \exp(\mathbf{x}_5\boldsymbol{\beta})$
4	5	1	$\mathbf{x}_4$	add to pool, $B_4 = B_5 + \exp(\mathbf{x}_4\boldsymbol{\beta})$
3	5	1	$\mathbf{x}_3$	add to pool, $B_3 = B_4 + \exp(\mathbf{x}_3\boldsymbol{\beta})$
2	3	0	$\mathbf{x}_2$	add to pool, $B_2 = B_3 + \exp(\mathbf{x}_2\boldsymbol{\beta})$
1	1	1	$\mathbf{x}_1$	add to pool, $B_1 = B_2 + \exp(\mathbf{x}_1\boldsymbol{\beta})$

With Stata, this is easy to do with

```
tempvar negt B
generate double `negt' = -t
sort `negt' d
generate double `B' = sum(exp(`xb'))
```

where `xb' is the variable that contains $\mathbf{x}_j\boldsymbol{\beta}$ (we get it using `mleval`).

Calculating A_i is no more difficult:

```
tempvar A
by 'negt' d: generate double 'A' = cond(_n==_N, sum('xb'), .) if d==1
```

We need one more ingredient, d_i, the number of people who die at each time:

```
tempvar sumd
by 'negt' d: generate double 'sumd' = cond(_n==_N, sum(d), .)
```

Now that we have 'A', 'B', and 'sumd', calculating the log-likelihood value is easy:

```
tempvar L last
generate double 'L' = 'A' - 'sumd'*ln('B')
by 'negt' d: generate byte 'last' = (_n==_N & d==1)
mlsum 'lnf' = 'L' if 'last'
```

The if 'last' on the mlsum statement is crucial. By default, mlsum assumes that every observation should contribute to the likelihood value. If there are any missing values among the contributions, mlsum assumes that means the likelihood cannot be calculated at the given value of β. It fills in the overall log likelihood 'lnf' with a missing value, and then later, when ml sees that result, it takes the appropriate action. We purposely have lots of missing values in 'L' because we have contributions to the likelihood only when there is a failure. Thus we need to mark the observations that should contribute to the likelihood and tell mleval about it. We were, we admit, tempted to write

```
generate double 'L' = 'A' - 'sumd'*ln('B')
mlsum 'lnf' = 'L' if !missing('L')
```

because then we would sum the contributions that do exist. That would have been a bad idea. Perhaps, given β, one of the contributions that should exist could not be calculated, meaning that the overall likelihood really could not be calculated at β. We want to be sure that, should that event arise, mlsum sets the overall $\ln L$ in 'lnf' to missing so that ml takes the appropriate action. Therefore, we carefully marked which values should be filled in and specified that on the mlsum statement.

Based on the above, our method-d0 evaluator is

```
———————————————————————————————————— begin mycox1_d0.ado ————————
program mycox1_d0
        version 18
        args todo b lnf
        local t "$ML_y1"                   // $ML_y1 is t
        local d "$ML_y2"                   // $ML_y2 is d
        tempvar xb B A sumd negt last L
        mleval `xb' = `b'
        quietly generate double `negt' = -`t'
        local by "`negt' `d'"
        local byby "by `by'"
        sort `by'
        quietly {
                generate double `B' = sum(exp(`xb'))
                `byby': generate double `A' = cond(_n==_N, sum(`xb'), .) if `d'==1
                `byby': generate `sumd' = cond(_n==_N, sum(`d'), .)
                `byby': generate byte `last' = (_n==_N & `d' == 1)
                generate double `L' = `A' - `sumd'*ln(`B') if `last'
                mlsum `lnf' = `L' if `last'
        }
end
———————————————————————————————————————— end mycox1_d0.ado ————————
```

▷ **Example**

Using `cancer.dta` and `mycox1_d0`, we fit a Cox regression of time of death on drug type and age:

```
. sysuse cancer, clear
(Patient survival in drug trial)
. ml model d0 mycox1_d0 (studytime died = i.drug age, nocons)
. ml maximize
Initial:       Log likelihood = -99.911448
Alternative:   Log likelihood = -99.606964
Rescale:       Log likelihood = -99.101334
Iteration 0:   Log likelihood = -99.101334
Iteration 1:   Log likelihood = -82.649192
Iteration 2:   Log likelihood = -81.737189
Iteration 3:   Log likelihood =  -81.65276
Iteration 4:   Log likelihood = -81.652567
Iteration 5:   Log likelihood = -81.652567
```

```
                                            Number of obs =      48
                                            Wald chi2(3)  =   28.33
Log likelihood = -81.652567                 Prob > chi2   =  0.0000
```

	Coefficient	Std. err.	z	P>|z|	[95% conf. interval]	
drug						
Other	-1.71156	.4943639	-3.46	0.001	-2.680495	-.7426241
NA	-2.956384	.6557432	-4.51	0.000	-4.241617	-1.671151
age	.11184	.0365789	3.06	0.002	.0401467	.1835333

◁

❑ Technical note

Our program is not as efficient as it might be. Every time our likelihood evaluator is called, it re-sorts the data. In the example above, for instance, `ml` called our evaluator 116 times, which we determined using `ml count` (see appendix B). Clearly, sorting the data just once would save considerable execution time.

Instead of sorting the data within the likelihood evaluator, we could sort the data just once before calling `ml model` and `ml maximize`. We also take this opportunity to include weights according to the formula for the log likelihood based on Breslow's approximation as shown in [ST] **stcox**:

```
───────────────────────────── begin mycox2_d0.ado ─────────────
program mycox2_d0
        version 18
        args todo b lnf
        local negt "$ML_y1"              // $ML_y1 is -t
        local d "$ML_y2"                 // $ML_y2 is d
        tempvar xb B A sumd last L
        mleval 'xb' = 'b'
        // data assumed already sorted by 'negt' and 'd'
        local byby "by 'negt' 'd'"
        local wxb "$ML_w*'xb'"
        quietly {
                generate double 'B' = sum($ML_w*exp('xb'))
                'byby': generate double 'A' = cond(_n==_N, sum('wxb'), .) if 'd'==1
                'byby': generate 'sumd' = cond(_n==_N, sum($ML_w*'d'), .)
                'byby': generate byte 'last' = (_n==_N & 'd' == 1)
                generate double 'L' = 'A' - 'sumd'*ln('B') if 'last'
                mlsum 'lnf' = 'L' if 'last', noweight
        }
end
──────────────────────────────── end mycox2_d0.ado ─────────────
```

```
. sysuse cancer
(Patient survival in drug trial)

. generate negtime = -studytime

. sort negtime died

. ml model d0 mycox2_d0 (negtime died = i.drug age, nocons)

. ml maximize, nolog
                                         Number of obs =      48
                                         Wald chi2(3)  =   28.33
Log likelihood = -81.652567             Prob > chi2   =  0.0000
```

	Coefficient	Std. err.	z	P>|z|	[95% conf. interval]
drug					
Other	-1.71156	.4943639	-3.46	0.001	-2.680495 -.7426242
NA	-2.956384	.6557432	-4.51	0.000	-4.241617 -1.671151
age	.11184	.0365789	3.06	0.002	.0401467 .1835333

We are unsure whether to encourage this. The do-file is only slightly faster because, in the original version, Stata did not actually sort the data 116 times. `sort` is smart, and when told to sort, it looks at the sort markers to determine if the dataset is already sorted. Thus the original code sorted the data once and made `sort` verify that it was sorted 115 times.

Still, there are situations where there is a considerable gain to making calculations once and only once. You can do that, but be aware that you make your do-file more dangerous to use.

❏

❏ **Technical note**

If we were using our evaluator as a subroutine in writing a new estimation command, we could gain the efficiency safely. We would be calling `ml` from inside another ado-file and so have complete control. We could create 'negt' and sort on it before the `ml model` statement and then use 'negt' in the `ml model` statement. The user could not forget to sort the data or mistakenly use time rather than negative time because we would be the ones doing that behind the scenes.

We did this in `mycox`, the estimation command discussed in section 17.5. A listing of `mycox` is shown in appendix E.5.

❏

8 Debugging likelihood evaluators

Things probably will not proceed as smoothly for you as they have for us in the previous chapters. You may write an evaluator and observe the following:

```
. ml model lf myprobit_lf (foreign = mpg weight)

. ml maximize
Initial:      Log likelihood =      -<inf>  (could not be evaluated)
—Break—
r(1);
```

We are the ones who pressed *Break* because we got tired of watching nothing happen. You might see

```
. ml model lf myprobit_lf (foreign = mpg weight)

. ml maximize
invalid syntax
r(198);
```

or

```
. ml model lf myprobit_lf (foreign = mpg weight)

. ml maximize
Initial:      Log likelihood =            0
Alternative:  Log likelihood =            0
Rescale:      Log likelihood =            0
could not calculate numerical derivatives—discontinuous region with missing
values encountered
r(430);
```

ml has some capabilities that will help you solve these problems.

8.1   ml check

The mistake we made in the above examples was not typing ml check between the ml model and ml maximize statements. If you are lucky—if your program works—here is what you might see:

```
. sysuse auto
(1978 automobile data)
. ml model lf myprobit_lf (foreign = mpg weight)
. ml check
```

Test 1: Calling myprobit_lf to check if it computes log likelihood and does
        not alter coefficient vector ...
        Passed.

Test 2: Calling myprobit_lf again to check if the same log likelihood value is
        returned ...
        Passed.

Test 3: Calling myprobit_lf to check if 1st derivatives are computed ...
        Test not relevant for type lf evaluators.

Test 4: Calling myprobit_lf again to check if the same 1st derivatives are
        returned ...
        Test not relevant for type lf evaluators.

Test 5: Calling myprobit_lf to check if 2nd derivatives are computed ...
        Test not relevant for type lf evaluators.

Test 6: Calling myprobit_lf again to check if the same 2nd derivatives are
        returned ...
        Test not relevant for type lf evaluators.
_____

Searching for alternate values for the coefficient vector to verify that
myprobit_lf returns different results when fed a different coefficient vector
...
Initial: Log likelihood =   -<inf> (could not be evaluated)
Searching for feasible values: +

Feasible: Log likelihood = -173.96812
Improving initial values: +.....+...
Improvement: Log likelihood = -108.86053

Continuing with tests ...
_____

Test 7: Calling myprobit_lf to check log likelihood at the new values ...
        Passed.

Test 8: Calling myprobit_lf requesting 1st derivatives at the new values ...
        Test not relevant for type lf evaluators.

Test 9: Calling myprobit_lf requesting 2nd derivatives at the new values ...
        Test not relevant for type lf evaluators.
_____

                    myprobit_lf HAS PASSED ALL TESTS
_____

Test 10: Does myprobit_lf produce unanticipated output? This is a minor issue.
         Stata has been running myprobit_lf with all output suppressed. This
         time, Stata will not suppress the output. If you see any
         unanticipated output, you need to place quietly in front of some of
         the commands in myprobit_lf.
_____
                                                     ─── begin execution
_____
                                                     ─── end execution
```

If your program does not work, you will probably see something like the following:
```

```
. sysuse auto
(1978 automobile data)
. ml model lf myprobit_bad_lf (foreign = mpg weight)
. ml check

Test 1: Calling myprobit_bad_lf to check if it computes log likelihood and
        does not alter coefficient vector ...
        Failed.
        myprobit_bad_lf returned error 132.

Here is a trace of its execution:
```

```
-> myprobit_bad_lf __000008 __000009
                    - 'begin'
                    = capture noisily version 18: myprobit_bad_lf __000008 __000009
                    ─────────────────────────────────── begin myprobit_bad_lf ───
                    - version 18
                    - args lnf xb
                    - quietly replace 'lnf' = ln(normal( 'xb')) if $ML_y1 == 1
                    = quietly replace __000008 = ln(normal( __000009)) if
> foreign == 1
                    - quietly replace 'lnf' = ln(normal(-'xb') if $ML_y1 == 0
                    = quietly replace __000008 = ln(normal(-__000009) if
> foreign == 0
too few ')' or ']'
                    ─────────────────────────────────── end myprobit_bad_lf ───
                    - 'end'
                    = set trace off
```

```
Fix myprobit_bad_lf.
r(132);
```

We left out a closing parenthesis in our program. The problem with our program is not that it altered the coefficient vector; it was just that, while attempting to test whether our program altered the coefficient vector, our program generated an error, so ml check backed up and showed us a trace of its execution.

ml check can be used with any method of likelihood evaluator.

ml check will save you hours of time.

8.2 Using the debug methods

All the evaluator methods that require analytic first or second derivatives have a corresponding debug method. Methods that require first derivatives, namely, lf1 and d1, have accompanying debug methods lf1debug and d1debug. Second-derivative-based methods lf2 and d2 have accompanying debug methods lf2debug and d2debug. These debug methods verify the accuracy of your analytic derivatives. Like with lf, lf0, and d0 methods, they compute all the required derivatives numerically when fitting the model. However, they also request the derivatives from your evaluator program and then compare those derivatives with the ones calculated numerically. A large discrepancy between the numeric and analytic derivatives is an indication that you have either incorrectly taken a derivative of your likelihood function or else made a programming error when implementing it in your program.

Remember, `ml check` does not verify that your derivatives are correct; it just looks for mechanical errors. That does not mean you should not bother with `ml check`—just do not expect `ml check` to catch substantive errors.

If you have written a method-d1 evaluator, we recommend that you type

```
. ml model d1debug ...
. ml check
. ml maximize
```

Similarly, if you have written a method-lf2 evaluator, we recommend that you type

```
. ml model lf2debug ...
. ml check
. ml maximize
```

We recommend using the debug methods along with `ml check`. They do not interact—`ml check` will produce no more information than it would if it were called with a regular evaluator method.

To show what can happen, we have written a method-lf1 evaluator for the Weibull model and purposely programmed the derivatives incorrectly. There are no mechanical errors in our program, however.

```
. sysuse cancer
(Patient survival in drug trial)
. ml model lf1debug myweibull_bad_lf1 (studytime died = i.drug age) ()
. ml check

Test 1: Calling myweibull_bad_lf1 to check if it computes log likelihood and
        does not alter coefficient vector ...
        Passed.

  (output omitted)

Test 3: Calling myweibull_bad_lf1 to check if 1st derivatives are computed ...
        Passed.

  (note that ml check did not notice that the derivatives are wrong!)

  (output omitted)
───────────────────────────────────────────────────────────────────────────

               myweibull_bad_lf1 HAS PASSED ALL TESTS
───────────────────────────────────────────────────────────────────────────

  (output omitted)

You should check that the derivatives are right.

Use the optimizer to obtain estimates.

The output will include a report comparing analytic to numeric derivatives. Do
not be concerned if your analytic derivatives differ from the numeric ones in
early iterations.

The analytic gradient will differ from the numeric one in early iterations,
then the mreldif() difference should become less than 1e-6 in the middle
iterations, and the difference will increase again in the final iterations as
the gradient goes to zero.
```

`ml check` even reminds us to check that the derivatives are right. We are already using method `lf1debug`, so we will discover our problem when we type `ml maximize`.

8.2.1 First derivatives

We have proceeded too quickly in our examples in the previous chapters. In real life, the probit and Weibull method-lf1 evaluators we wrote in chapter 6 and the d1 evaluators we wrote in chapter 7 would not have worked the first time we tried. Without a doubt, we would have made an error in producing the gradient vector, either substantively because we took derivatives incorrectly or mechanically in how we programmed the evaluator.

The recipe for success is to start with a working method-lf0 or method-d0 evaluator, make the requisite changes to convert it to a lf1 or d1 evaluator, and then specify method lf1debug or d1debug in the `ml model` statement.

For example, here is what happens when we check the gradient calculations in `myweibull_lf1` using method lf1debug:

```
. sysuse cancer
(Patient survival in drug trial)
. ml model lf1debug myweibull_lf1 (studytime died = i.drug age) ()
. ml max
Initial:      Log likelihood =       -744
Alternative:  Log likelihood = -356.14276
Rescale:      Log likelihood = -200.80201
Rescale eq:   Log likelihood = -136.69232
─────────────────────────────────────── begin derivative-comparison report
lf1debug: mreldif(gradient vectors) = 3.98e-10
lf1debug: Warning: Evaluator did not compute Hessian matrix.
─────────────────────────────────────── end derivative-comparison report
Iteration 0:  Log likelihood = -136.69232  (not concave)
─────────────────────────────────────── begin derivative-comparison report
lf1debug: mreldif(gradient vectors) = 1.45e-10
lf1debug: Warning: Evaluator did not compute Hessian matrix.
─────────────────────────────────────── end derivative-comparison report
Iteration 1:  Log likelihood = -124.11726
─────────────────────────────────────── begin derivative-comparison report
lf1debug: mreldif(gradient vectors) = 8.03e-09
lf1debug: Warning: Evaluator did not compute Hessian matrix.
─────────────────────────────────────── end derivative-comparison report
Iteration 2:  Log likelihood =  -113.9591
─────────────────────────────────────── begin derivative-comparison report
lf1debug: mreldif(gradient vectors) = 1.74e-09
lf1debug: Warning: Evaluator did not compute Hessian matrix.
─────────────────────────────────────── end derivative-comparison report
Iteration 3:  Log likelihood = -110.30683
─────────────────────────────────────── begin derivative-comparison report
lf1debug: mreldif(gradient vectors) = 8.50e-07
lf1debug: Warning: Evaluator did not compute Hessian matrix.
─────────────────────────────────────── end derivative-comparison report
Iteration 4:  Log likelihood = -110.26748
─────────────────────────────────────── begin derivative-comparison report
lf1debug: mreldif(gradient vectors) = .0000372
lf1debug: Warning: Evaluator did not compute Hessian matrix.
─────────────────────────────────────── end derivative-comparison report
```

```
Iteration 5:  Log likelihood = -110.26736
─────────────────────────────────── begin derivative-comparison report
lf1debug: mreldif(gradient vectors) = .0000381
lf1debug: Warning: Evaluator did not compute Hessian matrix.
─────────────────────────────────── end derivative-comparison report
Iteration 6:  Log likelihood = -110.26736
```

```
                                               Number of obs =      48
                                               Wald chi2(3)  =   35.25
Log likelihood = -110.26736                    Prob > chi2   =  0.0000
```

		Coefficient	Std. err.	z	P>|z|	[95% conf. interval]	
eq1							
	drug						
	Other	1.012966	.2903917	3.49	0.000	.4438086	1.582123
	NA	1.45917	.2821195	5.17	0.000	.9062261	2.012114
	age	-.0671728	.0205688	-3.27	0.001	-.1074868	-.0268587
	_cons	6.060723	1.152845	5.26	0.000	3.801188	8.320259
eq2							
	_cons	.5573333	.1402154	3.97	0.000	.2825162	.8321504

The difference between methods `lf1` and `lf1debug` is that `lf1debug` provides a derivative comparison report interspersed in the iteration log. For instance, between iterations 1 and 2 we see the following:

```
Iteration 1:  Log likelihood = -124.11726
─────────────────────────────────── begin derivative-comparison report
lf1debug: mreldif(gradient vectors) = 8.03e-09
lf1debug: Warning: Evaluator did not compute Hessian matrix.
─────────────────────────────────── end derivative-comparison report
Iteration 2:  Log likelihood =  -113.9591
```

Whenever `ml` needs first derivatives, method `lf1debug` calls the likelihood evaluator just as method `lf1` would, but method `lf1debug` also calculates the derivatives numerically. It then displays a comparison of the derivatives calculated both ways and returns to `ml` the numerically calculated derivatives. Thus `lf1debug` proceeds to a correct answer, even if the analytical derivative calculation is incorrect.

`mreldif()` in the derivative-comparison report stands for maximum relative difference; this would be easier to understand if we could see the gradient vector. Method `lf1debug` shows us the vectors as well as the `mreldif()` if we specify `ml maximize`'s `gradient` option:

```
. sysuse cancer
(Patient survival in drug trial)

. ml model lf1debug myweibull_lf1 (studytime died = i.drug age) ()

. ml max, gradient
Initial:     Log likelihood =       -744
Trying nonzero initial values: +.
Alternative: Log likelihood = -356.14276
Rescaling entire vector: .++.
Rescale:     Log likelihood = -200.80201
Rescaling equations: +++++++++++++++++++++++++++++++++++.
Sign reverse: +++++++++++++++++++++++++++++++++++.+++.
Rescaling equations: ....
Rescale eq:  Log likelihood = -136.69232
```

```
Iteration 0:
─────────────────────────────── begin derivative-comparison report
lf1debug: mreldif(gradient vectors) = 3.98e-10

lf1debug: Evaluator calculated gradient:
        eq1:       eq1:       eq1:       eq1:       eq1:       eq2:
        1b.        2.         3.
        drug       drug       drug       age        _cons      _cons
r1        0  -.7819348   1.243378   -632.982   -10.67705   2.580612

lf1debug: Numerically calculated gradient (used for stepping):
        eq1:       eq1:       eq1:       eq1:       eq1:       eq2:
        1b.        2.         3.
        drug       drug       drug       age        _cons      _cons
r1        0  -.7819348   1.243378   -632.982   -10.67705   2.580612

lf1debug: Relative differences:
        eq1:       eq1:       eq1:       eq1:       eq1:       eq2:
        1b.        2.         3.
        drug       drug       drug       age        _cons      _cons
r1        0   3.34e-10   3.98e-10   3.55e-10   3.51e-10   3.73e-10

lf1debug: Warning: Evaluator did not compute Hessian matrix.
─────────────────────────────── end derivative-comparison report
                              Log likelihood = -136.69232
                                          (not concave)
```

```
Iteration 1:
───────────────────────────────────────── begin derivative-comparison report
lf1debug: mreldif(gradient vectors) = 1.45e-10
lf1debug: Evaluator calculated gradient:
          eq1:      eq1:      eq1:      eq1:      eq1:       eq2:
          1b.       2.        3.
          drug      drug      drug      age       _cons      _cons
r1          0    12.10675  11.93993  1358.373  25.85834   7.363412
lf1debug: Numerically calculated gradient (used for stepping):
          eq1:      eq1:      eq1:      eq1:      eq1:       eq2:
          1b.       2.        3.
          drug      drug      drug      age       _cons      _cons
r1          0    12.10675  11.93993  1358.373  25.85834   7.363412
lf1debug: Relative differences:
          eq1:      eq1:      eq1:      eq1:      eq1:       eq2:
          1b.       2.        3.
          drug      drug      drug      age       _cons      _cons
r1          0    1.08e-11  1.45e-10  8.50e-11  7.95e-11   9.18e-11
lf1debug: Warning: Evaluator did not compute Hessian matrix.
───────────────────────────────────────── end derivative-comparison report
                                        Log likelihood = -124.11726

Iteration 2:
───────────────────────────────────────── begin derivative-comparison report
lf1debug: mreldif(gradient vectors) = 8.03e-09
lf1debug: Evaluator calculated gradient:
          eq1:       eq1:       eq1:       eq1:       eq1:        eq2:
          1b.        2.         3.
          drug       drug       drug       age        _cons       _cons
r1          0     -1.833031  -1.784294  -35.01956  -.7653024  -24.53372
lf1debug: Numerically calculated gradient (used for stepping):
          eq1:       eq1:       eq1:       eq1:       eq1:        eq2:
          1b.        2.         3.
          drug       drug       drug       age        _cons       _cons
r1          0     -1.833031  -1.784294  -35.01956  -.7653024  -24.53372
lf1debug: Relative differences:
          eq1:       eq1:       eq1:       eq1:       eq1:        eq2:
          1b.        2.         3.
          drug       drug       drug       age        _cons       _cons
r1          0      2.86e-10   3.24e-10   8.03e-09   2.89e-09   2.41e-10
lf1debug: Warning: Evaluator did not compute Hessian matrix.
───────────────────────────────────────── end derivative-comparison report
                                        Log likelihood = -113.9591
```

```
Iteration 3:
————————————————————————————————— begin derivative-comparison report
lf1debug: mreldif(gradient vectors) = 1.74e-09

lf1debug: Evaluator calculated gradient:
            eq1:        eq1:        eq1:        eq1:        eq1:        eq2:
            1b.         2.          3.
            drug        drug        drug        age         _cons       _cons
r1            0    -.0318114    -.074351    66.03781    1.153271    -2.112529

lf1debug: Numerically calculated gradient (used for stepping):
            eq1:        eq1:        eq1:        eq1:        eq1:        eq2:
            1b.         2.          3.
            drug        drug        drug        age         _cons       _cons
r1            0    -.0318114    -.074351    66.03781    1.153271    -2.112529

lf1debug: Relative differences:
            eq1:        eq1:        eq1:        eq1:        eq1:        eq2:
            1b.         2.          3.
            drug        drug        drug        age         _cons       _cons
r1            0     3.30e-10    4.85e-10    1.74e-09    9.51e-10    3.06e-10

lf1debug: Warning: Evaluator did not compute Hessian matrix.
————————————————————————————————————— end derivative-comparison report
                                        Log likelihood = -110.30683
```

```
Iteration 4:
————————————————————————————————— begin derivative-comparison report
lf1debug: mreldif(gradient vectors) = 8.50e-07

lf1debug: Evaluator calculated gradient:
            eq1:        eq1:        eq1:        eq1:        eq1:        eq2:
            1b.         2.          3.
            drug        drug        drug        age         _cons       _cons
r1            0     .0032121    .0005401    5.701895    .1003147    -.0921864

lf1debug: Numerically calculated gradient (used for stepping):
            eq1:        eq1:        eq1:        eq1:        eq1:        eq2:
            1b.         2.          3.
            drug        drug        drug        age         _cons       _cons
r1            0     .0032121    .0005401     5.7019    .1003148    -.0921865

lf1debug: Relative differences:
            eq1:        eq1:        eq1:        eq1:        eq1:        eq2:
            1b.         2.          3.
            drug        drug        drug        age         _cons       _cons
r1            0     1.93e-08    1.94e-08    8.50e-07    9.12e-08    7.84e-08

lf1debug: Warning: Evaluator did not compute Hessian matrix.
————————————————————————————————————— end derivative-comparison report
                                        Log likelihood = -110.26748
```

```
Iteration 5:
─────────────────────────────────────── begin derivative-comparison report
lf1debug: mreldif(gradient vectors) = .0000372
lf1debug: Evaluator calculated gradient:
          eq1:        eq1:        eq1:        eq1:        eq1:        eq2:
          1b.         2.          3.
          drug        drug        drug        age         _cons       _cons
r1          0     .0000175    9.64e-06    .0237448    .0004192    -.0003496
lf1debug: Numerically calculated gradient (used for stepping):
          eq1:        eq1:        eq1:        eq1:        eq1:        eq2:
          1b.         2.          3.
          drug        drug        drug        age         _cons       _cons
r1          0     .0000176    9.77e-06    .0237829    .0004199    -.0003505
lf1debug: Relative differences:
          eq1:        eq1:        eq1:        eq1:        eq1:        eq2:
          1b.         2.          3.
          drug        drug        drug        age         _cons       _cons
r1          0    1.30e-07    1.30e-07    .0000372    6.71e-07    9.03e-07
lf1debug: Warning: Evaluator did not compute Hessian matrix.
────────────────────────────────────────── end derivative-comparison report
                                              Log likelihood = -110.26736

Iteration 6:
─────────────────────────────────────── begin derivative-comparison report
lf1debug: mreldif(gradient vectors) = .0000381
lf1debug: Evaluator calculated gradient:
          eq1:        eq1:        eq1:        eq1:        eq1:        eq2:
          1b.         2.          3.
          drug        drug        drug        age         _cons       _cons
r1          0    -1.30e-07   -1.30e-07   -.0000377   -6.64e-07    8.97e-07
lf1debug: Numerically calculated gradient (used for stepping):
          eq1:        eq1:        eq1:        eq1:        eq1:        eq2:
          1b.         2.          3.
          drug        drug        drug        age         _cons       _cons
r1          0     3.13e-10    2.20e-10    3.83e-07    6.78e-09   -5.55e-09
lf1debug: Relative differences:
          eq1:        eq1:        eq1:        eq1:        eq1:        eq2:
          1b.         2.          3.
          drug        drug        drug        age         _cons       _cons
r1          0    1.30e-07    1.30e-07    .0000381    6.71e-07    9.03e-07
lf1debug: Warning: Evaluator did not compute Hessian matrix.
────────────────────────────────────────── end derivative-comparison report
                                              Log likelihood = -110.26736
```

```
                                               Number of obs =      48
                                               Wald chi2(3)  =   35.25
         Log likelihood = -110.26736           Prob > chi2   =  0.0000
```

	Coefficient	Std. err.	z	P>|z|	[95% conf. interval]	
eq1						
drug						
Other	1.012966	.2903917	3.49	0.000	.4438086	1.582123
NA	1.45917	.2821195	5.17	0.000	.9062261	2.012114
age	-.0671728	.0205688	-3.27	0.001	-.1074868	-.0268587
_cons	6.060723	1.152845	5.26	0.000	3.801188	8.320259
eq2						
_cons	.5573333	.1402154	3.97	0.000	.2825162	.8321504

When we did not specify the `gradient` option, we saw the following reported just before the `Iteration 1` line:

```
                                          begin derivative-comparison report
lf1debug: mreldif(gradient vectors) = 1.45e-10
lf1debug: Warning: Evaluator did not compute Hessian matrix.
                                          end derivative-comparison report
Iteration 1:  Log likelihood = -124.11726
```

With the `gradient` option, we now see

```
Iteration 1:
                                          begin derivative-comparison report
lf1debug: mreldif(gradient vectors) = 1.45e-10
lf1debug: Evaluator calculated gradient:
          eq1:     eq1:     eq1:     eq1:     eq1:     eq2:
          1b.       2.       3.
          drug     drug     drug     age     _cons    _cons
r1          0   12.10675  11.93993  1358.373  25.85834  7.363412
lf1debug: Numerically calculated gradient (used for stepping):
          eq1:     eq1:     eq1:     eq1:     eq1:     eq2:
          1b.       2.       3.
          drug     drug     drug     age     _cons    _cons
r1          0   12.10675  11.93993  1358.373  25.85834  7.363412
lf1debug: Relative differences:
          eq1:     eq1:     eq1:     eq1:     eq1:     eq2:
          1b.       2.       3.
          drug     drug     drug     age     _cons    _cons
r1          0   1.08e-11  1.45e-10  8.50e-11  7.95e-11  9.18e-11
lf1debug: Warning: Evaluator did not compute Hessian matrix.
                                          end derivative-comparison report
                              Log likelihood = -124.11726
```

We have the same `mreldif()` of 1.45e–10, but now we also see the gradient vector element by element. `mreldif()` is the maximum of the relative differences calculated element by element, where relative difference is defined as $|g_i - n_i|/(|n_i| + 1)$, g_i is the element of the gradient vector `myweibull_d1` calculated, and n_i is the numerically calculated gradient value. We can safely ignore the warning regarding computation of the Hessian matrix because we are working with an `lf1` evaluator, not an `lf2` one.

Specifying the `gradient` option can be useful when the calculated gradient value is suspect. Here `myweibull_lf1` works, and the reported `mreldif()`s are typical:

Iteration 0:   3.98e–10
Iteration 1:   1.45e–10
Iteration 2:   8.03e–09
Iteration 3:   1.74e–09
Iteration 4:   8.50e–07
Iteration 5:   3.72e–05
Iteration 6:   3.81e–05

What is being revealed here is not the accuracy of `myweibull_lf1` but that of the numerical approximation. Here they are initially good, remain good through the next few iterations, and then become poor near the end. They end up poor because we are measuring error as a relative difference and the derivatives themselves are going to zero as the algorithm converges. In other cases, you might see the initial relative differences be somewhat larger than those in the middle, though if you have programmed the analytic derivatives correctly, at some point the relative differences should become relatively small (1e–8 or smaller).

The purpose of the output, however, is not to test the accuracy of `ml`'s numeric derivative calculator but to verify that the analytic derivatives are in rough agreement with them. If they are, we can safely assume that the analytic derivatives are correct.

To demonstrate what you might see when the analytic derivatives are somehow wrong, we will use `myweibull_bad_lf1`, a modified copy of `myweibull_lf1`. We changed the line that reads

```
replace 'g2' = 'd' - 'R'*'p'*('M'-'d')
```

to read

```
replace 'g2' = 'd' + 'R'*'p'*('M'-'d')
```

and so introduce a rather typical sign error. The result of fitting the model with method `lf1debug` is

```
. sysuse cancer
(Patient survival in drug trial)

. ml model lf1debug myweibull_bad_lf1 (studytime died = i.drug age) ()

. ml max

Initial:      Log likelihood =        -744
Alternative:  Log likelihood = -356.14276
Rescale:      Log likelihood = -200.80201
Rescale eq:   Log likelihood = -136.69232
                                              ─────── begin derivative-comparison report
lf1debug: mreldif(gradient vectors) = .9407374
lf1debug: Warning: Evaluator did not compute Hessian matrix.
                                              ─────── end derivative-comparison report
Iteration 0:  Log likelihood = -136.69232   (not concave)
                                              ─────── begin derivative-comparison report
lf1debug: mreldif(gradient vectors) = .8496778
lf1debug: Warning: Evaluator did not compute Hessian matrix.
                                              ─────── end derivative-comparison report
Iteration 1:  Log likelihood = -124.11726
                                              ─────── begin derivative-comparison report
lf1debug: mreldif(gradient vectors) = 1.268853
lf1debug: Warning: Evaluator did not compute Hessian matrix.
                                              ─────── end derivative-comparison report
Iteration 2:  Log likelihood =  -113.9591
                                              ─────── begin derivative-comparison report
lf1debug: mreldif(gradient vectors) = 1.017086
lf1debug: Warning: Evaluator did not compute Hessian matrix.
                                              ─────── end derivative-comparison report
Iteration 3:  Log likelihood = -110.30683
                                              ─────── begin derivative-comparison report
lf1debug: mreldif(gradient vectors) = .9856113
lf1debug: Warning: Evaluator did not compute Hessian matrix.
                                              ─────── end derivative-comparison report
Iteration 4:  Log likelihood = -110.26748
                                              ─────── begin derivative-comparison report
lf1debug: mreldif(gradient vectors) = .9841326
lf1debug: Warning: Evaluator did not compute Hessian matrix.
                                              ─────── end derivative-comparison report
Iteration 5:  Log likelihood = -110.26736
                                              ─────── begin derivative-comparison report
lf1debug: mreldif(gradient vectors) = .984127
lf1debug: Warning: Evaluator did not compute Hessian matrix.
                                              ─────── end derivative-comparison report
Iteration 6:  Log likelihood = -110.26736
```

Number of obs	=	48
Wald chi2(3)	=	35.25
Log likelihood = -110.26736					Prob > chi2	=	0.0000

	Coefficient	Std. err.	z	P>|z|	[95% conf. interval]	
eq1						
drug						
Other	1.012966	.2903917	3.49	0.000	.4438086	1.582123
NA	1.45917	.2821195	5.17	0.000	.9062261	2.012114
age	-.0671728	.0205688	-3.27	0.001	-.1074868	-.0268587
_cons	6.060723	1.152845	5.26	0.000	3.801188	8.320259
eq2						
_cons	.5573333	.1402154	3.97	0.000	.2825162	.8321504

First, despite the error in `myweibull_bad_lf1`, the overall model is still correctly fit because method `lf1debug` used the numerically calculated derivatives. The log, however, reveals that our calculation of the derivatives has problems:

```
---------------------------------------------------- begin derivative-comparison report
lf1debug: mreldif(gradient vectors) = .9407374
lf1debug: Warning: Evaluator did not compute Hessian matrix.
---------------------------------------------------- end derivative-comparison report
Iteration 0:  Log likelihood = -136.69232  (not concave)
---------------------------------------------------- begin derivative-comparison report
lf1debug: mreldif(gradient vectors) = .8496778
lf1debug: Warning: Evaluator did not compute Hessian matrix.
---------------------------------------------------- end derivative-comparison report
Iteration 1:  Log likelihood = -124.11726
```

The `mreldif()`s we now observe are

Iteration 0:	0.94
Iteration 1:	0.85
Iteration 2:	1.27
Iteration 3:	1.02
Iteration 4:	0.99
Iteration 5:	0.98
Iteration 6:	0.98

Specifying the `gradient` option allows us to spot that the problem arises in calculating the derivative of the second equation. We can easily see this when we look at the relative differences of the gradient elements in the iteration log:

```
Iteration 1:
─────────────────────────────────────── begin derivative-comparison report
lf1debug: mreldif(gradient vectors) = .8496778

lf1debug: Evaluator calculated gradient:
          eq1:      eq1:      eq1:      eq1:      eq1:       eq2:
          1b.        2.        3.
          drug      drug      drug      age      _cons      _cons
     r1      0  12.10675  11.93993  1358.373  25.85834  54.63659

lf1debug: Numerically calculated gradient (used for stepping):
          eq1:      eq1:      eq1:      eq1:      eq1:       eq2:
          1b.        2.        3.
          drug      drug      drug      age      _cons      _cons
     r1      0  12.10675  11.93993  1358.373  25.85834  7.363412

lf1debug: Relative differences:
          eq1:      eq1:      eq1:      eq1:      eq1:       eq2:
          1b.        2.        3.
          drug      drug      drug      age      _cons      _cons
     r1      0  1.08e-11  1.45e-10  8.50e-11  7.95e-11  .8496778
```

Note the large value for `eq2:_cons` and the small values for `eq1`. This tells us that the derivatives for `eq2` are either incorrectly derived or badly coded but that the `eq1` derivatives are correct.

8.2.2 Second derivatives

As mentioned in the previous section, in real life the probit and Weibull evaluators we wrote in chapters 6 and 7 would not have worked the first time we tried. Without a doubt, we would have made an error in producing the Hessian, either substantively because we took the derivatives incorrectly or mechanically in how we programmed the evaluator.

Similar to starting with a derivative-free evaluator and then programming and testing our first derivatives, here we start with a working method-`lf1` or method-`d1` evaluator and make the changes necessary to convert it to method-`lf2` or method-`d2`. This time we specify method-`lf2debug` or method-`d2debug` in the `ml model` statement. They are just like methods `lf2` and `d2` except that they check the analytical derivatives against derivatives calculated numerically.

Here is the result of using method `d2debug` to check the Hessian calculations in `myprobit_d2`:

```
. sysuse auto
(1978 automobile data)

. ml model d2debug myprobit_d2 (foreign = mpg weight)

. ml max
Initial:        Log likelihood = -51.292891
Alternative:    Log likelihood = -45.055272
Rescale:        Log likelihood = -45.055272
──────────────────────────────────────── begin derivative-comparison report
d2debug: mreldif(gradient vectors) = .0002445
d2debug: mreldif(Hessian matrices) = .000014
──────────────────────────────────────── end derivative-comparison report
Iteration 0:   Log likelihood = -45.055272
──────────────────────────────────────── begin derivative-comparison report
d2debug: mreldif(gradient vectors) = 9.12e-10
d2debug: mreldif(Hessian matrices) = .0027612
──────────────────────────────────────── end derivative-comparison report
Iteration 1:   Log likelihood = -27.908441
──────────────────────────────────────── begin derivative-comparison report
d2debug: mreldif(gradient vectors) = 3.19e-09
d2debug: mreldif(Hessian matrices) = 1.21e-07
──────────────────────────────────────── end derivative-comparison report
Iteration 2:   Log likelihood = -26.863584
──────────────────────────────────────── begin derivative-comparison report
d2debug: mreldif(gradient vectors) = 2.20e-06
d2debug: mreldif(Hessian matrices) = 2.58e-08
──────────────────────────────────────── end derivative-comparison report
Iteration 3:   Log likelihood = -26.844207
──────────────────────────────────────── begin derivative-comparison report
d2debug: mreldif(gradient vectors) = .0003472
d2debug: mreldif(Hessian matrices) = 5.65e-09
──────────────────────────────────────── end derivative-comparison report
Iteration 4:   Log likelihood = -26.844189
──────────────────────────────────────── begin derivative-comparison report
d2debug: mreldif(gradient vectors) = .0003653
d2debug: mreldif(Hessian matrices) = 9.57e-09
──────────────────────────────────────── end derivative-comparison report
Iteration 5:   Log likelihood = -26.844189

                                        Number of obs =      74
                                        Wald chi2(2)  =   20.75
Log likelihood = -26.844189             Prob > chi2   =  0.0000
```

foreign	Coefficient	Std. err.	z	P>|z|	[95% conf. interval]	
mpg	-.1039503	.0515689	-2.02	0.044	-.2050236	-.0028771
weight	-.0023355	.0005661	-4.13	0.000	-.003445	-.0012261
_cons	8.275464	2.554144	3.24	0.001	3.269433	13.2815

The difference between methods `d2` and `d2debug` is that method `d2debug` provides a derivative comparison report interspersed in the iteration log. For instance, before iteration 1, we see the following:

```
─────────────────────────────────────── begin derivative-comparison report
d2debug: mreldif(gradient vectors) = 9.12e-10
d2debug: mreldif(Hessian matrices) = .0027612
─────────────────────────────────────── end derivative-comparison report
Iteration 1:  Log likelihood = -27.908441
```

Whenever `ml` needs derivatives, method `d2debug` calls the likelihood evaluator just as method `d2` would, but method `d2debug` also calculates the derivatives numerically. It then displays a comparison of the derivatives calculated both ways and then returns to `ml` the numerically calculated derivatives. Thus `d2debug` proceeds to a correct answer, even if an analytical derivative calculation (first or second) is incorrect.

`d2debug` reports comparisons for both first and second derivatives. The issues concerning comparing first derivatives are the same as for `d1debug`—see section 8.2.1; it is the value of the `mreldif()` in the middle iterations that matters.

Concerning second derivatives, it is the comparison in the last few iterations that matters. The numeric second derivatives `d2debug` calculates can be poor—very poor—in early and middle iterations:

Iteration 0:	1.40e–05
Iteration 1:	2.76e–03
Iteration 2:	1.21e–07
Iteration 3:	2.58e–08
Iteration 4:	5.65e–09
Iteration 5:	9.57e–09

That the numerical derivatives are poor in early iterations does not matter; by the last iteration, the numerical derivative will be accurate, and that is the one you should use to compare. Remember, the purpose of the output is not to test the accuracy of `ml`'s numeric derivative calculator but to verify that the analytic derivatives roughly agree with the numeric derivatives. If they do, we can safely assume that the analytic derivatives are correct.

To demonstrate what you might see when the analytic derivatives are somehow wrong, suppose our probit evaluator is `myprobit_bad_d2`. This is a copy of `myprobit_d2` in which the line that reads

```
    mlmatsum 'lnf' 'H' = -'g1'*('g1'+'xb'), eq(1)
```

was modified to read

```
    tempvar h
    generate double 'h' = 'g1'*('g1'+'xb') if $ML_y1 == 1
    replace    'h' = -'g1'*('g1'-'xb') if $ML_y1 == 0
    mlmatsum 'lnf' 'H' = -'h', eq(1)
```

With **d2debug**, our overall results would still be correct because **d2debug** uses the numeric derivatives (first and second), not the analytic values we calculate. But the `mreldif()`s for the Hessian are

Iteration 0:   3.581
Iteration 1:   21.72
Iteration 2:   45.15
Iteration 3:   48.94
Iteration 4:   46.73
Iteration 5:   46.73

Clearly our analytic Hessian calculation is incorrect. Whenever the `mreldif()` value is greater than 10^{-6} in the last iteration, you should suspect problems.

When using **d2debug**, you can specify **ml maximize**'s `hessian` option to see the analytically calculated Hessian matrix compared with the numerically calculated Hessian matrix, or specify `gradient` to see a comparison of gradient vectors, or both.

8.3   ml trace

Between **ml check** and methods **d1debug** and **d2debug**, you should be able to find and fix most problems.

Even so, there may be a bug in your code that appears only rarely because it is in a part of the code that is rarely executed. The way we typically find such bugs is to **set trace on** and then review a log of the output.

Do not do this with `ml` because the trace will contain a trace of `ml`'s code, too, and there is a lot of it. Instead, type `ml trace on`. That will trace the execution of your program only:

```
. sysuse auto, clear
(1978 automobile data)
. ml model lf myprobit_lf (foreign = mpg weight)
. ml trace on              // trace the evaluator
. ml maximize
-> myprobit_lf __000005 __000006
                  - 'begin'
                  = capture noisily version 18: myprobit_lf __000005 __000006
                  ─────────────────────────────────────────── begin myprobit_lf ───
                  - version 18
                  - args lnfj xb
                  - quietly replace 'lnfj' = ln(normal( 'xb')) if $ML_y1 == 1
                  = quietly replace __000005 = ln(normal( __000006))
> if foreign == 1
                  - quietly replace 'lnfj' = ln(normal(-'xb')) if $ML_y1 == 0
                  = quietly replace __000005 = ln(normal(-__000006))
> if foreign == 0
                  ─────────────────────────────────────────── end myprobit_lf ───
                  - 'end'
                  = set trace off
Initial:     Log likelihood = -51.292891
-> myprobit_lf __000005 __000006
                  - 'begin'
                  = capture noisily version 18: myprobit_lf __000005 __000006
                  ─────────────────────────────────────────── begin myprobit_lf ───
                  - version 18
                  - args lnfj xb
                  - quietly replace 'lnfj' = ln(normal( 'xb')) if $ML_y1 == 1
                  = quietly replace __000005 = ln(normal( __000006))
> if foreign == 1
                  - quietly replace 'lnfj' = ln(normal(-'xb')) if $ML_y1 == 0
                  = quietly replace __000005 = ln(normal(-__000006))
> if foreign == 0
                  ─────────────────────────────────────────── end myprobit_lf ───
                  - 'end'
                  = set trace off
      (output omitted )
```

You may type `ml trace off` later, but that is seldom necessary because the next time you issue an `ml model` statement, trace is turned off when the previous problem is cleared.

9 Setting initial values

Typically, but not always, you can get away without specifying initial values. In the examples so far in this book, we have typed

```
. ml model ...
. ml maximize
```

and not bothered at all with initial values. We have been able to do that because `ml maximize` is pretty smart about coming up with initial values on its own. By default,

1. `ml maximize` assumes an initial value vector $\boldsymbol{\beta}_0 = (0, 0, \ldots, 0)$.

2. If $\ln L(\boldsymbol{\beta}_0)$ can be evaluated, `ml maximize` tries to improve the initial values deterministically by rescaling $\boldsymbol{\beta}_0$, overall and equation by equation.

   If $\ln L(\boldsymbol{\beta}_0)$ cannot be evaluated,

   a. `ml maximize` uses a pseudorandom algorithm to hunt for a $\boldsymbol{\beta}_0$ vector that will allow $\ln L(\boldsymbol{\beta}_0)$ to be evaluated and, once found,

   b. `ml maximize` tries to improve the initial values deterministically by rescaling $\boldsymbol{\beta}_0$, overall and equation by equation.

`ml maximize` does all of this before attempting to maximize the function using the standard techniques.

Actually, `ml maximize` does not do all of this. It calls `ml search` to do all of this. There are good reasons to invoke `ml search` yourself, however, and there are other ways to set the initial values, too:

1. Invoking `ml search` does a more thorough job of setting initial values. When `ml search` is called by `ml maximize`, it tries to be quick and find something that is merely good enough.

2. `ml plot` graphs slices of the likelihood function (one parameter at a time) and resets the parameter to the value corresponding to the maximum of the slice. `ml plot` is great entertainment, too.

3. You can also use `ml init`. This is a method of last resort when you are working interactively, but programmers often use this method because they are willing to perform extra work to obtain good initial values, especially because it is faster than using either of the above techniques.

You may combine the three methods above. If you know that 5 is a good value for one of the parameters, you can use `ml init` to set it. You can then use `ml search` to grope around from there for even better values—`ml search` cannot make things worse. Then you can use `ml plot` to examine another of the parameters that you know is likely to cause problems and so set it to a reasonable value. You can then use `ml search` again to improve things. If you kept working like this, you could dispense with the maximization step altogether.

9.1 ml search

The syntax of `ml search` is

ml <u>sea</u>rch $\big[\,[\,/\,]\,eqname\,[\,:\,]$ $\#_{lb}$ $\#_{ub}\big]$ $\big[\,[\,/\,]\,eqname\,[\,:\,]$ $\#_{lb}$ $\#_{ub}\big]$ $\big[\,\ldots\,\big]$ $\big[\,,$
        <u>r</u>epeat($\#$) nolog <u>trace</u> <u>restart</u> <u>norescale</u>$\big]$

`ml search` looks for better (or feasible) initial values. By feasible, we mean values of the parameters for which the log likelihood can be calculated. For instance, in the log-likelihood function

$$\ln \ell_j = \ln \phi \left(\frac{y_j - \mu_j}{\sigma} \right) - \ln \sigma \tag{9.1}$$

any starting value with $\sigma \le 0$ is not feasible.

The simplest syntax for `ml search` is

    . ml search

You can also follow `ml search` with the name of an equation and two numbers, such as

    . ml search eq1 -3 3

and `ml search` will restrict the search to those bounds. Using the log likelihood in (9.1), suppose that we type

    . ml model lf mynormal_lf (mu: y = x1 x2) /sigma

It would make sense—but not be necessary—to restrict `ml search` to positive and reasonable values for σ:

    . ml search /sigma .1 5

You specify bounds for equations, not individual parameters. That means you do not specify bounds for the coefficient on, say, `x1`; rather, you specify bounds for the equation `mu`. In this example, if `y` varied between 1 and 10, it would make sense to specify

    . ml search /sigma .1 5 mu 1 10

In practice, it is seldom worth the trouble to specify the bounds because `ml search` comes up with bounds on its own by probing the likelihood function.

`ml search` defines initial values on the basis of equations by setting the intercept (coefficient on _cons) to numbers randomly chosen within the range and setting the remaining coefficients to zero. If the equation has no intercept, a number c within the range is chosen, and the coefficients are then set from the regression $c = \mathbf{x}\boldsymbol{\beta}_0 + e$.

`ml search` can be issued repeatedly.

Using ml search

We strongly recommend that you use `ml search`. You can specify bounds for the search, but typically you need only type

```
. ml search
```

We recommend that you use `ml search`, even if you specify initial values (which you do with `ml init`, as we will explain later):

```
. ml model lf mynormal_lf (mu: mpg = weight displ) /sigma
. ml init /sigma = 3
. ml search
```

If you use `ml search`, any previous starting values will be taken as a suggestion, so it would probably be better to type

```
. ml model lf mynormal_lf (mu: mpg = weight displ) /sigma
. ml search /sigma 1 10
```

and so restrict the search for `/sigma` to the range $1 < \sigma < 10$ rather than suggesting $\sigma = 3$ and leaving the range at $-\infty < \sigma < \infty$. Of course, we could do both, but the suggestion is not worth nearly as much as the bounds.

Not specifying an equation does not mean the equation is not searched; it means just that no special bounds are specified for the search.

Determining equation names

Remember, you do not have to specify bounds, but the search will be more efficient if you do. You specify the bounds in terms of the equations, not the individual parameters.

`ml model` labels equations as eq1, eq2, ... if you do not specify otherwise. You can name an equation by coding "(*eqname*: ...)" when you specify the equation. If you are confused, you can type `ml query` any time after the `ml model` statement to find out what you have defined and where you are. Suppose that we set up a problem by typing

```
. sysuse auto
(1978 automobile data)
. ml model lf myprobit_lf (foreign = mpg weight)
```

Then `ml query` will report

```
. ml query
Settings for moptimize()
Version:                                    1.00
Evaluator
    Type:                                   lf
    Function:                               myprobit_lf
Dependent variable:                         foreign
Equation 1
    Predictor 1:                            mpg
    Predictor 2:                            weight
                                            _cons

    (output omitted)
Starting values
    Coefficient values:
        1:                                  0
        2:                                  0
        3:                                  0
    Function value:                         .
Current status
    Coefficient values:                     <unknown>
    Function value:                         .
    Converged:                              No
```

Our equation is called `eq1`. Next let's search over the restricted range $(-3, 3)$:

```
. ml search eq1 -3 3
Initial:      Log likelihood = -51.292891
Alternative:  Log likelihood = -45.055272
Rescale:      Log likelihood = -45.055272
```

If we now do an `ml query`, we see

```
. ml query
Settings for moptimize()
Version:                                    1.00
Evaluator
    Type:                                   lf
    Function:                               myprobit_lf
Dependent variable:                         foreign
Equation 1
    Predictor 1:                            mpg
    Predictor 2:                            weight
                                            _cons

    (output omitted)
Starting values
    Coefficient values:
        1:                                  0
        2:                                  0
        3:                                  -.5
    Function value:                         -45.0552724
```

```
Current status
   Coefficient values:
      1:                                 0
      2:                                 0
      3:                                -.5
   Function value:                  -45.0552724
   Converged:                       No
```

`ml` remembers the search bounds we specified. If we simply type `ml search`, the $(-3, 3)$ bounds are assumed.

9.2   ml plot

The basic syntax of `ml plot` is

`ml plot` [*eqname:*] *name* [*#* [*#* [*#*]]]

`ml plot` graphically helps fill in initial values. In practice, this command is more cute than useful because `ml search` works so well. Still, `ml plot` focuses on individual parameters, not on the overall equation as with `ml search`, and that can sometimes be useful.

`ml plot` graphs the likelihood function for one parameter at a time and then replaces the value of the parameter according to the maximum value of the likelihood function. These graphs are cuts of the likelihood function—they are not graphs of the profile-likelihood function.

`ml plot` is not an alternative to `ml search`; you can use them both, in any order, and repeatedly. We recommend that you use `ml search` first and then try `ml plot`. This way, `ml plot` is at least starting with feasible values.

The simplest syntax is to type `ml plot` followed by a coefficient name:

```
. ml plot _cons
. ml plot /sigma
. ml plot eq2:_cons
. ml plot eq1:foreign
```

When the name of the equation is not explicitly supplied, the first equation, whatever the name, is assumed (thus _cons probably means eq1:_cons).

For instance, after setting up a model,

```
. sysuse auto
(1978 automobile data)
. ml model lf myprobit_lf (foreign = mpg weight)
```

we might type the following, which results in the plot in figure 9.1.

```
. ml plot _cons
              reset _cons =          -.5  (was              0)
              Log likelihood = -45.055272  (was -51.292891)
```

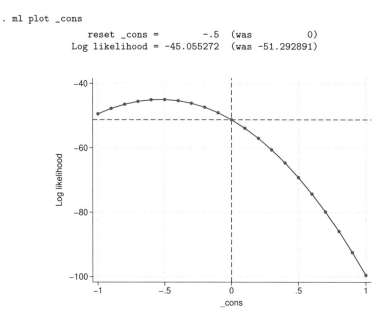

Figure 9.1. Probit log likelihood versus the constant term.

By default, `ml plot` graphs 21 values of the likelihood within ±1 of the current value
of the parameter. Rather than displaying a graph, `ml plot` might present you with a
message:

```
. ml plot _cons
(all values are missing)
```

This could mean that

1. The current value for the coefficient on `_cons` and all values within ±1 of the
   current value are infeasible, so you need to find a better range.

2. The current value for the coefficient on `_cons` is infeasible, and the range is too
   wide. Remember, `ml plot` graphs 21 points; perhaps values within ±1/100 of the
   range would be feasible.

That is why we recommend that you first use `ml search`.

You can also type `ml plot` followed by the name of a parameter and then a number.
For example, the following results in the plot in figure 9.2.

```
. ml plot weight .01
            Keeping weight =           0
            Log likelihood = -45.055272
```

Figure 9.2. Probit log likelihood versus the weight coefficient.

Here we graph ± 0.01 around the current value of the coefficient on weight. If we specified two numbers, such as ml plot mpg -.5 1, we would graph the range $[\beta - 0.5, \beta + 1]$ where β would be the current value of the coefficient on mpg. If we typed a third number, that many points (plus 1) would be graphed.

ml maintains the current value of the initial-value vector $\boldsymbol{\beta}_0$. $\boldsymbol{\beta}_0$ is set to zero when we issue an ml model statement, and thereafter other routines can update it. Each time we perform an ml search, $\boldsymbol{\beta}_0$ is updated. Each time we use ml plot, $\boldsymbol{\beta}_0$ is updated if a better value for the parameter is found. Thus you may use ml search and ml plot together. The current value of $\boldsymbol{\beta}_0$ is reported by ml query.

9.3   ml init

The syntax of ml init is

ml init $\{$ $\big[$ eqname: $\big]$ name=# | /eqname=# $\}$ $\big[\ldots\big]$

ml init # $\big[$ # $\ldots\big]$, copy

ml init matname $\big[$, skip copy$\big]$

ml init allows you to specify initial values explicitly. ml init without arguments resets the parameter vector to contain all zeros.

Using ml init

`ml init` can be issued not at all, once, or repeatedly.

To specify initial values, type the coefficient's name, an equal sign, and the value to which it should be set. For instance,

```
. ml model lf myprobit_lf (foreign = mpg weight)
. ml init mpg=1.5 _cons=-.5
```

More than one coefficient may be specified and not all coefficients need be specified. Initial values for omitted coefficients remain unchanged.

In multiple-equation models, specify the equation name and a colon to identify the coefficient:

```
. ml model lf myprog (y1 = x1 x2 x3) (y2 = x1 x4) /sigma1 /sigma2
. ml init eq1:_cons=5 eq2:_cons=3
```

Here the first two equations are named `eq1` and `eq2` because we did not supply our own equation names in the `ml model` statement. When you omit the equation name in the `ml init` statement, as in

```
. ml init x1=3.14
```

Stata assumes that you are referring to the first equation.

Also, when you set the coefficient on the constant, you can refer either to *eqname:*_cons or to */eqname*—they mean the same thing:

```
. ml init /sigma1=3 sigma2:_cons=4
```

You can even use the slash notation with equations that have more than just a constant. The line

```
. ml init eq1:_cons=5 eq2:_cons=3
```

could just as well be specified

```
. ml init /eq1=5 /eq2=3
```

In the rare instance when you wish to specify initial values for all parameters, you can type them one after the other:

```
. ml init 5 7 2 9  3 2 1  1  1, copy
```

In most such cases, however, you would have a vector containing the values, and it is easier to use the syntax described next.

Loading initial values from vectors

You can obtain initial values from row vectors (matrices). This is rarely done, and this feature is included for programmers who want to obtain initial values from some other estimator.

```
. regress y1 x2 x3
. matrix b0 = e(b)
. ml model lf myprog (y1 = x1 x2 x3) (y2 = x1 x4) /sigma1 /sigma2
. ml init b0
```

By default, values are copied from the specified vector by name; it does not matter that the vector obtained from `regress` is in a different order than that used by `ml`. In this case, the initial values were applied to the first equation because the vector `b0` fetched from `regress` did not include equation names; if we had set the equation names, the vector could have applied to the second equation. The vector you specify need not contain the same number of elements as there are parameters in your model. Parameters that do not appear in your vector are initialized to zero. If there are elements in your vector whose names do not match parameters in your model, `ml init` will exit with an error message unless you specify the `skip` option.

If you would rather specify a vector that does not have column names or that contains initial values for every parameter in your model and in the order in which they get declared in the `ml model` statement, you can use the `copy` option of `ml init`. In the example above, the `ml model` statement implies that there are a total of nine individual parameters, so if you use the `copy` option with `ml init`, the vector you specify must have nine columns.

Our use of the name `b0` has no special significance in the above example; we could just as well have called it `initvec` or anything else.

10   Interactive maximization

You type `ml maximize` to obtain estimates once you have issued the `ml model` statement. The process can be as short as

```
. ml model ...
. ml maximize
```

but we recommend

```
. ml model ...
. ml check
. ml search
. ml maximize
```

10.1   The iteration log

If you type `ml maximize`, you will see something like

```
. ml maximize
Initial:      Log likelihood =       -744
Alternative:  Log likelihood = -356.14276
Rescale:      Log likelihood = -200.80201
Rescale eq:   Log likelihood = -136.69232
Iteration 0:  Log likelihood = -136.69232  (not concave)
Iteration 1:  Log likelihood = -124.13044
Iteration 2:  Log likelihood = -113.99049
...
```

The first four lines of the output really have to do with `ml search`. The first line, `Initial`, reports the log likelihood at the initial values β_0. Here the log likelihood is -744, but it is often reported as `-<inf>`, meaning that the likelihood could not be evaluated at β_0. That would not cause problems. The next line, `Alternative`, reports the result of a random search for an alternative β_0.

ml then takes the better of the two. `Rescale` reports the results of rescaling the entire β_0 vector—that is, searching for a scalar c such that $\ln L(c\beta_0) > \ln L(\beta_0)$. `Rescale eq` then repeats the search for c, equation by equation.

At this point, maximization has not even begun: starting values are being improved, and already we have improved the log likelihood from -744 to -136.7.

The goal is to find a β_0 so that the chosen optimization algorithm has a reasonable chance of converging to the maximum.

At iteration 0, the maximization process really starts, and note that iteration 0 reports an opening log likelihood that is equal to where the searcher left off.

The purpose of an iteration is to take $\boldsymbol{\beta}_i$ and produce a better value $\boldsymbol{\beta}_{i+1}$, where i is the iteration number. Iteration 0 takes $\boldsymbol{\beta}_0$ and produces $\boldsymbol{\beta}_1$.

An iteration is the calculation of a new direction vector. When `ml` reports an iteration, it is reporting $\ln L(\boldsymbol{\beta}_i)$, the value received. The iteration calculates a direction vector, $\mathbf{d}_i$, based on the function at that point, where $\mathbf{d}_i$ is some function of the gradient and Hessian (or an alternative to the Hessian when using the BHHH, DFP, or BFGS algorithms). The iteration then steps in that direction until the log likelihood ceases to improve. $\boldsymbol{\beta}_{i+1}$ is defined as $\boldsymbol{\beta}_i + s\mathbf{d}_i$, where s is the step size taken. The process then repeats.

Thus the derivatives are calculated once per iteration. Between iterations, log-likelihood evaluations are made.

10.2   Pressing the Break key

You can press *Break* while `ml` is iterating:

```
. ml maximize
Initial:      Log likelihood =       -744
Alternative:  Log likelihood = -356.14276
Rescale:      Log likelihood = -200.80201
Rescale eq:   Log likelihood = -136.69232
Iteration 0:  Log likelihood = -136.69232  (not concave)
Iteration 1:  Log likelihood = -124.13044
Iteration 2:  Log likelihood = -113.99049
Iteration 3:  Log likelihood = -110.30732
—Break—
r(1);
```

Nothing is lost; you can pick up right from where you left off:

```
. ml maximize
Initial:      Log likelihood = -110.26748
Rescale:      Log likelihood = -110.26748
Rescale eq:   Log likelihood = -110.26748
Iteration 0:  Log likelihood = -110.26748
Iteration 1:  Log likelihood = -110.26736
Iteration 2:  Log likelihood = -110.26736
(estimation output appears)
```

The iteration numbers change, but note that the initial log-likelihood value is close to the last iteration number just before you pressed *Break*.

In between, you can use ml query to obtain a summary of where you are:

```
. ml query
Settings for moptimize()
Version:                                     1.00
Evaluator
    Type:                                    d1
    Function:                                myweibullb_d1
Dependent variables:
    1:                                       studytime
    2:                                       died
Equation 1
    Predictor 1:                             1b.drug
    Predictor 2:                             2.drug
    Predictor 3:                             3.drug
    Predictor 4:                             age
                                             _cons
Equation 2
    Predictors:                              <none>
                                             _cons

User-defined information:                    <none>

Weights:                                     <none>

Optimization technique:                      nr
Singular H method:                           m-marquardt

Convergence
    Maximum iterations:                      300
    Tolerance for the function value:        1.0000e-07
    Tolerance for the parameters:            1.0000e-06
    Tolerance for the scaled gradient:       1.0000e-05
    Tolerance for the quasi-scaled gradient: .
    Warning:                                 on

Iteration log
    Value ID:                                log likelihood
    Trace dots:                              off
    Trace value:                             on
    Trace tol:                               off
    Trace step:                              off
    Trace coefficients:                      off
    Trace gradient:                          off
    Trace Hessian:                           off

Constraints:                                 Yes

Search for starting values:                  off

Numerical derivative step size lower bounds: off

Starting values
    Coefficient values:
        1:                                   0
        2:                                   0
        3:                                   0
        4:                                   0
        5:                                   4
        6:                                   -.25
    Function value:                          -136.692316
```

```
Current status
    Coefficient values:
        1:                            0
        2:                            1.01373004
        3:                            1.46009119
        4:                            -.067167153
        5:                            6.05907233
        6:                            .558621113
    Function value:                   -110.307324
    Converged:                        No
```

Or you can type `ml report`:

```
. ml report
Current coefficient vector:
      eq1:    eq1:    eq1:    eq1:    eq1:    eq2:
      1b.      2.      3.
      drug    drug    drug    age   _cons   _cons
r1       0       0       0       0       4    -.25
Value of log likelihood function = -136.69232
Gradient vector (length = 633.079):
      eq1:       eq1:       eq1:       eq1:       eq1:       eq2:
      1b.        2.         3.
      drug       drug       drug       age      _cons      _cons
r1       0  -.7819348  1.243378  -632.982  -10.67705   2.580612
```

You can even use any of the methods for setting initial values—`ml search`, `ml plot`, or `ml init`—to attempt to improve the parameter vector before continuing with the maximization by typing `ml maximize`.

10.3 Maximizing difficult likelihood functions

When you have a difficult likelihood function, specify `ml maximize`'s `difficult` option. The `difficult` option varies how a direction is found when the negative Hessian, $-\mathbf{H}$, cannot be inverted and thus the usual direction calculation $\mathbf{d} = \mathbf{g}(-\mathbf{H})^{-1}$ cannot be calculated. `ml` flags such instances by mentioning that the likelihood is "not concave":

```
. ml maximize
Initial:       Log likelihood =      -744
Alternative:   Log likelihood = -356.14276
Rescale:       Log likelihood = -200.80201
Rescale eq:    Log likelihood = -136.69232
Iteration 0:   Log likelihood = -136.69232  (not concave)
Iteration 1:   Log likelihood = -124.13044
Iteration 2:   Log likelihood = -113.99049
...
```

`ml`'s usual solution is to mix a little steepest ascent into the calculation, which works well enough in most cases.

When you specify `difficult`, `ml` goes through more work. First, `ml` computes the eigenvalues of $-\mathbf{H}$ and then, for the part of the orthogonal subspace where the eigenvalues are negative or small positive numbers, `ml` uses steepest ascent. In the other subspace, `ml` uses a regular Newton–Raphson step.

A "not concave" message does not necessarily indicate problems. It can be particularly maddening when `ml` reports "not concave" iteration after iteration as it crawls toward a solution. It is only a problem if `ml` reports "not concave" at the last iteration. The following log has "not concave" messages along the way and at the end:

```
. ml maximize
Initial:       Log likelihood =   -2702.155
Alternative:   Log likelihood =   -2702.155
Rescale:       Log likelihood = -2315.2187
Rescale eq:    Log likelihood = -2220.5288
Iteration 0:   Log likelihood = -2220.5288  (not concave)
Iteration 1:   Log likelihood = -2152.5211  (not concave)
Iteration 2:   Log likelihood = -2141.3916  (not concave)
Iteration 3:   Log likelihood = -2140.7298  (not concave)
Iteration 4:   Log likelihood = -2139.5807  (not concave)
Iteration 5:   Log likelihood = -2139.0407  (not concave)
(output omitted)
Iteration 40:  Log likelihood = -2138.0806  (not concave)
Iteration 41:  Log likelihood = -2138.0803  (not concave)
Iteration 42:  Log likelihood = -2138.0801  (not concave)
Iteration 43:  Log likelihood = -2138.0799  (not concave)
(results presented; output omitted)
```

Specifying `difficult` sometimes helps:

```
. ml maximize, difficult
Initial:       Log likelihood =   -2702.155
Alternative:   Log likelihood =   -2702.155
Rescale:       Log likelihood = -2315.2187
Rescale eq:    Log likelihood = -2220.5288
Iteration 0:   Log likelihood = -2220.5288  (not concave)
Iteration 1:   Log likelihood = -2143.5267  (not concave)
Iteration 2:   Log likelihood = -2138.2358
Iteration 3:   Log likelihood = -2137.9355
Iteration 4:   Log likelihood = -2137.8971
Iteration 5:   Log likelihood = -2137.8969
(results presented; output omitted)
```

Here `difficult` not only sped estimation but solved the problem of nonconcavity at the final iteration. Note that the final log likelihood without `difficult` was -2138.1, which is less than -2137.9. The difference may seem small, but it was enough to place the final results on a concave portion of the likelihood function.

Usually, however, specifying `difficult` just slows things down.

11 Final results

Once you have typed `ml maximize` and obtained the results, you can do anything Stata usually allows after estimation. That means you can use `test` to perform hypothesis tests, `predict` and `predictnl` to obtain predicted values, and so on.

11.1 Graphing convergence

A special feature of `ml` is that, once estimation output has been produced, you can graph the convergence by typing `ml graph`. Typing the following produces the output shown in figure 11.1.

```
. sysuse cancer
(Patient survival in drug trial)
. ml model lf1 myweibull_lf1 (studytime died = i.drug age) ()
. ml maximize, nolog
```

```
                                              Number of obs =       48
                                              Wald chi2(3)  =   35.25
Log likelihood = -110.26736                   Prob > chi2   = 0.0000
```

	Coefficient	Std. err.	z	P>|z|	[95% conf. interval]	
eq1						
drug						
Other	1.012966	.2903917	3.49	0.000	.4438086	1.582123
NA	1.45917	.2821195	5.17	0.000	.9062261	2.012114
age	-.0671728	.0205688	-3.27	0.001	-.1074868	-.0268587
_cons	6.060723	1.152845	5.26	0.000	3.801188	8.320259
eq2						
_cons	.5573333	.1402154	3.97	0.000	.2825163	.8321504

. ml graph

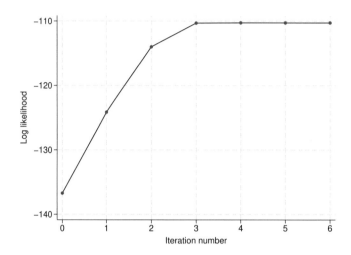

Figure 11.1. Weibull log-likelihood values by iteration.

Given how Stata's optimizers work, for smooth likelihood functions the graph may jump around at first, but once it gets close to the maximum, it should move smoothly and with smaller and smaller steps toward it; see section 1.4. This graph shows that the process has truly converged.

The basic syntax of ml graph is

ml graph [#]

specifies the number of iterations to be graphed, counted from the last. For example, ml graph 5 would graph the last five iterations. ml graph keeps track of the last 20 iterations, sufficient for determining convergence.

11.2 Redisplaying output

Once `ml maximize` has displayed results, you can use `ml display` to redisplay them.

```
. ml display
```

```
                                                Number of obs =       48
                                                Wald chi2(3)  =    35.25
        Log likelihood = -110.26736             Prob > chi2   =  0.0000
```

	Coefficient	Std. err.	z	P>|z|	[95% conf. interval]	
eq1						
drug						
Other	1.012966	.2903917	3.49	0.000	.4438086	1.582123
NA	1.45917	.2821195	5.17	0.000	.9062261	2.012114
age	-.0671728	.0205688	-3.27	0.001	-.1074868	-.0268587
_cons	6.060723	1.152845	5.26	0.000	3.801188	8.320259
eq2						
_cons	.5573333	.1402154	3.97	0.000	.2825163	.8321504

The syntax of `ml display` is

`ml display` [`, noheader nofootnote level(#) first neq(#) showeqns plus`
  `nocnsreport noomitted vsquish noemptycells baselevels`
  `allbaselevels cformat(%fmt) pformat(%fmt) sformat(%fmt)`
  `nolstretch coeflegend` *eform_option*]

`noheader` suppresses the display of the header above the coefficient table. In the non-svy context, the header displays the final log-likelihood (criterion) value, the number of observations, and the model significance test. In the `svy` context, the header displays the number of observations, the estimate of the population size, and the model significance test.

`nofootnote` suppresses the footnote display below the coefficient table. The footnote displays a warning if the model fit did not converge within the specified number of iterations. Use `ml footnote` to display the warning if 1) you add to the coefficient table using the `plus` option or 2) you have your own footnotes and want the warning to be last.

`level(#)` is the standard confidence-level option. It specifies the confidence level, as a percentage, for confidence intervals of the coefficients. The default is `level(95)` or as set by `set level`; see [U] **20.8 Specifying the width of confidence intervals**.

`first` displays a coefficient table reporting results for the first equation only, and the report makes it appear that the first equation is the only equation. This option is used by programmers who estimate ancillary parameters in the second and subsequent equations and who wish to report the values of such parameters themselves.

`neq(#)` is an alternative to `first`. `neq(#)` displays a coefficient table reporting results for the first # equations. This option is used by programmers who estimate ancillary parameters in the #+1 and subsequent equations and who wish to report the values of such parameters themselves.

`showeqns` is a seldom-used option that displays the equation names in the coefficient table. `ml display` uses the numbers stored in `e(k_eq)` and `e(k_aux)` to determine how to display the coefficient table. `e(k_eq)` identifies the number of equations, and `e(k_aux)` identifies how many of these are for ancillary parameters. The `first` option is implied when `showeqns` is not specified and all but the first equation are for ancillary parameters.

`plus` displays the coefficient table, but rather than ending the table in a line of dashes, ends it in dashes–plus-sign–dashes. This is so that programmers can write additional display code to add more results to the table and make it appear as if the combined result is one table. Programmers typically specify `plus` with the `first` or `neq()` option. `plus` implies `nofootnote`.

`nocnsreport` suppresses the display of constraints above the coefficient table. This option is ignored if constraints were not used to fit the model.

`noomitted` specifies that variables that were omitted because of collinearity not be displayed. The default is to include in the table any variables omitted because of collinearity and to label them as "(omitted)".

`vsquish` specifies that the blank space separating factor-variable terms or time-series–operated variables from other variables in the model be suppressed.

`noemptycells` specifies that empty cells for interactions of factor variables not be displayed. The default is to include in the table interaction cells that do not occur in the estimation sample and to label them as "(empty)".

`baselevels` and `allbaselevels` control whether the base levels of factor variables and interactions are displayed. The default is to exclude from the table all base categories.

> `baselevels` specifies that base levels be reported for factor variables and for interactions whose bases cannot be inferred from their component factor variables.

> `allbaselevels` specifies that all base levels of factor variables and interactions be reported.

`cformat(%fmt)` specifies how to format coefficients, standard errors, and confidence limits in the coefficient table.

`pformat(%fmt)` specifies how to format p-values in the coefficient table.

`sformat(%fmt)` specifies how to format test statistics in the coefficient table.

`nolstretch` specifies that the width of the coefficient table not be automatically widened to accommodate longer variable names. The default, `lstretch`, is to automatically widen the coefficient table up to the width of the Results window. Specifying `lstretch` or `nolstretch` overrides the settings given by `set lstretch`. If `set lstretch` has not been set, the default is `lstretch`.

`coeflegend` specifies that the legend of the coefficients and how to specify them in an expression be displayed rather than the coefficient table.

eform_option: <u>ef</u>orm(*string*), <u>ef</u>orm, hr, shr, <u>irr</u>, or, and <u>rrr</u> display the coefficient table in exponentiated form: for each coefficient, $\exp(b)$ rather than b is displayed, and standard errors and confidence intervals are transformed. Display of the intercept, if any, is suppressed. *string* is the table header that will be displayed above the transformed coefficients and must be 11 characters or shorter in length—for example, `eform("Odds ratio")`. The options `eform`, `hr`, `shr`, `irr`, `or`, and `rrr` provide a default *string* equivalent to "exp(b)", "Haz. ratio", "SHR", "IRR", "Odds ratio", and "RRR", respectively. These options may not be combined.

`ml display` looks at `e(k_eform)` to determine how many equations are affected by an *eform_option*; by default, only the first equation is affected. Type `ereturn list, all` to view `e(k_eform)`; see [P] **ereturn**.

12 Writing do-files to maximize likelihoods

There are two reasons to maximize likelihood functions:

1. You need to fit the model for a particular analysis or just a few analyses.

2. You need the estimator around constantly because you use it all the time.

We recommend using do-files in case 1 and ado-files (covered in the next chapter) in case 2.

12.1 The structure of a do-file

We organize our do-files differently depending on the problem, but we always start the same way:

```
───────────────────────────────────── begin outline.do ──────────
clear
capture program drop myprog
program myprog
        version 18
        args ...
        ...
end
use sampledata
ml model ... myprog ...
                                       ───────── end outline.do ──────────
```

Then, working interactively, we can type

```
. do outline
. ml check
. ml search
. ml max
```

Typically, `ml check` detects problems, but this structure makes it easy to try again. We can edit our do-file and then start all over by typing `do outline`. We do not actually name the file `outline.do`, of course. At this stage, we typically give it some very short name: `it.do` is our favorite. Then we can just tell Stata to `do it`.

We put the loading of some sample data and the `ml model` statement in the do-file. The remaining `ml` commands we need to type are short; we just did not want to be forever retyping things like

```
. ml model d2 myweibull_d2 (lneta: studytime died = i.drug age) /lngamma
```

while we are debugging our code.

12.2   Putting the do-file into production

Once we have our evaluator working, we modify our do-file and rename it:

```
──────────────────────────────── begin readytouse.do ───────────
capture program drop myprog
program myprog
        version 18
        args ...
        ...
end
capture program drop model
program model
        version 18
        ml model method myprog (lneta: '0') /lngamma
end
──────────────────────────────── end readytouse.do ───────────
```

The changes we made were as follows:

1. We removed the `clear` at the top; we no longer want to eliminate the data in memory. Now we can load the data and run the do-file in any order.

2. We removed the "`use` *sampledata*" and the hardcoded `ml model` statement.

3. We substituted in their place a new program—`model`, a two-line program we wrote to make it easy to type the `ml model` statement.

We previously imagined that model statements such as

```
. ml model d2 myweibull_d2 (lneta: studytime died = i.drug age) /lngamma
```

were reasonable for this estimator. Some of what we need to type will always be the same:

```
. ml model d2 myweibull_d2 (lneta: ...) /lngamma
```

To make it easier for us to use the estimator, we code the fixed part into a new program:

```
program model
        version 18
        ml model d2 myweibull_d2 (lneta: '0') /lngamma
end
```

'0' is just Stata jargon for "what the user typed goes here". So with `model`, we can type things like

```
. model studytime died = i.drug age
```

With *readytouse*.`do` in place in our analysis do-file, we can code things like

```
────────────────────────────────────────── begin analysis.do ──────────
run readytouse
use realdata
model studytime died = i.drug age
ml search
ml max
──────────────────────────────────────────── end analysis.do ──────────
```

13 Writing ado-files to maximize likelihoods

When you find yourself writing or editing do-files for fitting a specific model to data regularly, it might be worth the effort to write an ado-file that will fit the model for you. Such a program is commonly called an estimation command.

This chapter discusses writing estimation commands. Chapter 14 discusses how to accommodate survey data analysis in your existing ml–based estimation commands. Other example estimation commands are presented in chapter 17.

13.1 Writing estimation commands

The term *estimation command* has a special meaning in Stata, because it implies several properties:

- Estimation commands share a common syntax.

- You can, at any time, review the previous estimates by typing the estimation command without arguments. This is known as *replay* in Stata parlance.

- Estimation commands share a common output format.

- You can obtain the variance–covariance matrix of the estimators by typing estat vce after estimation.

- You can restrict later commands to the estimation subsample by adding the suffix if e(sample), for example, summarize ... if e(sample).

- You can obtain predictions and the standard error of the linear prediction using predict. You can use predictnl for nonstandard predictions and their standard errors.

- You can refer to coefficients and standard errors in expressions (such as with generate) by typing _b[*name*] or [*eqname*]_b[*name*] and _se[*name*] or [*eqname*]_se[*name*].

- You can perform tests on the estimated parameters using test (Wald test of linear hypotheses), testnl (Wald test of nonlinear hypotheses), and lrtest (likelihood-ratio tests). You can obtain point estimates and confidence intervals of linear and nonlinear functions of the estimated parameters using lincom and nlcom, respectively.

Except for the first two bulleted items, `ml` satisfies all these requirements; you do not
have to do anything more. `ml`'s syntax is different from that of other estimation commands (first bullet), and you cannot type `ml` without arguments to redisplay results
(second bullet), but you can type `ml display` to redisplay results, so the second viola-
tion is only a technicality.

Convenience of syntax is the main reason to package an `ml` estimator into an ado-file.
Rather than typing

```
. do such-and-such (to load the evaluator)
. ml model lf mynormal_lf (mpg = foreign weight displacement) () if age>20,
> vce(robust)
. ml search
. ml maximize
```

it would be more convenient to simply type

```
. mynormal0 mpg foreign weight displacement if age>20, vce(robust)
```

Creating such an ado-file would be easy:

─────────────────────────────────────── begin mynormal0.ado ───────────
```
program mynormal0
        version 18
        syntax varlist [if] [in] [, vce(passthru) ]
        gettoken y rhs : varlist
        ml model lf mynormal_lf (`y' = `rhs') () `if' `in', `vce'
        ml maximize
end
```
─────────────────────────────────────── end mynormal0.ado ───────────

─────────────────────────────────────── begin mynormal_lf.ado ───────────
```
program mynormal_lf
        version 18
        args lnf mu lnsigma
        quietly replace `lnf' = ln(normalden($ML_y1,`mu',exp(`lnsigma')))
end
```
─────────────────────────────────────── end mynormal_lf.ado ───────────

The `mynormal0` command works, although it does not really meet our standards for an
estimation command. Typing `mynormal0` without arguments will not redisplay results,
although it would not be difficult to incorporate that feature. Moreover, this command
does not allow us to use factor-variable notation when specifying our variables.

The point is that `mynormal0` works and is fairly simple, and it may work well enough
for your purposes. We are about to go to considerable effort to solve what are really
minor criticisms. We want to add every conceivable option, and we want to make the
program bulletproof. Such work is justified only if you or others plan on using the
resulting command often.

In the following, we do not mention how to write your estimation command so that it
will allow the use of Stata's by *varlist*: prefix. If you are going to use the outline for an
estimation command detailed in section 13.3 and want to allow by, see the *Combining
byable(recall) with byable(onecall)* section in [P] **byable**.

13.2   The standard estimation-command outline

The outline for any estimation command—not just those written with `ml`—is

```
──────────────────────────────────────────── begin newcmd.ado ──────────
program newcmd, eclass sortpreserve
        version 18
        local options "display-options"
        if replay() {
                if ("`e(cmd)'" != "newcmd") error 301
                syntax [, `options']
        }
        else {
                syntax ... [, estimation-options `options' ]

                // mark the estimation subsample
                marksample touse

                // obtain the estimation results
                ...
                ereturn local cmd "newcmd"
        }
        display estimation results respecting display options
end
──────────────────────────────────────────── end newcmd.ado ──────────
```

The idea behind this outline is as follows. Estimation commands should

- Redisplay results when typed without arguments.

- Compute and display results when typed with arguments.

Variations on the outline are acceptable, and the one we most often use is

```
──────────────────────────────────────────── begin newcmd.ado ──────────
program newcmd
        version 18
        if replay() {
                if ("`e(cmd)'" != "newcmd") error 301
                Replay `0'
        }
        else    Estimate `0'
end
program Estimate, eclass sortpreserve
        syntax ... [, estimation-options display-options ]

        // mark the estimation subsample
        marksample touse

        // obtain the estimation results
        ...
        ereturn local cmd "newcmd"

        Replay, display-options
end
program Replay
        syntax [, display-options ]
        display estimation results respecting the display options
end
──────────────────────────────────────────── end newcmd.ado ──────────
```

The names `Estimate` and `Replay` are actually used for the subroutine names; there is no possibility of conflict, even if other ado-files use the same names, because the subroutines are local to their ado-file. (We do not bother to restate `version 18` in the subroutines `Estimate` and `Replay`. The subroutines can be called only by *newcmd*, and *newcmd* is already declared to be `version 18`.)

13.3   Outline for estimation commands using ml

The outline for a new estimation command implemented with `ml` is

```
────────────────────────────────────────── begin newcmd.ado ──────────
program newcmd
        version 18
        if replay() {
                if ("'e(cmd)'" != "newcmd") error 301
                Replay '0'
        }
        else    Estimate '0'
end

program Estimate, eclass sortpreserve
        syntax varlist(fv) [if] [in] [fweight] [,           ///
                other-estimation-options                    ///
                display-options                             ///
        ]

        // mark the estimation subsample
        marksample touse

        // perform estimation using ml
        ml model method newcmd_ll ...                       ///
                ['weight''exp'] if 'touse',                 ///
                other-estimation-options                    ///
                maximize
        ereturn local cmd "newcmd"

        Replay, display-options
end

program Replay
        syntax [, display-options ]
        ml display, display-options
end
────────────────────────────────────────── end newcmd.ado ──────────


────────────────────────────────────────── begin newcmd_ll.ado ──────────
program newcmd_ll
        version 18
        args ...
        ...
end
────────────────────────────────────────── end newcmd_ll.ado ──────────
```

An estimation command implemented in terms of `ml` requires two ado-files:

newcmd.ado	The estimation command the user sees
newcmd_ll.ado	The secret subroutine (the evaluator) `ml` calls

The evaluator must be in a separate ado-file because otherwise `ml` could not find it.

We recommend that you give the secret subroutine the name formed by adding _11 as a suffix to the name of the new estimation command *newcmd*: *newcmd*_11. No user will ever have to type this name, nor would we want the user to type it, and placing an underscore in the name pretty much ensures that. The standard ending _11 reminds you that it is a log-likelihood evaluator. Using the name of the command reminds you which ado-file this subroutine goes with.

For instance, if you planned on calling your new estimation command `mynormal`, then we would recommend that you name the likelihood evaluator `mynormal_11.ado`. For this book, we decided to use the method as a suffix; this facilitates having different copies of likelihood evaluators for a given model (for example, `mynormal_1f.ado` and `mynormal_1f2.ado`).

In `newcmd.ado`, we did not show how to handle *other-estimation-options* or *display-options*. There are really two types of estimation options: those that are specific to the estimator you are implementing and those that control or modify how `ml` maximizes the log-likelihood function. Later, we will see how to accept both types of estimation options as well as display options that control, for example, the level of the confidence intervals in the output table and the number formats used to display coefficients and statistics.

13.4   Using ml in noninteractive mode

There are no `ml search`, `ml init`, or `ml maximize` commands. We use only `ml model` and then, later, `ml display`.

`ml model`, it turns out, has a `maximize` option that causes `ml model` to issue the `ml maximize` command for us. It also has a `search()` option corresponding to `ml search`, an `init()` option corresponding to `ml init`, and many other options.

The `ml` system has two modes of operation: interactive mode, which we have used in the previous chapters, and noninteractive mode, which we want to use now. When you specify `ml model`'s `maximize` option, you turn off interactive mode. The `ml model` command proceeds from definition to final solution.

Turning off interactive mode is desirable in this context. We wish to use `ml` as a subroutine, and we do not want the user to know. If we used `ml` in interactive mode— if we coded the `ml model` and `ml maximize` statements separately—then if the user pressed *Break* at the wrong time, `ml` would leave behind traces of the incomplete maximization. Also in interactive mode, `ml` temporarily adds variables to the dataset, but these variables are not official temporary variables as issued by the `tempvar` command. Instead, `ml` gets rid of them when the `ml maximize` command successfully completes. These variables could be left behind were the user to press *Break*.

In noninteractive mode, `ml` uses real temporary variables and arranges that all traces of itself disappear no matter how things work out.

Noninteractive mode is turned on—interactive mode is switched off—when we specify `ml model`'s `maximize` option.

▷ **Example**

The following is our first draft of an estimation command for linear regression. `mynormal1` has all the properties listed in section 13.1.

```
                                            begin mynormal1.ado
program mynormal1
        version 18
        if replay() {
                if ('"'e(cmd)'"' != "mynormal1") error 301
                Replay '0'
        }
        else    Estimate '0'
end
program Estimate, eclass sortpreserve
        syntax varlist(fv) [if] [in] [,      ///
                vce(passthru)                ///
                *                            /// -Replay- options
        ]
        // check syntax
        gettoken lhs rhs : varlist
        // parse Replay options
        _get_diopts diopts, 'options'
        // mark the estimation sample
        marksample touse
        // fit the full model
        ml model lf2 mynormal_lf2            ///
                (mu: 'lhs' = 'rhs')          ///
                /lnsigma                     ///
                if 'touse',                  ///
                'vce'                        ///
                maximize
        ereturn local cmd mynormal1
        Replay , 'diopts'
end
program Replay
        syntax [, *]
        ml display , 'options'
end
                                            end mynormal1.ado
```

◁

13.4.1 Parsing with help from _get_diopts

Estimation commands allow the user to specify reporting options such as `level()` for the size of confidence levels, `cformat()` to control how coefficients are formatted, and `vsquish` to control the amount of blank space around time-series and factor-variable operators. The full list of reporting options is in [R] **Estimation options**. We specified `*` with `syntax` to allow the user to specify those reporting options. Any options the user specifies that do not match an option listed in the `syntax` statement are placed in 'options'.

The tool we need to parse 'options' to ensure it contains valid reporting options is the Stata command _get_diopts. This command inspects the list of options you pass to it, picks out any reporting options in it, and performs syntax checking on those options.

There are two ways to use _get_diopts. In the first way, you code

```
syntax varlist(fv) ... [, *]
_get_diopts diopts, 'options'
```

In this version, local macro 'diopts' will contain the valid reporting options, if any. If the user specifies any other options not explicitly declared in the `syntax` statement, then _get_diopts will exit with a syntax error to the user. This is how we will use _get_diopts in most of our examples.

The second way to use _get_diopts is to code

```
syntax varlist(fv) ... [, *]
_get_diopts diopts otheropts, 'options'
```

Here, instead of issuing a syntax error, _get_diopts will place any unrecognized options in 'otheropts'. Your code can then do further syntax parsing on 'otheropts'.

In `mynormal1`, we use _get_diopts to parse our reporting options. We then pass those reporting options on to our `Replay` program, which actually displays the output. We could skip _get_diopts and just pass 'options' to `Replay` because `ml display` accepts reporting options and will exit with an error message if the user specifies an invalid reporting option. That is bad practice. Suppose you are implementing a complex estimator that requires more than a split second of time to finish. Then the user would not know about the invalid reporting option until after estimation is complete. You want to catch errors as quickly as possible.

In `Replay`, we do not bother to parse 'options' for valid reporting options using _get_diopts because `ml display` does that anyway.

13.5   Advice

Experience has taught us that developing new commands is best done in steps. For example, we advocate the following when developing an estimation command using `ml`:

1. Derive the log-likelihood function from your probability model.

2. Write a likelihood evaluator that calculates the log-likelihood values and, optionally, its derivatives. We recommend method `lf` whenever possible; otherwise, start with method `d0`.

3. Interactively test this likelihood evaluator using the debugging utilities described in chapter 8.

4. Use the outline from section 13.3 to write a first draft for your estimation command.

5. Test your first draft for correctness and ensure that it has all the properties expected of an estimation command. Put these tests into a certification script. For more information on writing certification scripts, see Gould (2001).

6. When adding new features to your estimation command, add one option at a time, test each new option, and add the new tests to the certification script.

For the rest of this chapter, we will give some detailed advice based on our experience writing estimation commands. We will build each piece of advice into our estimation command for linear regression.

13.5.1   Syntax

Keep the syntax simple. Remember, you have to parse it and the user has to type it. Making the syntax unnecessarily complicated serves nobody's interest.

Many models are just single equations with one or more "dependent" variables. In that case, we recommend that the syntax be

cmd varlist $\begin{bmatrix} if \end{bmatrix}$ $\begin{bmatrix} in \end{bmatrix}$ $\begin{bmatrix} weight \end{bmatrix}$ $\begin{bmatrix} , \ \ldots \end{bmatrix}$

No extraneous characters are placed between the dependent and independent variables. This syntax is easy to parse and use. If there is one dependent variable, then the outline is

```
syntax varlist(fv) [fweight pweight] [if] [in] [, ...]
gettoken lhs rhs : varlist
...
ml model ... ('lhs'='rhs') ...
```

This approach works, even if there is really more than one equation but the subsequent equations concern ancillary parameters such as the variance of the residuals in linear regression. The `ml model` statement would then read

```
ml model ... ('lhs'='rhs') /lnsigma ...
```

Including the specification (fv) with `syntax varlist` indicates that users can include factor-variable notation in the *varlist* they specify. Typically, the dependent variable is not allowed to contain factor-variable notation. To ensure that the user does not specify such a dependent variable, you can use the `_fv_check_depvar` utility. Programming

```
_fv_check_depvar 'lhs'
```

will check for factor variables and interactions in `'lhs'` and issue an error message if factor variables or interactions are present. In general, you will not need to do any additional work to handle factor variables because `ml` does the rest automatically.

If there are two dependent variables (for example, in Stata's `intreg` command), the outline becomes

```
syntax varlist(min=2 fv) [fweight pweight] [if] [in] [, ... ]
gettoken lhs1 rhs : varlist
gettoken lhs2 rhs : rhs
_fv_check_depvar 'lhs1' 'lhs2', k(2)
...
ml model ... ('lhs1' 'lhs2' = 'rhs') ...
```

The option `k(2)` to `_fv_check_depvar` indicates that we have two dependent variables.

Many multiple-equation models, such as Stata's `heckman` command, lend themselves to the second equation being specified as a required option. In those cases, the outline (with one dependent variable) becomes

```
syntax varlist(fv) [fweight pweight] [if] [in], SECond(varlist) [...]
gettoken lhs rhs : varlist
_fv_check_varlist 'lhs'
...
ml model ... ('lhs' = 'rhs') ('second') ...
```

Above, we specified the independent variables for the second equation in the required option `second()`, which could be minimally abbreviated `sec()`. The name of the option, of course, would be your choice.

▷ Example

We will add another option called `hetero(`*varlist*`)` to our estimation command. This new option will allow the user to specify variables to model heteroskedasticity in the errors. Because it is so easy, we will also allow *varlist* to contain factor-variable notation. By default, `mynormal2` will be a linear regression model with a constant error variance (homoskedastic errors).

Here is the ado-file for the next iteration of our command, which we call `mynormal2`:

```
————————————————————————————— begin mynormal2.ado ———————
program mynormal2
        version 18
        if replay() {
                if ('""'e(cmd)'"' != "mynormal2") error 301
                Replay '0'
        }
        else    Estimate '0'
end
program Estimate, eclass sortpreserve
        syntax varlist(fv) [if] [in] [,      ///
                vce(passthru)                ///
                HETero(varlist fv)           /// <- NEW
                *                            /// -Replay- options
        ]
        // check syntax
        gettoken lhs rhs : varlist
        _fv_check_depvar 'lhs'
        // parse Replay options
        _get_diopts diopts, 'options'
        // mark the estimation sample
        marksample touse

        // fit the full model
        ml model lf2 mynormal_lf2             ///
                (mu: 'lhs' = 'rhs')           ///
                (lnsigma: 'hetero')           /// <- NEW
                if 'touse',                   ///
                'vce'                         ///
                maximize

        ereturn local cmd mynormal2

        Replay , 'diopts'
end
program Replay
        syntax [, *]
        ml display , 'options'
end
————————————————————————————————— end mynormal2.ado ———————
```

◁

13.5.2   Estimation subsample

In your code, you may need to make calculations on the data. If so, you must ensure that you make them on the same data that `ml` uses, that is, the estimation subsample.

You should identify that sample early in your code and then restrict your attention to it. In the standard Stata syntax case, this is easy:

```
syntax varlist(fv) [fweight pweight] [if] [in] [, ... ]
marksample touse
```

Thereafter, you code `if 'touse'` on the end of your Stata commands. `marksample` creates a temporary variable—a variable that you thereafter refer to in single quotes—that contains 1 if the observation is to be used and 0 otherwise.

If you obtain a second equation from an option, account for missing values in it as well:

```
syntax varlist(fv) [fweight pweight] [if] [in], SECond(varlist fv) [, ... ]
marksample touse
markout 'touse' 'second'
```

`markout` takes the `'touse'` variable created by `marksample` and sets elements in it to zero in observations where the subsequent specified variables are missing.

We must also allow for the possibility that the user specified `vce(cluster varname)` to request a robust variance calculation that does not assume observations within each cluster are independent but only that observations in different clusters are independent. In general, we need to restrict our calculations to those observations for which the cluster variable does not contain missing values.

The easiest way to find out whether the user specified `vce(cluster varname)` is to use the `_vce_parse` utility:

```
syntax ... , [ vce(passthru) ... ]
marksample touse
...
_vce_parse 'touse', opt(Robust opg oim) argopt(CLuster) : , 'vce'
```

With this syntax, `_vce_parse` will scan local macro `'vce'` and see if the user specified `vce(cluster varname)`; we capitalized the C and L in `CLuster` to indicate that `cluster` can be abbreviated as `cl` (or `clu`, `clus`, ...). If the user did specify a cluster variable, then `_vce_parse` will mark out our `touse` variable if *varname* contains missing values. For more information on `_vce_parse`, type in `help _vce_parse`.

Once we have correctly identified the estimation subsample, there is no reason for `ml` to go to the work of doing it again. We can specify our `ml model` statement as

```
ml model ... if 'touse', missing ...
```

`ml model`'s `missing` option literally means that missing values among the dependent and independent variable are okay, so they will not be marked out from the estimation sample. In fact, we have already marked such missing values out, so we are merely saving time. We are also ensuring that, if we have marked the estimation subsample incorrectly, `ml` will use the same incorrect sample and that will probably cause the estimation to go violently wrong. Thus such a bug would not go undiscovered.

▷ **Example**

mynormal3 uses markout to identify missing values in the hetero() option and the cluster variable (if specified), and it also uses the missing option in the ml model statement.

Here are the differences between mynormal2 and mynormal3.

```
──────────────────────────────────────── fragment mynormal3.ado ───────
program mynormal3
        version 18
        if replay() {
                if (`"`e(cmd)'"' != "mynormal3") error 301
                Replay `0'
        }
        else    Estimate `0'
end

program Estimate, eclass sortpreserve
        ...
        // mark the estimation sample
        marksample touse
        markout `touse' `hetero'                    // <- NEW
        _vce_parse `touse', opt(Robust opg oim)     /// <- NEW
                argopt(CLuster): , `vce'            // <- NEW
        // fit the full model
        ml model lf2 mynormal_lf2                   ///
                ...                                 ///
                missing                             /// <- NEW
                maximize

        ereturn local cmd mynormal3

        Replay , `diopts'
end

program Replay
        ...
end
──────────────────────────────────────── end mynormal3.ado ───────
```

Using auto.dta and mynormal3, we fit the regression of mileage (in miles per gallon) on vehicle weight and engine displacement. We also use an indicator variable to model heteroskedasticity between foreign and domestic vehicles.

```
. sysuse auto
(1978 automobile data)

. mynormal3 mpg weight displacement, hetero(foreign)
Initial:      Log likelihood =     -<inf>  (could not be evaluated)
Feasible:     Log likelihood = -6441.8168
Rescale:      Log likelihood = -368.12496
Rescale eq:   Log likelihood =  -257.3948
Iteration 0:  Log likelihood =  -257.3948  (not concave)
Iteration 1:  Log likelihood = -240.09029  (not concave)
Iteration 2:  Log likelihood = -229.23871  (not concave)
Iteration 3:  Log likelihood = -222.61255  (not concave)
Iteration 4:  Log likelihood = -215.68284  (not concave)
Iteration 5:  Log likelihood = -207.35199
Iteration 6:  Log likelihood = -186.51002
Iteration 7:  Log likelihood = -183.74152
Iteration 8:  Log likelihood = -183.71457
Iteration 9:  Log likelihood = -183.71454
Iteration 10: Log likelihood = -183.71454
```

		Number of obs =	74
		Wald chi2(2) =	202.12
Log likelihood = -183.71454		Prob > chi2 =	0.0000

mpg	Coefficient	Std. err.	z	P>|z|	[95% conf. interval]	
mu						
weight	-.0059578	.000831	-7.17	0.000	-.0075865	-.004329
displacement	.0000592	.0068563	0.01	0.993	-.0133789	.0134974
_cons	39.4995	1.597494	24.73	0.000	36.36847	42.63053
lnsigma						
foreign	.824055	.1810665	4.55	0.000	.4691712	1.178939
_cons	.8187011	.0982573	8.33	0.000	.6261204	1.011282

◁

❏ Technical note

For speed, if you implement your command using an lf- or a d-family evaluator, the most important option you can add to the ml model statement is nopreserve. The way ml ordinarily works with these evaluators is to preserve the data and drop the irrelevant observations prior to optimization. Dropping irrelevant observations allows the evaluator to be written sloppily and still produce correct results. The data are automatically restored following estimation.

If you specify nopreserve, you must manage your estimation subsample carefully, and that means you must write your evaluator to respect $ML_samp. The speed advantages in doing so are considerable.

We recommend that during development, you do not specify nopreserve. Once you have completed your code, add the nopreserve option and then test it carefully against the version without the option.

To use `nopreserve` in `mynormal3`, you would modify the `ml model` statement to read

```
ml model lf2 mynormal_lf2                        ///
        (mu: 'lhs' = 'rhs')                      ///
        (lnsigma: 'hetero')                      ///
        if 'touse',                              ///
        'vce'                                    ///
        missing                                  ///
        nopreserve                               /// <- NEW
        maximize
```

❑

13.5.3   Parsing with help from mlopts

`mlopts` is an official Stata command that will assist in checking syntax and parsing estimation commands that use `ml`.

The problem `mlopts` solves is this: you are about to write a new estimation command, and doubtlessly, it is going to allow options:

`mynewcmd` *varlist* $\big[$*if*$\big]$ ... $\big[$, *options* $\big]$

Among those options, you would like to allow all the standard maximization options, such as `iterate(#)`, `tolerance(#)`, and `trace`. Handling these options, once parsed, is easy; you merely need to include them on the `ml model` statement. The problem is that even you, the programmer, probably do not know what all these options are (see appendix B), and even if you do, separating these options from the options unique to `mynewcmd` will be work.

That is where `mlopts` helps. `mlopts` takes the options specified by the user of your command and separates them into two lists: the standard ones that you will want to pass along to `ml model`, and all the rest.

There are two ways to use `mlopts`. In the first, you code

```
syntax varlist(fv) ... [, myopts *]
mlopts std, 'options'
```

In the second, you code

```
syntax varlist(fv) ... [, myopts *]
mlopts std mine, 'options'
```

In both cases, the standard maximization options, such as `tolerance(#)` and `trace`, are placed in local macro 'std'. In the first case, if the user specifies an invalid option, `mlopts` will issue a syntax error. In the second case, any other options the user may have specified that are not explicitly declared in *myopts* are placed in 'mine'. You will then have to parse 'mine' and decide what to do with options specified in it.

We use the second way in our examples because we want to allow for both maximization and reporting (display) options. Local macro 'options' will contain both

types of options. We use `mlopts` to pick out the maximization options and then use
`_get_diopts` to pick out the reporting options. If the user specifies an invalid option,
`_get_diopts` will issue the error message for us.

We will show you an example but first, let's get the official documentation out of
the way.

Syntax

`mlopts` *mlopts* [*rest*] [, *options_to_parse*]

Description

The options specified in *options_to_parse* will be separated into two groups and returned
in local macros specified by the caller. The names of the local macros will be *mlopts*
and *rest*.

mlopts is required and will contain those options that the `ml` command recognizes as
   maximize options; see appendix B. Options returned in *mlopts* are

   <u>tr</u>ace <u>grad</u>ient <u>hess</u>ian showstep <u>techn</u>ique(nr | bhhh | dfp | bfgs)
   <u>iter</u>ate(#) <u>tol</u>erance(#) <u>ltol</u>erance(#) <u>nrtol</u>erance(#)
   qtolerance(#) <u>nonrtol</u>erance <u>showtol</u>erance <u>difficult</u>
   constraints(*clist* | *matname*) <u>col</u>linear <u>nocnsnotes</u>
   vce(oim | opg | <u>r</u>obust)

rest is optional and, if specified, will contain all other options that were not returned in
   mlopts. If *rest* is not supplied, any extra options will cause an error.

Stored results

`mlopts` stores the following in `s()`:

Macros
s(constraints)	contents of `constraints()` option
s(collinear)	`collinear`, if option was specified
s(technique)	contents of `technique()` option

▷ Example

In `mynormal4`, we use `mlopts` to retrieve the *maximize* options and pass them on to
the `ml model` statement. We also added the `nolog` and `noconstant` options; `nolog` suppresses the iteration log from being displayed, and `noconstant` suppresses the constant
term (intercept) in the regression equation.

Here are the differences between `mynormal3` and `mynormal4`.

```
─────────────────────────────── fragment mynormal4.ado ───────────
program mynormal4
        version 18
        if replay() {
                if ('"'e(cmd)'"' != "mynormal4") error 301
                Replay '0'
        }
        else    Estimate '0'
end
program Estimate, eclass sortpreserve
        syntax varlist(fv) [if] [in] [,      ///
                vce(passthru)                ///
                noLOg                        /// <- NEW
                noCONStant                   /// <- NEW
                ...                          ///
                *                            ///
        ]
        // check syntax
        mlopts mlopts other, 'options'            // <- NEW
        // parse Replay options
        _get_diopts diopts, 'other'               // <- CHANGED
        gettoken lhs rhs : varlist
        _fv_check_depvar 'lhs'
        ...
        // fit the full model
        ml model lf2 mynormal_lf2                      ///
                (mu: 'lhs' = 'rhs', 'constant')        /// <- CHANGED
                (lnsigma: 'hetero')                    ///
                if 'touse',                            ///
                'vce'                                  ///
                'log'                                  /// <- NEW
                'mlopts'                               /// <- NEW
                ...
        ereturn local cmd mynormal4
        Replay , level('level')
end
program Replay
        ...
end
─────────────────────────────── end mynormal4.ado ───────────
```

We can now request the outer product of the gradients variance estimates by specifying the `vce(opg)` option to be automatically processed by `mlopts`. We also use the `noconstant` option to directly estimate the intercept for both foreign and domestic vehicles. Finally, we change the coefficient number format using `cformat()`, which is processed by `_get_diopts`.

```
. sysuse auto
(1978 automobile data)

. generate domestic = foreign==0

. mynormal4 mpg foreign domestic weight displacement,
>  vce(opg) noconstant nolog cformat(%9.3f)
Initial:     Log likelihood =      -<inf>  (could not be evaluated)
Feasible:    Log likelihood = -6441.8168
Rescale:     Log likelihood = -368.12496
Rescale eq:  Log likelihood =  -257.3948
                                                Number of obs =       74
                                                Wald chi2(4)  = 1854.07
Log likelihood = -194.16368                     Prob > chi2   =  0.0000
```

		OPG				
mpg	Coefficient	std. err.	z	P>|z|	[95% conf. interval]	
mu						
foreign	40.247	2.620	15.36	0.000	35.112	45.382
domestic	41.848	3.295	12.70	0.000	35.391	48.305
weight	-0.007	0.002	-4.07	0.000	-0.010	-0.004
displacement	0.002	0.014	0.14	0.889	-0.025	0.029
lnsigma						
_cons	1.205	0.067	17.95	0.000	1.073	1.336

◁

13.5.4 Weights

We recommend that your estimation commands allow frequency and sampling weights, if they can. For sampling weights, the issues are the same as for allowing robust estimates of variance. There is no issue for method-lf evaluators, pweights are not possible with method-d0 evaluators, and method-lf1 and method-lf2 evaluators must return equation-level scores for pweights to be allowed (see page 102).

If you allow weights, you should code

```
syntax ... [fweight pweight] ...
...
if "`weight'" != "" {
        local wgt "[`weight'`exp']"
rbr
```

The idea here is to define the macro `wgt` so that weights are more convenient to use. Then you can code the ml model statement as

```
. ml model ... `wgt' if `touse' ...
```

because `wgt` either substitutes to nothing or substitutes to [fweight=*exp*] or [pweight=*exp*].

▷ **Example**

We incorporated fweights and pweights into mynormal5.

Here are the differences between mynormal4 and mynormal5:

──────────────────────────────── fragment mynormal5.ado ────────

```
program mynormal5
        version 18
        if replay() {
                if (`"`e(cmd)'"' != "mynormal5") error 301
                Replay `0'
        }
        else    Estimate `0'
end
program Estimate, eclass sortpreserve
        syntax varlist(fv) [if] [in]            ///
                [fweight pweight] [,            /// <- NEW
                noLOg                           /// -ml model- options
                ...                             ///
        ]
        ...
        if "`weight'" != "" {                             // NEW block
                local wgt "[`weight'`exp']"
        }
        // mark the estimation sample
        marksample touse
        markout `touse' `hetero'
        _vce_parse `touse', opt(Robust oim opg)     ///
                argopt(CLuster): `wgt' , `vce'      //  <- CHANGED
        ...
        // fit the full model
        ml model lf2 mynormal_lf2               ///
                (mu: `lhs' = `rhs', `constant')  ///
                (lnsigma: `hetero')              ///
                `wgt' if `touse',                /// <- CHANGED
                `vce'                            ///
                ...
        ereturn local cmd mynormal5
        Replay , `diopts'
end
program Replay
        ...
end
```

──────────────────────────────── end mynormal5.ado ────────

◁

13.5.5 Constant-only model

We recommend adopting a two-step procedure for obtaining estimates:

```
display as txt _n "Fitting constant-only model:"
ml model ..., maximize ...
display as txt _n "Fitting full model:"
ml model ..., maximize continue ...
```

The idea here is first to fit the constant-only model and then to proceed to fit the full model. `continue` specifies that `ml` is to report a likelihood-ratio test based on the previous `ml` results in place of the Wald test. There are two reasons why we recommend this two-step procedure with the `continue` option:

1. Users want to see a likelihood-ratio test reported, if it is relevant. The two-step procedure, with the `continue` option used at the second step, will accomplish this.

2. Estimation for many commonly used models proceeds more smoothly if you use the coefficients from the constant-only fit as starting values.

With some likelihood functions, you can obtain a solution for the constant-only model analytically. In that case, you can supply the point estimates using the `init()` option and the log-likelihood value using the `lf0()` option (not to be confused with evaluator method `lf0`). The outline then becomes

```
obtain analytic solution
display as txt _n "Fitting full model:"
ml model ..., maximize lf0(...) init(...) ...
```

In both outlines above, we would also specify the `search(off)` option, but that does not matter as far as reporting likelihood-ratio tests is concerned. We would specify `search(off)` because it is unlikely that `ml search` will improve on these starting values; searching would be a waste of time. However, do not specify `search(off)` if the constant-only model involves setting any coefficient on `_cons` to zero in any equation or if some equations have no `_cons`.

The likelihood-ratio test is inappropriate in some circumstances. It would be inappropriate for clustered data or when robust variance estimates are being computed; see [SVY] **Survey**. The likelihood-ratio test might also be inappropriate if the option `constraints()` is specified, depending on the constraints.

In any case, `ml` watches for these problems and reports the Wald test, even if you specify the `continue` option. In the last case, Wald tests are always reported when the `constraints()` option is specified because it is too difficult to determine whether the likelihood-ratio test is appropriate.

When fitting the constant-only model, you need not bother honoring the user's request for any type of robust variance estimates. Calculating robust variance estimates takes additional time, and doing so does not affect the likelihood-ratio test or the point estimates that are in turn used as starting values for the full model. Only for the full model do you need to compute robust variance estimates if requested by the user.

▷ Example

Below we modify `mynormal5` to create `mynormal6` that follows the two-step procedure outlined above. `mynormal6` will first fit the constant-only model as long as option `noconstant` is not specified. Also, we add a `nolrtest` option to go back to reporting the Wald test in place of the new likelihood-ratio test. We do this to show that we can.

Here are the differences between `mynormal5` and `mynormal6`:

────────────────────────────── fragment mynormal6.ado ──────────

```
program mynormal6
        version 18
        if replay() {
                if (`"`e(cmd)'"' != "mynormal6") error 301
                Replay `0'
        }
        else    Estimate `0'
end

program Estimate, eclass sortpreserve
        syntax varlist(fv) [if] [in]            ///
                [fweight pweight] [,            ///
                ...                             ///
                noLRTEST                        /// <- NEW
                *                               ///
        ]
        // check syntax
        mlopts mlopts, `options'
        ...
        if "`log'" != "" {
                local qui quietly
        }
        ...
        if "`constant'" == "" {                         // NEW block
                `qui' di as txt _n "Fitting constant-only model:"
                ml model lf mynormal_lf                 ///
                        (mu: `lhs' = )                  ///
                        (lnsigma: `hetero')             ///
                        `wgt' if `touse',               ///
                        `log'                           ///
                        `mlopts'                        ///
                        nocnsnotes                      ///
                        missing                         ///
                        maximize
                if "`lrtest'" == "" {
                        local contin continue search(off)
                }
                else {
                        tempname b0
                        mat `b0' = e(b)
                        local contin init(`b0') search(off)
                }
                `qui' di as txt _n "Fitting full model:"
        }
        // fit the full model
        ml model lf2 mynormal_lf2                       ///
                (mu: `lhs' = `rhs', `constant')         ///
                (lnsigma: `hetero')                     ///
                `wgt' if `touse',                       ///
                `vce'                                   ///
                ...                                     ///
                `contin'                                /// <- NEW
                missing                                 ///
                maximize
        ereturn local cmd mynormal6
```

```
        Replay , 'diopts'
end

program Replay
        ...
end
```
———————————————————————————————— end mynormal6.ado ————————

Using `auto.dta` and `mynormal6`, we fit a regression model and see the value of the likelihood-ratio test statistic for the coefficients in the regression equation.

```
. sysuse auto, clear
(1978 automobile data)

. mynormal6 mpg foreign weight displacement

Fitting constant-only model:

Initial:      Log likelihood =     -<inf>  (could not be evaluated)
Feasible:     Log likelihood = -6441.8168
Rescale:      Log likelihood = -368.12496
Rescale eq:   Log likelihood =  -257.3948
Iteration 0:  Log likelihood =  -257.3948
Iteration 1:  Log likelihood = -253.97778
Iteration 2:  Log likelihood = -234.74755
Iteration 3:  Log likelihood = -234.39642
Iteration 4:  Log likelihood = -234.39434
Iteration 5:  Log likelihood = -234.39434

Fitting full model:

Iteration 0:  Log likelihood = -234.39434  (not concave)
Iteration 1:  Log likelihood = -221.82805  (not concave)
Iteration 2:  Log likelihood = -218.56082
Iteration 3:  Log likelihood = -195.60215
Iteration 4:  Log likelihood = -194.16762
Iteration 5:  Log likelihood = -194.16368
Iteration 6:  Log likelihood = -194.16368
```

					Number of obs =	74
					LR chi2(3) =	80.46
Log likelihood = -194.16368					Prob > chi2 =	0.0000

mpg	Coefficient	Std. err.	z	P>|z|	[95% conf. interval]	
mu						
foreign	-1.600631	1.083131	-1.48	0.139	-3.72353	.5222673
weight	-.0067745	.0011346	-5.97	0.000	-.0089982	-.0045508
displacement	.0019286	.0097942	0.20	0.844	-.0172676	.0211249
_cons	41.84795	2.286289	18.30	0.000	37.3669	46.32899
lnsigma						
_cons	1.204895	.0821995	14.66	0.000	1.043787	1.366003

◁

13.5.6   Initial values

Previously, we did not bother with supplying `ml` with initial values; `ml` is pretty good about filling them in for us. For estimation commands, however, we recommend that you take the effort to produce good starting values.

Mostly, `ml search` does a good job, and if randomly it does not, the interactive `ml` user can type `ml search` yet again. In programming a canned estimator, do not depend on randomness working in your favor.

Produce starting values that you know are good. For most models, it is adequate to produce known-to-be-good starting values for the constant-only model, obtain the constant-only estimates, and then proceed from there to fitting the full model:

```
produce good starting values for constant-only model
display as txt _n "Fitting constant-only model:"
ml model ..., maximize init(...) search(off) ...
display as txt _n "Fitting full model:"
ml model ..., maximize continue search(off) ...
```

It would be acceptable to remove the `search(off)` from the first `ml model` statement. If, however, your starting values really are good, you will want to leave the `search(off)` in place because otherwise you will just be wasting computer time.

If you had questionable starting values, in addition to removing `search(off)` you might be tempted to add, say, `repeat(5)` and thus allow `ml search` a better opportunity to improve the values. In general, we do not do this because, if randomly searching really does improve things, then we are going to have to explain to users why they got one iteration log one time and a different iteration log another. We would also be worried that randomness may not work in our favor next time. Work harder on solving the problem of initial values, because randomly searching really is a waste of computer time.

Notice our very different attitude when programming an estimator for our own use and writing an estimation command. When we are just using `ml` interactively, we seldom bother with initial values. We use `ml search`, and that is safe because we are monitoring results. When we write an estimation command, users may not anticipate problems and therefore not watch for them. Because the estimator is a command, they will assume it works, and we therefore recommend performing the work necessary to meet their expectations.

▷ **Example**

The maximum likelihood estimators of the mean, μ, and variance, σ^2, for the normal distribution are

$$\widehat{\mu} = \overline{y} = \frac{1}{N} \sum_{j=1}^{N} y_j \tag{13.1}$$

$$\widehat{\sigma}^2 = \frac{1}{N} \sum_{j=1}^{N} (y_j - \overline{y})^2 \tag{13.2}$$

mynormal7 uses summarize to compute the initial values according to (13.1) and (13.2) for the constant-only linear regression model.

When weights are supplied, we define the 'awgt' macro for initial value calculations.

```
local awgt "[aw'exp']"
```

We use aweights because summarize does not allow pweights but, when used with aweights, reproduces the correct mean and variance calculations for pweights.

Here are the differences between mynormal6 and mynormal7:

——————————————————————————— fragment mynormal7.ado ———————

```
program mynormal7
        version 18
        if replay() {
                if ('"'e(cmd)'"' != "mynormal7") error 301
                Replay '0'
        }
        else    Estimate '0'
end
program Estimate, eclass sortpreserve
        ...
        if "'weight'" != "" {
                local wgt "['weight''exp']"
                // for initial value calculations
                if "'weight'" == "pweight" {                // <-NEW block
                        local awgt "[aw'exp']"
                }
                else    local awgt "'wgt'"
        }
        ...
        if "'constant'" == "" {
                // initial values                          -- NEW block
                quietly sum 'lhs' 'awgt' if 'touse'
                local mean = r(mean)
                local lnsd = ln(r(sd))+ln((r(N)-1)/r(N))/2
                local initopt init(/mu='mean' /lnsigma='lnsd') search(off)
                'qui' di as txt _n "Fitting constant-only model:"
```

```
             ml model lf mynormal_lf                      ///
                     ...                                  ///
                    'mlopts'                              ///
                    'initopt'                             /// <- NEW
                    nocnsnotes                            ///
                    missing                               ///
                    maximize
                 ...
         }
         ...
         ereturn local cmd mynormal7
         Replay , 'diopts'
   end
   program Replay
         ...
   end
```
————————————————————————————— end mynormal7.ado ——————————

In the following, we use `auto.dta` and `mynormal7` to fit a linear regression. It seems that `ml` performed an extra iteration when fitting the constant-only model, even though `mynormal7` passed it values calculated using the analytical solution. Actually, `ml` took one iteration to verify that it converged; `Iteration 0` is displayed to show the value of the log-likelihood function just before optimization begins in earnest.

```
. sysuse auto
(1978 automobile data)

. mynormal7 mpg foreign weight displacement

Fitting constant-only model:

Iteration 0:  Log likelihood = -234.39434
Iteration 1:  Log likelihood = -234.39434

Fitting full model:

Iteration 0:  Log likelihood = -234.39434  (not concave)
Iteration 1:  Log likelihood = -221.82805  (not concave)
Iteration 2:  Log likelihood = -218.56082
Iteration 3:  Log likelihood = -195.60215
Iteration 4:  Log likelihood = -194.16762
Iteration 5:  Log likelihood = -194.16368
Iteration 6:  Log likelihood = -194.16368

                                         Number of obs =      74
                                         LR chi2(3)    =   80.46
Log likelihood = -194.16368              Prob > chi2   =  0.0000
```

mpg	Coefficient	Std. err.	z	P>|z|	[95% conf. interval]	
mu						
foreign	-1.600631	1.083131	-1.48	0.139	-3.72353	.5222673
weight	-.0067745	.0011346	-5.97	0.000	-.0089982	-.0045508
displacement	.0019286	.0097942	0.20	0.844	-.0172676	.0211249
_cons	41.84795	2.286289	18.30	0.000	37.3669	46.32899
lnsigma						
_cons	1.204895	.0821995	14.66	0.000	1.043787	1.366003

13.5.7   Storing results in e()

`ml` automatically stores the standard things in `e()` for you. You need to add `e(cmd)` so that the rest of Stata will know that these results were produced by a full-fledged estimation command:

```
ereturn local cmd "newcmd"
```

You may want to store other things in `e()` for use by your `Replay` subroutine. Store them before setting `e(cmd)` so that, should the user press *Break* while you are still setting values in `e()`, Stata will know that the estimation is not complete.

Do not store anything in `e()` until after you have issued the last `ml model` command. `ml model` wipes out whatever is stored in `e()` every time it is run.

13.5.8   Displaying ancillary parameters

Ancillary parameters are often estimated in a transformed metric; for example, rather than estimating σ, we estimate $\ln \sigma$. The output would look much better if the estimation output table labeled the parameter with `/lnsigma` rather than `lnsigma:_cons`. It would also be an improvement if it displayed the respective point and interval estimates of σ.

`ml display` will label $\ln \hat{\sigma}$ with `/lnsigma` for us if we identify it as an ancillary parameter, which we do by filling in `e(k_aux)`. `ml display` assumes that `e(k_aux)` contains the number of ancillary (auxiliary) parameters in the model and that those parameters make up the last elements of the vector of point estimates. `e(k_aux)` will be effective only if it contains an integer smaller than the number of equations (`ml` stores the number of equations in `e(k_eq)`).

To take advantage of `e(k_aux)` in `ml display`, you must code your likelihood evaluator so that the ancillary parameters make up the last elements of the vector of coefficients, meaning that they are specified last in the `ml model` statement:

```
ml model lf mynormal_lf (y = x variables) /lnsigma , ...
```

Then somewhere after the `ml model` statement but before storing the name of the estimation command in `e(cmd)`, we must store the number of ancillary parameters in `e(k_aux)` by coding

```
ereturn scalar k_aux = 1
```

When `ml display` is used to display the output table, it looks at `e(k_aux)`, assumes that the last equation contains an ancillary parameter, and labels it appropriately in the output table.

Now that we know how to label our ancillary parameters, we might also want to display these parameters in another metric. To do this, use the diparm() option of ml model. ml model accepts multiple diparm(*diparm_args*) options, and a simplified syntax diagram for *diparm_args* is

$$eqname \; \big[\, , \; \big[\texttt{exp} \, | \, \texttt{tanh} \, | \, \texttt{invlogit} \big] \; \underline{\texttt{lab}}\texttt{el}(string) \; \underline{\texttt{p}}\texttt{rob} \; \texttt{dof}(\#) \; \underline{\texttt{l}}\texttt{evel}(\#) \big]$$

where *eqname* is the equation name for the ancillary parameter. Basically, you specify the transformation (exp, tanh, or invlogit) for the ancillary parameter and use the label() option to give the transformation its own title in the output table.

▷ **Example**

When performing regression with homoskedastic errors, mynormal8 will use e(k_aux) to display $\ln \widehat{\sigma}$ with label /lnsigma and the diparm() option to display $\widehat{\sigma}$ with label sigma.

Here are the differences between mynormal7 and mynormal8:

```
 ──────────────────────────────────────── fragment mynormal8.ado ───────────
program mynormal8
        version 18
        if replay() {
                if ('"'e(cmd)'"' != "mynormal8") error 301
                Replay '0'
        }
        else    Estimate '0'
end

program Estimate, eclass sortpreserve
        ...
        if "'log'" != "" {
                local qui quietly
        }
        if '"'hetero'"' == "" {                        // NEW block
                local diparm diparm(lnsigma, exp label("sigma"))
        }
        // mark the estimation sample
        marksample touse
        markout 'touse' 'hetero'
        ...
        if "'hetero'" == "" {                          // NEW block
                ereturn scalar k_aux = 1
        }
        else    ereturn scalar k_aux = 0
        ereturn local cmd mynormal8
        Replay , 'diopts'
end

program Replay
        ...
end
 ──────────────────────────────────────── end mynormal8.ado ───────────
```

The output table looks much nicer with these new features:

```
. sysuse auto
(1978 automobile data)
. mynormal8 mpg foreign weight displacement, nolog
                                            Number of obs =      74
                                            LR chi2(3)    =   80.46
Log likelihood = -194.16368                 Prob > chi2   =  0.0000
```

mpg	Coefficient	Std. err.	z	P>|z|	[95% conf. interval]	
foreign	-1.600631	1.083131	-1.48	0.139	-3.72353	.5222673
weight	-.0067745	.0011346	-5.97	0.000	-.0089982	-.0045508
displacement	.0019286	.0097942	0.20	0.844	-.0172676	.0211249
_cons	41.84795	2.286289	18.30	0.000	37.3669	46.32899
/lnsigma	1.204895	.0821995	14.66	0.000	1.043787	1.366003
sigma	3.336409	.2742511			2.839951	3.919653

◁

❑ **Technical note**

The `diparm()` option has many more features and correspondingly a much more complicated syntax than we have shown above. Most importantly, any transformation is available, including user-defined transformations. To learn about them, type

```
. help _diparm
```

Note the underscore in front of `diparm`. The contents of the `diparm()` option are in fact processed by the `_diparm` command.

❑

13.5.9 Exponentiated coefficients

For some models, exponentiating the coefficients from the fit assists interpretation. The `eform(string)` option of `ml display` will exponentiate the coefficients, use the delta method to compute the respective standard errors, transform the respective confidence intervals, and label the column of transformed coefficients with *string*.

In logistic regression, exponentiating the coefficients provides what are commonly known as odds ratios. Thus if we were working on an estimation command for logistic regression, we might add an option that will cause `eform("Odds ratio")` to be passed to `ml display`. In fact, this is such a common occurrence that `ml display` already has an option that does this, namely, `or`. Table 13.1 lists all the shortcut options such as `or`.

Table 13.1. Common shortcuts for the `eform(`*string*`)` option

Shortcut	Equivalent
eform	eform("exp(b)")
hr	eform("Haz. ratio")
shr	eform("SHR")
irr	eform("IRR")
or	eform("Odds ratio")
rrr	eform("RRR")

▷ **Example**

No one would want a linear-regression routine that reported exponentiated coefficients, but it is probably better to stay with a familiar example rather than introduce a new one. `mynormal9` has a new `eform` option.

Here are the differences between `mynormal8` and `mynormal9`:

```
──────────────────────────────────── fragment mynormal9.ado ─────────
program mynormal9
        version 18
        if replay() {
                if (`"`e(cmd)'"' != "mynormal9") error 301
                Replay `0'
        }
        else    Estimate `0'
end

program Estimate, eclass sortpreserve
        syntax varlist(fv) [if] [in]            ///
                ...                             ///
                EForm                           /// <- NEW
                *                               ///
        ]

        ...
        ereturn local cmd mynormal9

        Replay , level(`level') `eform'         // <- CHANGED
end

program Replay
        syntax [, EForm *]                      // <- CHANGED
        ml display, `eform' `options'           // <- CHANGED
end
──────────────────────────────────────────── end mynormal9.ado ─────────
```

Here we see the new **eform** option in action.

```
. sysuse auto
(1978 automobile data)
. mynormal9 mpg foreign weight displacement, nolog eform
                                            Number of obs =      74
                                            LR chi2(3)    =   80.46
Log likelihood = -194.16368                 Prob > chi2   = 0.0000
```

mpg	exp(b)	Std. err.	z	P>|z|	[95% conf. interval]	
foreign	.2017691	.2185425	-1.48	0.139	.0241486	1.685846
weight	.9932484	.0011269	-5.97	0.000	.9910421	.9954596
displacement	1.001931	.0098131	0.20	0.844	.9828806	1.02135
_cons	1.49e+18	3.42e+18	18.30	0.000	1.69e+16	1.32e+20
/lnsigma	1.204895	.0821995	14.66	0.000	1.043787	1.366003
sigma	3.336409	.2742511			2.839951	3.919653

Note: Estimates are transformed only in the first equation.

◁

13.5.10 Offsetting linear equations

Suppose that you need to include a variable in an equation, and the coefficient on this variable must be a fixed value of your choosing. One way of doing this is by using constraints: set the fixed value of the coefficient using the **constraint** command, and specify the constraint in the **constraints()** option.

Another way is to specify the variable as an offset for the equation. On page 57 and in the technical note in section 4.7, we discussed the syntax for an equation in the **ml model** statement. We purposely omitted two options to keep the discussion simple. The full syntax for an equation in the **ml model** statement is

$\big(\big[\, equation\_name: \,\big]\big[\, varlist_y\ =\big]\big[\, varlist_x \,\big]\big[\, ,\ \underline{\text{noconst}}\text{ant } \underline{\text{off}}\text{set}(varname)$

$\underline{\text{expo}}\text{sure}(varname)\,\big]\ \text{freeparm}\big)$

The two new options are as follows.

offset(*varname*) specifies that the equation is to be $\mathbf{x}\boldsymbol{\beta} + varname$; in other words, the equation is to include *varname* with coefficient constrained to be 1.

exposure(*varname*) is an alternative to **offset**(*varname*); it specifies that the equation is to be $\mathbf{x}\boldsymbol{\beta} + \ln(varname)$. The equation is to include $\ln(varname)$ with coefficient constrained to be 1.

See [R] **poisson** for examples using the **exposure()** option.

▷ **Example**

mynormal10 has the new offset(*varname*) and exposure(*varname*) options. The options affect only the regression equation, that is, $\theta_{1j} = \mu_j = \mathbf{x}_{1j}\boldsymbol{\beta}_1$.

Here are the differences between mynormal9 and mynormal10:

```
──────────────────────────────────── fragment mynormal10.ado ──────────
program mynormal10
        version 18
        if replay() {
                if ('"'e(cmd)'"' != "mynormal10") error 301
                Replay '0'
        }
        else    Estimate '0'
end

program Estimate, eclass sortpreserve
        syntax varlist(fv) [if] [in]            ///
                [fweight pweight] [,            ///
                ...                             ///
                noLRTEST                        ///
                OFFset(varname numeric)         /// <- NEW
                EXPosure(varname numeric)       /// <- NEW
                *                               ///
        ]

        ...
        if "'offset'" != "" {                           // NEW block
                local offopt "offset('offset')"
        }
        if "'exposure'" != "" {                         // NEW block
                local expopt "exposure('exposure')"
        }

        // mark the estimation sample
        marksample touse
        markout 'touse' 'hetero' 'offset' 'exposure'        // <- CHANGED
        ...

        if "'constant'" == "" {
                ...
                'qui' di as txt _n "Fitting constant-only model:"
                ml model lf mynormal_lf                  ///
                        (mu: 'lhs' =, 'offopt' 'expopt' )    /// <- CHANGED
                        (lnsigma: 'hetero')              ///
                        'wgt' if 'touse',                ///
                        ...
                ...
        }

        // fit the full model
        ml model lf2 mynormal_lf2                        ///
                (mu: 'lhs' = 'rhs',                      /// <- CHANGED
                        'constant' 'offopt' 'expopt'     /// <- CHANGED
                )                                        ///
                (lnsigma: 'hetero')                      ///
                'wgt' if 'touse',                        ///
                ...
        ...
end
```

```
program Replay
        ...
end
```
———————————————————————————————— end mynormal10.ado ——————————

Here we see the new `offset()` option in action.

```
. sysuse auto
(1978 automobile data)

. generate one = 1

. mynormal10 mpg foreign weight displacement, nolog offset(one)
                                                   Number of obs =       74
                                                   LR chi2(3)    =    80.46
Log likelihood = -194.16368                        Prob > chi2   =  0.0000
```

mpg	Coefficient	Std. err.	z	P>|z|	[95% conf. interval]	
foreign	-1.600631	1.083131	-1.48	0.139	-3.72353	.5222673
weight	-.0067745	.0011346	-5.97	0.000	-.0089982	-.0045508
displacement	.0019286	.0097942	0.20	0.844	-.0172676	.0211249
_cons	40.84795	2.286289	17.87	0.000	36.3669	45.32899
one	1	(offset)				
/lnsigma	1.204895	.0821995	14.66	0.000	1.043787	1.366003
sigma	3.336409	.2742511			2.839951	3.919653

We can see that it works by comparing the above output with

```
. mynormal10 mpg foreign weight displacement, nolog
  (output omitted )
```

The intercept (`_cons`) changed by 1.

◁

13.5.11   Program properties

Some commands use program properties to check for supported features in other commands; see [P] **program properties**. Program properties are associated with a program by using the `properties()` option of `program`. Associating your estimation command with program property *prop* advertises that your command meets the requirements that *prop* implies.

We discuss program properties for the `svy` prefix in section 14.1. Here we discuss the program properties that allow your estimation command to work with the `nestreg` prefix.

If command *mycmd* has the `swml` program property, `nestreg` assumes that the following items are all true:

- *mycmd* is an estimation command that follows standard Stata syntax and allows the `if` qualifier; see [U] **11 Language syntax** and [U] **11.1.3 if exp**.

In this case, the minimum requirement for standard Stata syntax is

$$mycmd \quad varlist \quad [\, if \,] \quad [\, , \quad options \,]$$

- *mycmd* stores the model coefficients and ancillary parameters in e(b).

- e(N) contains the estimation sample size, and e(sample) identifies the estimation sample after *mycmd*.

- For the likelihood-ratio test, the log-likelihood value is stored in e(ll) and the model degrees of freedom is stored in e(df_m).

- For the Wald test, the estimated covariance matrix for the coefficient estimates is stored in e(V).

These requirements are easily met if you follow the outline specified in section 13.3. Notice that ml satisfies all but the first bulleted item; it is up to you to make sure that your command satisfies bullet 1.

▷ **Example**

mynormal11 has the new swml program property.

Here are the differences between mynormal10 and mynormal11:

```
───────────────────────────────── fragment mynormal11.ado ─────────
program mynormal11, properties(swml)          // <- NEW property
        version 18
        if replay() {
                if (`"`e(cmd)'"' != "mynormal11") error 301
                Replay `0'
        }
        else    Estimate `0'
end
program Estimate, eclass sortpreserve
        ...
        ereturn local cmd mynormal11
        Replay , `eform' `diopts'
end
program Replay
        ...
end
────────────────────────────────── end mynormal11.ado ─────────
```

We revisit the regression example using auto.dta to see that the **nestreg** prefix is allowed with mynormal11.

```
. sysuse auto, clear
(1978 automobile data)
. nestreg, quietly lrtable: mynormal11 mpg (foreign weight) (displacement)
Block 1: foreign weight
Block 2: displacement
```

Block	LL	LR	df	Pr > LR	AIC	BIC
1	-194.1831	80.42	2	0.0000	396.3661	405.5824
2	-194.1637	0.04	1	0.8439	398.3274	409.8477

We conclude from the likelihood-ratio statistic that the coefficient on `displacement` (that is, Block 2) is not significant at the 5% level.

`mynormal11` contains every conceivable (non-`svy`) option and is bulletproof. The only features left to add are for the `svy` prefix command, discussed in chapter 14.

◁

14 Writing ado-files for survey data analysis

The way to write a survey estimation command is to 1) write a standard (nonsurvey) estimation command and then 2) make a few changes so that your new estimation command will work with the svy prefix; see [SVY] **svy**. This chapter details how to do that. We will assume that the first step is already complete—we will start with `mynormal11` from the end of chapter 13. Although `mynormal11` is a highly developed estimator, we could just as well start with `mynormal1`. The extra features mentioned in chapter 13 are not required for making your estimator work with the svy prefix.

Stata's svy prefix command provides the following variance estimation methods:

- svy brr: balanced repeated replication (BRR)
- svy bootstrap: bootstrap for complex survey data
- svy jackknife: jackknife for complex survey data
- svy linearized: linearization/Taylor expansion
- svy sdr: successive difference replication (SDR)

The programming changes required by BRR, bootstrap, jackknife, and SDR variance estimation methods are minimal. Linearization requires that your evaluator support the robust variance estimator; thus your likelihood evaluator must either be method `lf` or a member of the `lf` family.

14.1 Program properties

We introduced program properties in section 13.5.11; also see [P] **program properties**. The svy prefix command uses program properties to identify which of the variance estimation methods are supported by the prefixed estimation command. The program properties are `svyb` for BRR, the complex survey bootstrap, and SDR; `svyj` for the complex survey jackknife; and `svyr` for linearization.

The syntax for the svy prefix is

svy *vcetype* $\begin{bmatrix} , & svy\_options \end{bmatrix}$: *command varlist* $\begin{bmatrix} if \end{bmatrix}$ $\begin{bmatrix} in \end{bmatrix}$ $\begin{bmatrix} , & options \end{bmatrix}$

Here are the extra requirements for programs (ado-files) to support the `svy` prefix:

1. *command* must be `eclass`, allow `iweight`s, and accept the standard estimation syntax; see [P] **program**, [P] **syntax**, and [P] **mark**.

   command varlist [*if*] [*in*] [*weight*] [, *options*]

   varlist may contain multiple-equation specifications as in the `ml model` statement.

2. *command* must save the model coefficients and ancillary parameters in `e(b)` and the estimation sample size in `e(N)`, and it must identify the estimation subsample in `e(sample)`; see [P] **ereturn**.

3. `svy brr` requires that *command* have `svyb` as a program property. For example, the program definition for `regress` appears as

   ```
   program regress, properties(...  svyb ...)  ...
   ```

4. `svy bootstrap` requires that *command* have `svyb` as a program property.

5. `svy sdr` requires that *command* have `svyb` as a program property.

6. `svy jackknife` requires that *command* have `svyj` as a program property.

7. `svy linearized` has the following requirements:

   a. *command* must have `svyr` as a program property.

   b. `predict` after *command* must be able to generate scores using the following syntax:

   predict [*type*] *stub*∗ [*if*] [*in*], s̲cores

   This syntax implies that estimation results with k equations will cause `predict` to generate k new equation-level score variables. These new equation-level score variables are *stub1* for the first equation, *stub2* for the second equation, ..., and *stubk* for the last equation. Actually, `svy` does not strictly require that these new variables be named this way, but this is a good convention to follow.

8. *command* must save the model-based variance estimator for the coefficients and ancillary parameters in `e(V)`; see [SVY] **Variance estimation**.

Item 1 is satisfied by following the outline in section 13.3, and items 2 and 8 are handled by the `ml` command and by setting the appropriate program properties.

▷ **Example**

   `mynormal12` has the new program properties required by `svy`.

Here are the differences between `mynormal11` and `mynormal12`:

```
─────────────────────────────────────────────── fragment mynormal12.ado ───────────
program mynormal12, properties(svyb svyj svyr swml)
        // NEW properties        ^^^^^^^^^^^^^^^
        version 18
        if replay() {
                if (`"`e(cmd)'"' != "mynormal12") error 301
                Replay `0'
        }
        else    Estimate `0'
end

program Estimate, eclass sortpreserve
        syntax varlist(fv) [if] [in]            ///
                [fweight pweight iweight] [,     /// <- ADDED iweight
                ...                              ///
        ]
        ...
        ereturn local cmd mynormal12
        Replay , `eform' `diopts'
end

program Replay
        ...
end
─────────────────────────────────────────────────── end mynormal12.ado ───────────
```

Here we use `svy: mynormal12` with some survey data:

```
. use http://www.stata-press.com/data/r11/nhanes2a, clear
. svyset [pweight=leadwt], strata(stratid) psu(psuid) vce(jackknife)

Sampling weights: leadwt
             VCE: jackknife
             MSE: off
     Single unit: missing
         Strata 1: stratid
 Sampling unit 1: psuid
           FPC 1: <zero>

. svy, subpop(rural): mynormal12 loglead age female
(running mynormal12 on estimation sample)

Jackknife replications (62): ss.......10.........20.........30........ss.......
> ..50.........60.. done
s: Estimation sample is entirely outside the subpopulation of interest.

Survey data analysis

Number of strata = 29                  Number of obs    =       4,678
Number of PSUs   = 58                  Population size = 106,180,946
                                       Subpop. no. obs =       1,837
                                       Subpop. size    =  36,182,921
                                       Replications     =          58
                                       Design df        =          29
                                       F(2, 28)         =      128.08
                                       Prob > F         =      0.0000
```

loglead	Coefficient	Jackknife std. err.	t	P>|t|	[95% conf. interval]	
age	.0023002	.0008534	2.70	0.012	.0005549	.0040455
female	-.4111291	.0253709	-16.20	0.000	-.4630183	-.3592399
_cons	2.60715	.051626	50.50	0.000	2.501563	2.712737
/lnsigma	-.961362	.045809	-20.99	0.000	-1.055052	-.8676721

Note: 2 strata omitted because they contain no subpopulation members.
. estat effects, deff deft meff meft

loglead	Coefficient	Jackknife std. err.	DEFF	DEFT	MEFF	MEFT
age	.0023002	.0008534	1.75361	1.32424	2.5775	1.60546
female	-.4111291	.0253709	1.73034	1.31543	2.08058	1.44242
_cons	2.60715	.051626	3.04471	1.74491	3.27341	1.80926
/lnsigma	-.961362	.045809	4.60444	2.1458	7.70975	2.77664

◁

14.2   Writing your own predict command

Most estimation commands require a predict subroutine, but ml takes care of much of the work for you. Stata's predict command will pass control to the subroutine identified in e(predict). ml's predict subroutine is ml_p, and it accepts the standard options documented in [R] **predict**. Specifically, ml_p accepts the scores option and uses ml score to compute the equation-level scores, which is a significant part of the requirements for svy linearized.

ml score automatically computes equation-level scores for method-lf and lf-family likelihood evaluators; no extra work is required. ml score does not work with d-family evaluators, because observation-level log-likelihood values (and their derivatives) are not available.

Write your own predict subroutine when you need predict to generate variables that are functions of e(b) and your data. Keep in mind that your predict subroutine could call ml_p for computing linear predictions (xb), standard error of the linear predictions (stdp), equation-level scores, or any other standard prediction documented in [R] **predict**.

▷ **Example**

Suppose we wanted predict to generate a variable with the predicted values of $\ln \sigma$ or σ. We can use _predict to get the predicted values of $\ln \sigma$ (which is a linear prediction of the lnsigma equation). Then we just exponentiate those values to get the predicted value of σ. Here is a listing of mynormal_p.ado, a predict subroutine for our maximum likelihood regression command. `mynormal_p` calls `ml_p` to handle the situations when the `lnsigma` and `sigma` options are not specified.

```
───────────────────────────────────────── begin mynormal_p.ado ───────────
program mynormal_p
        syntax anything(id="newvarname") [if] [in] [, LNSigma SIGma * ]

        if "`lnsigma'" != "" {
                syntax newvarname [if] [in] [, LNSigma ]
                _predict `typlist' `varlist' `if' `in', equation(lnsigma)
                label variable `varlist' "predicted ln(sigma)"
                exit
        }
        if "`sigma'" != "" {
                syntax newvarname [if] [in] [, SIGma ]
                _predict `typlist' `varlist' `if' `in', equation(lnsigma)
                quietly replace `varlist' = exp(`varlist')
                label variable `varlist' "Predicted sigma"
                exit
        }
        // let -ml_p- handle all other cases
        ml_p `0'
end
──────────────────────────────────────── end mynormal_p.ado ───────────
```

`mynormal13` uses `mynormal_p` as its `predict` subroutine. Here are the differences between `mynormal12` and `mynormal13`:

```
───────────────────────────────────────── fragment mynormal13.ado ───────────
program mynormal13, properties(svyb svyj svyr swml)
        version 18
        if replay() {
                if (`"`e(cmd)'"' != "mynormal13") error 301
                Replay `0'
        }
        else    Estimate `0'
end
program Estimate, eclass sortpreserve
        syntax varlist(fv) [if] [in]          ///
                [fweight pweight iweight] [,   ///
                ...                            ///
        ]
        ...
        ereturn local predict mynormal_p      // <- NEW e(predict)
        ereturn local cmd mynormal13
        Replay , `eform' `diopts'
end
program Replay
        ...
end
──────────────────────────────────────── end mynormal13.ado ───────────
```

Here we use svy: mynormal13 with some survey data and predict to generate a variable containing the estimated value of σ:

```
. use http://www.stata-press.com/data/r11/nhanes2a

. svyset [pweight=leadwt], strata(stratid) psu(psuid)

Sampling weights: leadwt
            VCE: linearized
    Single unit: missing
       Strata 1: stratid
Sampling unit 1: psuid
          FPC 1: <zero>

. svy: mynormal13 loglead age female
(running mynormal13 on estimation sample)

Survey data analysis

Number of strata = 31                    Number of obs   =        4,948
Number of PSUs   = 62                    Population size = 112,915,016
                                         Design df       =           31
                                         F(2, 30)        =       625.61
                                         Prob > F        =       0.0000
```

loglead	Coefficient	Linearized std. err.	t	P>|t|	[95% conf. interval]	
age	.0024985	.0004326	5.78	0.000	.0016162	.0033807
female	-.3648611	.0101531	-35.94	0.000	-.3855684	-.3441537
_cons	2.663709	.0295542	90.13	0.000	2.603432	2.723985
/lnsigma	-.9892149	.0294101	-33.64	0.000	-1.049197	-.9292325
sigma	.3718685	.0109367			.3502188	.3948566

```
. predict shat, sigma

. describe shat
```

Variable name	Storage type	Display format	Value label	Variable label
shat	float	%9.0g		Predicted sigma

```
. summarize shat
```

Variable	Obs	Mean	Std. dev.	Min	Max
shat	4,948	.3718686	0	.3718686	.3718686

◁

15 Mata-based likelihood evaluators

Mata is a full-featured matrix programming language built into Stata. It is designed for linear algebra and other advanced mathematical computation. Many complicated statistical models are written using matrix notation, and Mata makes programming those models in Stata easy. In this chapter, we illustrate how to write likelihood-evaluator programs in Mata while continuing to use Stata's `ml` commands to control the process. In the next chapter, we show how advanced programmers can use the `moptimize()` suite of Mata functions instead of `ml`. This chapter assumes that you are already familiar with the basics of Mata and that you know how to write likelihood-evaluator programs using Stata's ado-language. For a good primer on Mata, see Gould (2018).

15.1 Introductory examples

The easiest way to see how likelihood evaluators are written in Mata is to compare a couple of them with evaluators that we have already discussed, which were written in ado-code.

15.1.1 The probit model

In section 5.2.1, we used the following method-`lf` likelihood evaluator for the probit model:

──────────────────────────── begin myprobit_lf.ado ───────────

```
program myprobit_lf
        version 18
        args lnfj xb
        quietly replace `lnfj' = ln(normal( `xb')) if $ML_y1 == 1
        quietly replace `lnfj' = ln(normal(-`xb')) if $ML_y1 == 0
end
```
──────────────────────────── end myprobit_lf.ado ───────────

To fit a probit model of `died` on `i.drug` and `age`, we would type

```
. ml model lf myprobit_lf (died = i.drug age)
. ml maximize
```

To implement the same model using a Mata-based likelihood evaluator, we create the following do-file:

```
────────────────────────────────────────── begin myprobit_lf_mata.do ──────────
version 18
sysuse cancer, clear
mata:
void myprobit_lf(transmorphic scalar ML, real rowvector b,
                 real colvector lnfj)
{
        real vector depvar, xb
        depvar = moptimize_util_depvar(ML, 1)
        xb     = moptimize_util_xb(ML, b, 1)
        lnfj   = depvar:*ln(normal(xb)) + (1:-depvar):*ln(normal(-xb))
}
end
ml model lf myprobit_lf() (died = i.drug age)
ml maximize
────────────────────────────────────────────── end myprobit_lf_mata.do ──────────
```

We wrote our evaluator and `ml` commands in a do-file so that we could define our Mata function `myprobit_lf`. We could type the function interactively, but that is error-prone, and if you type as badly as we do, it would take you several tries to type it in correctly. For most purposes, defining the Mata function in a do-file is the easiest alternative. Other alternatives include creating a Mata .mo file (see [M-3] **mata mosave**) or compiling the Mata code into a library file (see [M-3] **mata mlib**), which is most practical if you are creating an ado-file, as described in chapter 13.

Method-lf likelihood evaluators in Mata always receive three arguments. The first argument is a structure that `ml` uses to record information about the current problem; we refer to this structure as the "problem handle". If you include declarations for the arguments to your function, the problem handle must be declared a `transmorphic scalar`; we named ours ML. We named the second argument b; this holds the vector of coefficients at which we are to evaluate the likelihood function and is passed as a `real rowvector`. Unlike lf evaluators written in ado-code, which receive the values of linear combinations of the independent variables and coefficients, `ml` always passes the raw vector of coefficients to evaluators written in Mata. The third argument is a column vector we called `lnfj` that is to contain the observation-level log-likelihood values; it is analogous to the `lnfj` temporary variable passed to ado-based lf evaluators.

The functions `moptimize_util_depvar()` and `moptimize_util_xb()` are utilities that aid in implementing likelihood evaluators. The first of these requires us to pass the problem handle as well as an equation number. In return, we obtain a vector containing the dependent variable for the kth equation, where k is the second argument passed to `moptimize_util_depvar()`. The second function, `moptimize_util_xb()` is analogous to the `mleval` command used in ado-based evaluators. We pass to it the problem handle, coefficient vector, and an equation number, and it returns the linear combination (model parameter) for that equation. At the end of our function, we replace the elements of vector `lnfj` with the current log-likelihood values.

The final portion of our do-file uses `ml model` to set up our problem and then calls `ml maximize` to fit it. The only difference between this call to `ml model` and calls used when likelihood evaluators are written as ado-files is that here we followed the name of our Mata-based likelihood evaluator with parentheses. The () after the name indicates to `ml` that we have a Mata-based evaluator. All other features of `ml model` and, in fact, all other `ml` subcommands work identically whether the evaluator is ado based or Mata based.

Executing our do-file, we obtain

```
. version 18
. sysuse cancer, clear
(Patient survival in drug trial)
. mata:
─────────────────────────────────────── mata (type end to exit) ───────
: void myprobit_lf(transmorphic scalar ML, real rowvector b,
>                  real colvector lnfj)
> {
>         real vector depvar, xb
>         depvar = moptimize_util_depvar(ML, 1)
>         xb     = moptimize_util_xb(ML, b, 1)
>
>         lnfj = depvar:*ln(normal(xb)) + (1:-depvar):*ln(normal(-xb))
> }
: end
─────────────────────────────────────────────────────────────────────

. ml model lf myprobit_lf() (died = i.drug age)
. ml maximize
Initial:      Log likelihood = -33.271065
Alternative:  Log likelihood = -31.427839
Rescale:      Log likelihood = -31.424556
Iteration 0:  Log likelihood = -31.424556
Iteration 1:  Log likelihood = -21.883453
Iteration 2:  Log likelihood = -21.710901
Iteration 3:  Log likelihood = -21.710799
Iteration 4:  Log likelihood = -21.710799
```

		Number of obs =	48
		Wald chi2(3) =	13.58
Log likelihood = -21.710799		Prob > chi2 =	0.0035

died	Coefficient	Std. err.	z	P>|z|	[95% conf. interval]	
drug						
Other	-1.941275	.597057	-3.25	0.001	-3.111485	-.7710644
NA	-1.792924	.5845829	-3.07	0.002	-2.938686	-.6471626
age	.0666733	.0410666	1.62	0.104	-.0138159	.1471624
_cons	-2.044876	2.284283	-0.90	0.371	-6.521988	2.432235

Our results match those we obtained in section 5.2.1.

15.1.2   The Weibull model

In section 6.4.3, we wrote a method-lf2 likelihood evaluator for the Weibull model. Here we write a do-file that implements the likelihood evaluator in Mata:

```
                                                    begin myweibull_lf2_mata.do
version 18
sysuse cancer, clear
mata:
void myweibull_lf2(transmorphic scalar ML, real scalar todo,
                   real rowvector b, real colvector lnfj,
                   real matrix S, real matrix H)
{
        real scalar lgam, p
        real vector leta, t, d, M, R
        real matrix H11, H12, H22

        leta = moptimize_util_xb(ML, b, 1)
        lgam = moptimize_util_xb(ML, b, 2)
        t    = moptimize_util_depvar(ML, 1)
        d    = moptimize_util_depvar(ML, 2)

        p = exp(lgam)
        M = (t:*exp(-1*leta)):^p
        R = ln(t) - leta
        lnfj = -M + d:*(lgam:-leta + (p-1):*R)
        if (todo == 0) return

        S = J(rows(leta), 2, .)
        S[.,1] = p*(M - d)
        S[.,2] = d - p*R:*(M - d)
        if (todo == 1) return

        H11 = moptimize_util_matsum(ML, 1, 1, p^2 :*M, lnfj[1])
        H12 = moptimize_util_matsum(ML, 1, 2, -p*(M - d + p*R:*M), lnfj[1])
        H22 = moptimize_util_matsum(ML, 2, 2, p*R:*(p*R:*M + M - d), lnfj[1])
        H = -1*(H11, H12 \ H12', H22)
}
end

ml model lf2 myweibull_lf2() (lneta: studytime died = i.drug age) /lngamma
ml maximize
                                                    end myweibull_lf2_mata.do
```

Method-lf2 Mata-based likelihood evaluators are passed six arguments: the problem handle, a scalar that functions just like the ado-based todo parameter, the coefficient vector, a vector that we are to fill in with the observation-level log-likelihood values, a matrix to hold the observation-level scores, and a matrix to contain the Hessian of the overall log-likelihood function.

The first part of our program again uses the utility functions moptimize_util_xb() and moptimize_util_depvar() to obtain the model parameters and dependent variables. We then create several vectors containing intermediate calculations, analogous to our use of temporary variables in the ado-version of our evaluator program. We use these vectors to fill in lnfj with the log likelihood. We check todo and return from the function if ml does not want our evaluator to compute derivatives (todo = 0).

If todo is nonzero, we next compute the scores. Our model has two equations, so our matrix of scores will have two columns and several rows equal to the number of observations in the dataset. Because we want to fill in the matrix S one column at a time, we must first initialize it to have the correct dimensions. With that done, we then fill in the two columns with the two equations' scores. If `todo` equals one, we need not compute the Hessian, and in that case, we return from the function.

The last part of our code computes the Hessian of the overall log likelihood. We use the function `moptimize_util_matsum()` to aid us in those computations; that function is equivalent to the `mlmatsum` command. The first argument is again the problem handle, and the second and third arguments are the two equations for which we are computing the second derivative. The fourth argument represents the second derivative $\partial^2 \ln L/\partial\theta_i\theta_j$. The fifth argument is the variable we use to store the log likelihood; `moptimize_util_matsum()` will set this to missing if the Hessian cannot be computed.

Executing our do-file, we obtain

```
. version 18
. sysuse cancer, clear
(Patient survival in drug trial)
. mata:
─────────────────────────────────────────────── mata (type end to exit) ───
: void myweibull_lf2(transmorphic scalar ML, real scalar todo,
>                    real rowvector b, real colvector lnfj,
>                    real matrix S, real matrix H)
> {
>         real scalar lgam, p
>         real vector leta, t, d, M, R
>         real matrix H11, H12, H22
>
>         leta = moptimize_util_xb(ML, b, 1)
>         lgam = moptimize_util_xb(ML, b, 2)
>         t    = moptimize_util_depvar(ML, 1)
>         d    = moptimize_util_depvar(ML, 2)
>
>         p = exp(lgam)
>         M = (t:*exp(-1*leta)):^p
>         R = ln(t) - leta
>         lnfj = -M + d:*(lgam:-leta + (p-1):*R)
>         if (todo == 0) return
>
>         S = J(rows(leta), 2, .)
>         S[.,1] = p*(M - d)
>         S[.,2] = d - p*R:*(M - d)
>         if (todo == 1) return
>
>         H11 = moptimize_util_matsum(ML, 1, 1, p^2 :*M, lnfj[1])
>         H12 = moptimize_util_matsum(ML, 1, 2, -p*(M - d + p*R:*M), lnfj[1])
>         H22 = moptimize_util_matsum(ML, 2, 2, p*R:*(p*R:*M + M - d), lnfj[1])
>         H = -1*(H11, H12 \ H12', H22)
> }
: end
─────────────────────────────────────────────────────────────────────────────

. ml model lf2 myweibull_lf2() (lneta: studytime died = i.drug age) /lngamma
```

```
. ml maximize
Initial:        Log likelihood =        -744
Alternative:    Log likelihood = -356.14276
Rescale:        Log likelihood = -200.80201
Rescale eq:     Log likelihood = -136.69232
Iteration 0:    Log likelihood = -136.69232  (not concave)
Iteration 1:    Log likelihood = -124.13044
Iteration 2:    Log likelihood = -113.99047
Iteration 3:    Log likelihood = -110.30732
Iteration 4:    Log likelihood = -110.26748
Iteration 5:    Log likelihood = -110.26736
Iteration 6:    Log likelihood = -110.26736
```

```
                                          Number of obs =        48
                                          Wald chi2(3)  =     35.25
Log likelihood = -110.26736               Prob > chi2   =    0.0000
```

	Coefficient	Std. err.	z	P>|z|	[95% conf. interval]	
drug						
Other	1.012966	.2903917	3.49	0.000	.4438086	1.582123
NA	1.45917	.2821195	5.17	0.000	.9062261	2.012114
age	-.0671728	.0205688	-3.27	0.001	-.1074868	-.0268587
_cons	6.060723	1.152845	5.26	0.000	3.801188	8.320259
/lngamma	.5573333	.1402154	3.97	0.000	.2825163	.8321504

These results match those reported in section 6.4.3.

15.2   Evaluator function prototypes

We have seen that method-lf evaluators receive three arguments, while method-lf2
evaluators receive six. In this section, we outline the arguments required for each type
of likelihood evaluator written in Mata.

Method-lf evaluators

void *lfeval*(transmorphic scalar M, real rowvector b, real rowvector *lnfj*)

input:
M:	problem handle
b:	$1 \times k$ coefficient vector

output:
lnfj:	$N \times 1$ log-likelihood vector

Notes:

- *lfeval* is the name of your evaluator function.

- k is the number of coefficients in the model, and N is the number of observations in the estimation subsample accounting for if, in, and missing values.

- *lnfj* is to be filled in with the observation-level log-likelihood values.

- Unlike ado-based method-lf evaluators, you are passed the coefficient vector b. Use moptimize_util_xb to extract the θ probability model parameters.

- To define your model using ml model, you type

      ml model lf *lfeval*() ...

   You must include the parentheses after the name of your Mata-based evaluator.

lf-family evaluators

void *lfxeval*(transmorphic scalar M, real scalar *todo*, real rowvector b,
      real rowvector *lnfj*, real matrix S, real matrix H)

input:
M:	problem handle
todo:	scalar *todo* = 0, 1, or 2
b:	$1 \times k$ coefficient vector

output:
lnfj:	$N \times 1$ log-likelihood vector
S:	$N \times E$ matrix of scores
H:	$k \times k$ Hessian matrix

Notes:

- *lfxeval* is the name of your evaluator function.

- Your function must accept all six arguments even if you are writing a method-lf0 or method-lf1 evaluator. Simply ignore the S and H arguments in your program if you are not required to fill them in.

- If you are implementing a method-lf1 or method-lf2 evaluator and $todo = 1$, you must fill in the $N \times E$ matrix S with the observation-level scores, where E is the number of equations in the model. The columns in S are ordered in the same way you specified the equations in your ml model statement. The (s, t) element of S is to contain $\partial \ln \ell_s / \partial \theta_{ts}$, where $s = 1, \ldots, N$ indexes observations and $t = 1, \ldots, E$ indexes equations.

- If you are implementing a method-lf2 evaluator and $todo = 2$, you must fill in the $k \times k$ matrix H with the Hessian matrix of the overall log-likelihood function. The (s, t) element of H is to contain $\partial^2 \ln L / \partial \beta_s \partial \beta_t$, where $s = 1, \ldots, k$ and $t = 1, \ldots, k$ index coefficients.

- To define your model using ml model, you type

    ml model lf0 *lfxeval()* ...

  or

    ml model lf1 *lfxeval()* ...

  or

    ml model lf2 *lfxeval()* ...

  depending on whether you have implemented an evaluator of method lf0, lf1, or lf2, respectively. You must include the parentheses after the name of your evaluator.

d-family evaluators

    void *deval*(transmorphic scalar M, real scalar *todo*, real rowvector b,
          real scalar *lnf*, real rowvector g, real matrix H)

    input:
            M: problem handle
        todo: scalar $todo = 0$, 1, or 2
            b: $1 \times k$ coefficient vector
    output:
          lnf: scalar containing log likelihood
            g: $1 \times k$ gradient vector
            H: $k \times k$ Hessian matrix

Notes:

- *deval* is the name of your evaluator function.

- Your function must accept all six arguments even if you are writing a method-d0 or method-d1 evaluator. Simply ignore the g and H arguments in your program if you are not required to fill them in.

- If you are implementing a method-**d1** or method-**d2** evaluator and *todo* = 1, you must fill in the $1 \times k$ vector g with the gradient of the overall log-likelihood function. Element s of g is to contain $\partial \ln L / \partial \beta_s$, where $s = 1, \ldots, k$ indexes coefficients.

- If you are implementing a method-**lf2** evaluator and *todo* = 2, you must fill in the $k \times k$ matrix H with the Hessian matrix of the overall log-likelihood function. The (s, t) element of H is to contain $\partial^2 \ln L / \partial \beta_s \partial \beta_t$, where $s = 1, \ldots, k$ and $t = 1, \ldots, k$ index coefficients.

- To define your model using **ml model**, you type

      ml model d0 *deval()* ...

  or

      ml model d1 *deval()* ...

  or

      ml model d2 *deval()* ...

  depending on whether you have implemented an evaluator of method **d0**, **d1**, or **d2**, respectively. You must include the parentheses after the name of your evaluator.

15.3   Utilities

When writing ado-file–based likelihood evaluators, utilities like **mleval**, **mlvecsum**, and **mlmatsum** are often useful, as are the macros **$ML_y1**, **$ML_y2**, and so on. Analogous functions exist for Mata-based likelihood evaluators. Here we focus on those functions that have analogues in the ado-language. For other functions that may be of interest, see [M-5] **moptimize( )** .

Table 15.1 lists the most popular utilities used in ado-based likelihood evaluators and their corresponding Mata functions. We then briefly discuss each function in turn, using the same notation as in section 15.2.

Table 15.1. Ado-based and Mata-based utilities

Ado-code	Mata function
$ML_y1, $ML_y2, ...	moptimize_util_depvar()
mleval	moptimize_util_xb()
	moptimize_util_eq_indices()
mlsum	moptimize_util_sum()
mlvecsum	moptimize_util_vecsum()
mlmatsum	moptimize_util_matsum()
mlmatbysum	moptimize_util_matbysum()

Dependent variables

In ado-based likelihood evaluators, the names of the dependent variables are stored in macros \$ML_y1, \$ML_y2, and so on. You can use these macros in, for example, `generate` statements to refer to the model's dependent variables.

In Mata-based likelihood evaluators, if you program

```
y1 = moptimize_util_depvar(M, 1)
```

then `y1` will be a column vector containing the observations on the first dependent variable. This vector is automatically restricted to the estimation subsample, so you need not do extra work to handle any `if` or `in` restrictions specified in the `ml model`. In this example, M is the problem handle, and the second argument represents the equation for which you want the dependent variable.

Obtaining model parameters

For all ado-based likelihood-evaluator types other than `lf`, you use the `mleval` command to obtain the θ model parameters from 'b'. With Mata-based likelihood evaluators, you instead use the `moptimize_util_xb()` function. Like the name suggests, this function computes $\theta_j = \mathbf{X}_j \beta_j$. For example, the code

```
xj = moptimize_util_xb(M, b, j)
```

creates column vector `xj` that contains the model parameters for the jth equation evaluated at the coefficient vector b. As with all the utility functions documented in this section, `moptimize_util_xb()` automatically restricts calculations to the relevant estimation sample.

When working with scalar parameters such as variances in homoskedastic models, you could specify the `scalar` option to `mleval`. Obtaining the same result in Mata is a two-step process. For example, say that your `ml model` statement is

```
ml model myeval() (mpg = gear_ratio turn) /lnsigma /theta
```

The command

```
j = moptimize_util_eq_indices(M, 1)
```

will create $j = [1, 3]$. The first element of `j` indicates that the coefficients for the first equation start at position 1 of b. The second element indicates that the coefficients for the first equation end at position 3 of b. That is, b[1], b[2], and b[3] are the coefficients for the first equation (corresponding to the variables `mpg` and `gear_ratio` and the constant, respectively).

The command

```
j = moptimize_util_eq_indices(M, 2)
```

will set $j = [4, 4]$, indicating that the scalar `lnsigma` is stored in the fourth element of b. Because j is a 1×2 vector, you can use range subscripts to extract the relevant portion of b. Hence, here you would program

```
j = moptimize_util_eq_indices(M, 2)
sigma = exp(b[|j|])
```

and work with the scalar `sigma`. Finally,

```
j = moptimize_util_eq_indices(M, 3)
```

will set $j = [5, 5]$, allowing you to extract the scalar `theta` as the fifth element of b.

Summing individual or group-level log likelihoods

With d-family evaluators, you return the overall log-likelihood function. As we saw with panel-data regression and the Cox model, we obtain a log likelihood for a group of observations; summing those groups' log likelihoods yields the overall log likelihood. In the simple case of linear regression, the overall log likelihood is simply the sum of each observation's individual log likelihood. For ado-based likelihood evaluators, the `mlsum` command is used to assist in computing the overall log likelihood.

With Mata-based evaluators, you use the `moptimize_util_sum()` function. For example,

```
lnf = moptimize_util_sum(M, v)
```

will sum the vector v to obtain the overall log likelihood, applying weights if specified, restricting the summation to the estimation subsample, and verifying that v contains no missing values. If v does contain missing values, `moptimize_util_sum()` returns a missing value, indicating that the overall log likelihood cannot be computed at the current value of the coefficient vector.

In fact, `moptimize_util_sum()` does not check that the number of rows in v equals the number of observations. For panel-data and other group-level log likelihoods, you could create v to have a number of rows equal to the number of panels and groups. However, if you do this, you cannot specify weights during estimation.

Calculating the gradient vector

The tool `mlvecsum` can be used in ado-based likelihood functions to assist in calculating the gradient vector $\partial \ln L / \partial \beta_i$ when the likelihood function meets the linear-form restrictions. The corresponding Mata utility is `moptimize_util_vecsum()`. The format for calling this utility is

```
gi = moptimize_util_vecsum(M, i, s, lnf)
```

M is again the problem handle. i is the equation number for which you are computing the gradient and corresponds to the `eq()` option of `mlvecsum`. s is a column vector that contains $\partial \ln \ell_j / \partial \theta_{ij}$. Argument *lnf* is the name of the scalar that stores the overall log likelihood of your model. Like `mlvecsum`, if `moptimize_util_vecsum()` cannot compute the gradient vector, likely because of missing values in s, it will set *lnf* to missing to indicate to `ml` that the current value of the coefficient vector is invalid. Weights are handled automatically, and the sum is restricted to the estimation subsample.

Calculating the Hessian

In section 6.2.5, we discussed the command `mlmatsum` to compute the Hessian matrix of a likelihood function that satisfies the linear-form restrictions. The corresponding Mata utility is `moptimize_util_matsum()` and has the syntax

```
Hmn = moptimize_util_matsum(M, m, n, d, lnf)
```

M is again the problem handle. Arguments m and n refer to the two equations whose cross-partial derivative you are computing and correspond to the two arguments in the `eq()` option of `mlmatsum`. Argument d is a vector containing $\partial^2 \ln \ell_j / \partial \theta_m \partial \theta_n$. Argument *lnf* is again the name of the scalar that stores the overall log likelihood of your model. If `moptimize_util_matsum()` cannot compute the Hessian, it will set *lnf* to missing. Weights are handled automatically, and the sum is restricted to the estimation subsample.

In section 7.4.3, we discussed using the `mlmatbysum` command to facilitate computation of the Hessian of models that have panel- or group-level log-likelihood functions. Terms on the (block) diagonal of the Hessian matrix have the form

$$\mathbf{H}_m = \sum_{i=1}^{N} \left(\sum_{t=1}^{T_i} a_{it} \right) \left(\sum_{t=1}^{T_i} b_{it} \mathbf{x}'_{mit} \right) \left(\sum_{t=1}^{T_i} b_{it} \mathbf{x}_{mit} \right)$$

where m refers to the equation whose submatrix of the Hessian you are computing. To form $\mathbf{H}_m$ using `mlmatbysum`, you would use

```
mlmatbysum 'lnf' 'Hm' varname_a varname_b, by(byvar) eq(m, m)
```

To perform the same operation using `moptimize_util_matbysum()`, you would code

```
moptimize_init_by(M, "byvar")
Hm = moptimize_util_matbysum(M, m, a, b, lnf)
```

M is again the problem handle, m is the equation number, a and b are column vectors that correspond to *varname_a* and *varname_b*, respectively, and *lnf* is the scalar that holds the overall log likelihood. *lnf* will be set to missing if the calculation cannot be completed. Weights and the estimation subsample are taken into account.

Before calling `moptimize_util_matbysum()`, you must call `moptimize_init_by()` to declare the panel or group identifier variable. As with virtually all `moptimize*()` functions, the first argument you pass to `moptimize_init_by()` is the problem handle. Technically, you need to declare the `byvar` only once to fit your model, but the easiest thing to do is just include the function call inside your likelihood evaluator. That will redeclare your *byvar* every time your likelihood is evaluated, but `moptimize_init_by()` is so fast that you will not notice a difference.

Terms on the off-diagonal of the Hessian matrix have the form

$$\mathbf{H}_{mn} = \sum_{i=1}^{N} \left(\sum_{t=1}^{T_i} a_{it} \right) \left(\sum_{t=1}^{T_i} b_{it} \mathbf{x}'_{1it} \right) \left(\sum_{t=1}^{T_i} c_{it} \mathbf{x}_{2it} \right)$$

To compute a matrix like that with `mlmatbysum`, you would use

    `mlmatbysum 'lnf' 'Hm'` *varname_a varname_b varname_c*, `by(`*byvar*`) eq(`*m*`, `*n*`)`

The corresponding `moptimize_util_matbysum()` call is

    `Hmn = moptimize_util_matbysum(`*M*`, `*m*`, `*n*`, `*a*`, `*b*`, `*c*`, `*lnf*`)`

Instead of specifying a single equation (m) as for elements on the diagonal of the Hessian, here you must specify two equations (m and n), and you must specify c, a row vector corresponding to c_{it} in the formula above and *varname_c* in `mlmatbysum`. If you have not previously called `moptimize_init_by()` within your likelihood evaluator, you will need to do so before using `moptimize_util_matbysum()`.

15.4   Random-effects linear regression

In section 7.4, we discussed likelihood functions for panel data, using the random-effects linear regression model as an example. Here we replicate that model using a Mata-based likelihood-evaluator function; this example allows us to use every Mata function in table 15.1. We will first present our Mata-based evaluator function in pieces to aid discussion. You may find it useful to bookmark section 7.4 so that you can flip back and forth to compare ado-code and Mata code.

15.4.1   Calculating lnf

The first part of our function, used to compute the likelihood function, is

```
void myrereg_d2_mata(transmorphic scalar M, real scalar todo,
                     real rowvector b, real scalar lnf,
                     real rowvector g, real matrix H)
{
        (Variable declarations not shown for brevity.)
        xb  = moptimize_util_xb(M, b, 1)
        j   = moptimize_util_eq_indices(M, 2)
        s_u = exp(b[|j|])
        j   = moptimize_util_eq_indices(M, 3)
        s_e = exp(b[|j|])

        y = moptimize_util_depvar(M, 1)
        z = y - xb

        // Create view of panels and set up panel-level processing
        st_view(panvar = ., ., st_global("MY_panel"))
        paninfo = panelsetup(panvar, 1)
        npanels = panelstats(paninfo)[1]

        lnfj = J(npanels, 1, 0)
        for(i=1; i <= npanels; ++i) {
                zi      = panelsubmatrix(z, i, paninfo)
                S_z2    = sum(zi:^2)
                Sz_2    = sum(zi)^2
                T       = rows(zi)
                a       = s_u^2 / (T*s_u^2 + s_e^2)
                lnfj[i] = - 0.5 * (                          ///
                            (S_z2 - a*Sz_2) / s_e^2 +        ///
                            ln(T*s_u^2/s_e^2 + 1) +          ///
                            T*ln(2*pi()*s_e^2) )
        }
        lnf = moptimize_util_sum(M, lnfj)
        if ((todo == 0) | (lnf == .)) return
```

Following the prototype in section 15.2 for a d-family evaluator, we name our function `myrereg_d2_mata()` and make it accept the requisite arguments. We then use `moptimize_util_xb()` to obtain the first equation (θ_1) of our model. Using the procedure discussed in section 15.3 for extracting scalar parameters from b, we then fill in s_u and s_e with the standard deviations of the panel-level and idiosyncratic error terms. We use `moptimize_util_depvar()` to retrieve our dependent variable and create the vector z to contain the residuals.

When writing likelihoods for panel-data models, you should be familiar with Mata's `panelsetup()` and related utilities; see [M-5] **panelsetup( )** . These utilities allow you to set up your data so that they can be processed one panel at a time. Particularly useful in this application is the `panelsubmatrix()` function, which allows us to obtain from a data matrix only those rows that correspond to the panel we are considering.

Because the likelihood function exists only at the panel level, we set up vector `lnfj` to have several rows equal to the number of panels in the dataset. We then loop over each panel of the dataset, calculating the panel-level likelihood as in (7.3). We use `moptimize_util_sum()` to sum the elements of `lnfj` to obtain the overall log likelihood, taking advantage of the fact that this function will work with a vector even if the vector does not have its number of rows equal to the number of observations in the estimation sample. Finally, we check to see if `ml` expects us to compute the gradient or Hessian and exit if we are not required to do so.

15.4.2 Calculating g

Our code for computing the gradient is

```
dxb = J(rows(panvar), 1, 0)
du  = 0
de  = 0
for(i=1; i <= npanels; ++i) {
        zi    = panelsubmatrix(z, i, paninfo)
        T     = rows(zi)
        a     = s_u^2 / (T*s_u^2 + s_e^2)
        S_zi = sum(zi)
        S_z2 = sum(zi:^2)
        Sz_2 = sum(zi)^2
        dxb[|paninfo[i,1] \ paninfo[i,2]|] = (zi :- a*S_zi)/s_e^2
        du = du + a^2*Sz_2 / s_u^2 - T*a
        de = de + S_z2 / s_e^2 - a*Sz_2 / s_e^2 -   ///
                a^2*Sz_2 / s_u^2 - T + 1 -          ///
                a*s_e^2 / s_u^2
        if ((du == .) | (de == .)) {
                lnf = .
                break
        }
}
        dxb = moptimize_util_vecsum(M, 1, dxb, lnf)
        g = (dxb, du, de)
        if ((todo == 1) | (lnf == .)) return
```

Examining (7.5), we see that $\partial \ln L_i / \partial \mathbf{x}_{it} \beta$ is observation specific, while from (7.6) and (7.7), we see that the derivatives with respect to the variance components are panel specific. We therefore create the vector `dxb` to have as its number of rows the number of observations in the estimation sample, which is easily gleaned by looking at the length of a vector that we know contains a row for each observation in the sample. We initialize `du` and `de` to be scalar zero; we will add each panel's contribution to those elements of the gradient ourselves.

As in computing the log-likelihood function, we loop over each panel individually. We use range subscripting (see [M-2] **Subscripts**) to fill in at once all the elements of `dxb` corresponding to the current panel. The function `panelsetup()` returns a $P \times 2$ matrix, where P is the number of panels. The first element of the ith row contains the number of the first row of data corresponding to panel i and the second element contains the number of the last row corresponding to panel i, making range subscripting simple. We add the current panel's contributions to the gradients with respect to the variance components. Importantly, because we are not using an `moptimize()` utility function to do the sums, we must check for ourselves that the gradient is valid (does not contain missing values); if it is not valid, we set `lnf` to missing and exit. We use `moptimize_util_vecsum()` to sum the contributions to the gradient with respect to $\mathbf{x}'_{it}\beta$. Finally, we assemble the gradient vector and exit if we are done.

15.4.3 Calculating H

The computations for the Hessian matrix proceed equally as those for the gradient, though of course the formulas are more complicated. Our code is virtually a direct translation into Mata of the ado-code from section 7.4.3. The Mata code is longer because the ado-based `mlmatsum` and `mlmatbysum` commands allow you to specify expressions, while the Mata utilities `moptimize_util_matsum()` and `moptimize_util_matbysum()` require you to create your own vectors before calling them.

Our code for computing the Hessian is

```
// Compute Hessian
avec  = J(rows(z), 1, 0)
d2u   = J(npanels, 1, 0)
d2e   = J(npanels, 1, 0)
dude  = J(npanels, 1, 0)
dxbdu = J(rows(z), 1, 0)
dxbde = J(rows(z), 1, 0)
for(i=1; i<= npanels; ++i) {
        zi   = panelsubmatrix(z, i, paninfo)
        S_zi = sum(zi)
        S_z2 = sum(zi:^2)
        Sz_2 = sum(zi)^2
        T    = rows(zi)
        a    = s_u^2 / (T*s_u^2 + s_e^2)

        avec[paninfo[i,2]] = a
        d2u[i] = -(2*a^2*Sz_2 / s_u^2 -                  ///
                  4*s_e^2*a^3*Sz_2 / s_u^4 +             ///
                  2*T*a^2*s_e^2 / s_u^2)
        d2e[i] = -(2*(S_z2 - a*Sz_2) / s_e^2 -           ///
                  2*a^2*Sz_2 / s_u^2 -                   ///
                  4*a^3*Sz_2*s_e^2 / s_u^4 +             ///
                  2*a*s_e^2 / s_u^2 -                    ///
                  2*a^2*s_e^4 / s_u^4)
        dude[i] = -(4*a^3*S_zi^2*s_e^2 / s_u^4 -         ///
                   2*T*a^2*s_e^2 / s_u^2)
        dxbdu[|paninfo[i,1] \ paninfo[i,2]|] =           ///
                J(rows(zi), 1, -2*a^2*S_zi / s_u^2)
        dxbde[|paninfo[i,1] \ paninfo[i,2]|] =           ///
                -(2*(zi :- a*S_zi)/s_e^2 :-              ///
                2*a^2*S_zi / s_u^2)
}
d2u   = moptimize_util_matsum(M, 2, 2, d2u, lnf)
d2e   = moptimize_util_matsum(M, 3, 3, d2e, lnf)
dude  = moptimize_util_matsum(M, 2, 3, dude, lnf)
dxbdu = moptimize_util_matsum(M, 1, 2, dxbdu, lnf)
dxbde = moptimize_util_matsum(M, 1, 3, dxbde, lnf)

moptimize_init_by(M, st_global("MY_panel"))
d2xb1 = moptimize_util_matbysum(M, 1, avec, J(rows(z), 1, 1), lnf)

d2xb2 = moptimize_util_matsum(M, 1, 1, J(rows(z), 1, 1), lnf)
d2xb  = -(d2xb2 - d2xb1) / s_e^2
```

```
H = (d2xb,    dxbdu, dxbde \              ///
     dxbdu', d2u,   dude  \              ///
     dxbde', dude,  d2e)
```

We created d2u, d2e, and dude to have their numbers of rows equal to the number of panels because they only have one value per panel. The component $a_i = \sigma_u^2/(T_i\sigma_u^2 + \sigma_e^2)$ has only one value per panel, but because we use the moptimize_util_matbysum() function, we need our avec vector to have the same length as the estimation sample. (We will admit we did not realize that until after discovering that our first draft of the program reported a conformability error.)

15.4.4   Results at last

Executing our complete do-file, we obtain

```
. version 18

. use http://www.stata-press.com/data/ml5/tablef7-1, clear

. mata: mata set matastrict on

. mata:
───────────────────────────────────────────── mata (type end to exit) ───────
: void myrereg_d2_mata(transmorphic scalar M, real scalar todo,
>                       real rowvector b,     real scalar lnf,
>                       real rowvector g,     real matrix H)
> {
>         // Likelihood
>         real scalar      a, i, npanels, s_e, s_u, S_z2, Sz_2, T
>         real colvector   j, lnfj, xb, panvar, y, z
>         real matrix      paninfo
>
>         // Gradient
>         real scalar      de, du, S_zi
>         real colvector   dxb, zi
>
>         // Hessian
>         real colvector   avec, d2e, d2u, dude, dxbdu, dxbde
>         real colvector   d2xb1, d2xb2, d2xb
>
>         xb  = moptimize_util_xb(M, b, 1)
>         j   = moptimize_util_eq_indices(M, 2)
>         s_u = exp(b[|j|])
>         j   = moptimize_util_eq_indices(M, 3)
>         s_e = exp(b[|j|])
>
>         y = moptimize_util_depvar(M, 1)
>         z = y - xb
>
>         // Create view of panels and set up panel-level processing
>         st_view(panvar = ., ., st_global("MY_panel"))
>         paninfo = panelsetup(panvar, 1)
>         npanels = panelstats(paninfo)[1]
>
>         lnfj = J(npanels, 1, 0)
>         for(i=1; i <= npanels; ++i) {
>                 zi       = panelsubmatrix(z, i, paninfo)
>                 S_z2     = sum(zi:^2)
```

```
>                     Sz_2    = sum(zi)^2
>                     T       = rows(zi)
>                     a       = s_u^2 / (T*s_u^2 + s_e^2)
>                 lnfj[i] = -0.5 * (                              ///
>                              (S_z2 - a*Sz_2) / s_e^2 +          ///
>                               ln(T*s_u^2/s_e^2 + 1) +           ///
>                               T*ln(2*pi()*s_e^2) )
>             }
>             lnf = moptimize_util_sum(M, lnfj)
>             if ((todo == 0) | (lnf == .)) return
>
>             // Compute the gradient
>             dxb = J(rows(panvar), 1, 0)
>             du  = 0
>             de  = 0
>             for(i=1; i <= npanels; ++i) {
>                     zi   = panelsubmatrix(z, i, paninfo)
>                     T    = rows(zi)
>                     a    = s_u^2 / (T*s_u^2 + s_e^2)
>                     S_zi = sum(zi)
>                     S_z2 = sum(zi:^2)
>                     Sz_2 = sum(zi)^2
>                     dxb[|paninfo[i,1] \ paninfo[i,2]|] = (zi :- a*S_zi)/s_e^2
>                     du = du + a^2*Sz_2 / s_u^2 - T*a
>                     de = de + S_z2 / s_e^2 - a*Sz_2 / s_e^2 -   ///
>                             a^2*Sz_2 / s_u^2 - T + 1 -          ///
>                             a*s_e^2 / s_u^2
>                     if ((du == .) | (de == .)) {
>                             lnf = .
>                             break
>                     }
>             }
>             dxb = moptimize_util_vecsum(M, 1, dxb, lnf)
>             g = (dxb, du, de)
>             if ((todo == 1) | (lnf == .)) return
>
>             // Compute Hessian
>             avec  = J(rows(z), 1, 0)
>             d2u   = J(npanels, 1, 0)
>             d2e   = J(npanels, 1, 0)
>             dude  = J(npanels, 1, 0)
>             dxbdu = J(rows(z), 1, 0)
>             dxbde = J(rows(z), 1, 0)
>             for(i=1; i<= npanels; ++i) {
>                     zi   = panelsubmatrix(z, i, paninfo)
>                     S_zi = sum(zi)
>                     S_z2 = sum(zi:^2)
>                     Sz_2 = sum(zi)^2
>                     T    = rows(zi)
>                     a    = s_u^2 / (T*s_u^2 + s_e^2)
>                     avec[paninfo[i,2]] = a
>                     d2u[i]  = -(2*a^2*Sz_2 / s_u^2 -            ///
>                                 4*s_e^2*a^3*Sz_2 / s_u^4 +      ///
>                                 2*T*a^2*s_e^2 / s_u^2)
>                     d2e[i]  = -(2*(S_z2 - a*Sz_2) / s_e^2 -     ///
>                                 2*a^2*Sz_2 / s_u^2 -            ///
>                                 4*a^3*Sz_2*s_e^2 / s_u^4 +      ///
>                                 2*a*s_e^2 / s_u^2 -             ///
>                                 2*a^2*s_e^4 / s_u^4)
```

```
>                    dude[i] = -(4*a^3*S_zi^2*s_e^2 / s_u^4 -        ///
>                            2*T*a^2*s_e^2 / s_u^2)
>                    dxbdu[|paninfo[i,1] \ paninfo[i,2]|] =         ///
>                            J(rows(zi), 1, -2*a^2*S_zi / s_u^2)
>                    dxbde[|paninfo[i,1] \ paninfo[i,2]|] =         ///
>                            -(2*(zi :- a*S_zi)/s_e^2 :-            ///
>                            2*a^2*S_zi / s_u^2)
>            }
>        d2u   = moptimize_util_matsum(M, 2, 2, d2u, lnf)
>        d2e   = moptimize_util_matsum(M, 3, 3, d2e, lnf)
>        dude  = moptimize_util_matsum(M, 2, 3, dude, lnf)
>        dxbdu = moptimize_util_matsum(M, 1, 2, dxbdu, lnf)
>        dxbde = moptimize_util_matsum(M, 1, 3, dxbde, lnf)
>        moptimize_init_by(M, st_global("MY_panel"))
>        d2xb1 = moptimize_util_matbysum(M, 1, avec, J(rows(z), 1, 1), lnf)
>        d2xb2 = moptimize_util_matsum(M, 1, 1, J(rows(z), 1, 1), lnf)
>        d2xb  = -(d2xb2 - d2xb1) / s_e^2
>
>        H = (d2xb,   dxbdu, dxbde \          ///
>            dxbdu', d2u,   dude  \          ///
>            dxbde', dude,  d2e)
> }
: end
```

```
. global MY_panel i
. sort $MY_panel
. ml model d2 myrereg_d2_mata() (xb: lnC = lnQ lnPF lf) /lns_u /lns_e
. ml maximize
Initial:      Log likelihood = -615.46011
Alternative:  Log likelihood = -315.38194
Rescale:      Log likelihood =  -242.2031
Rescale eq:   Log likelihood = -111.43007
Iteration 0:  Log likelihood = -111.43007  (not concave)
Iteration 1:  Log likelihood = -40.001214  (not concave)
Iteration 2:  Log likelihood =  49.947998  (not concave)
Iteration 3:  Log likelihood =  71.902601  (not concave)
Iteration 4:  Log likelihood =  89.534248
Iteration 5:  Log likelihood =  110.25833
Iteration 6:  Log likelihood =  111.13119
Iteration 7:  Log likelihood =  112.87175
Iteration 8:  Log likelihood =  114.72075
Iteration 9:  Log likelihood =  114.72904
Iteration 10: Log likelihood =  114.72904
```

```
                                          Number of obs  =       90
                                          Wald chi2(3)   = 11468.30
Log likelihood = 114.72904                Prob > chi2    =   0.0000
```

lnC	Coefficient	Std. err.	z	P>|z|	[95% conf. interval]	
lnQ	.90531	.0253759	35.68	0.000	.8555742	.9550459
lnPF	.4233757	.013888	30.48	0.000	.3961557	.4505957
lf	-1.064456	.1962308	-5.42	0.000	-1.449061	-.6798508
_cons	9.618648	.2066218	46.55	0.000	9.213677	10.02362
/lns_u	-2.170818	.3026641	-7.17	0.000	-2.764028	-1.577607
/lns_e	-2.828404	.077319	-36.58	0.000	-2.979947	-2.676862

15.5   Ado-file considerations

In chapter 13 and chapter 14, we showed how to write estimation commands that use `ml` to maximize log-likelihood functions. Accompanying your estimation command file is a second file that implements the log-likelihood function. When you write your log-likelihood function in Mata, you will need to compile that function and save it as a Mata object file using `mata mosave` or include it in a Mata library using `mata mlib`. Mata object files in particular are easy to use, so we will show that method here. In section 17.8, we show how to create a Mata library with multiple functions that are needed to implement an estimation command.

You may have seen or written an ado-file that includes Mata code to implement some functionality. Here is a short ado-file that implements the command `simple_mean`:

```
                                            ─── begin simple_mean.ado ───
// A very simple command
program simple_mean
        syntax varlist(min=1 max=1)
        mata: _simple_mean("`varlist'")
end
mata:mata set matastrict on
mata:
void _simple_mean(string scalar varname) {
        real scalar      m
        m = mean(st_data(., varname))
        printf("The mean of %s is %f.\n", varname, m)
}
end
exit
                                            ─── end simple_mean.ado ───
```

Our command accepts a single variable and displays its sample mean. Try it out if you like.

Unfortunately, we cannot use the same program structure and define our Mata-based log-likelihood evaluator at the end of our ado-file. When Mata functions are defined in an ado-file, they are visible only to other code within that same ado-file. There is no way to call our function `_simple_mean()` outside of using `simple_mean`. The upshot is that if we implement our log-likelihood evaluator in our ado-file, neither `ml` nor the Mata code underlying `ml` can access it.

The solution we will pursue here is to compile and save our function as a `.mo` file using `mata mosave`. We will use a do-file to define and save our log-likelihood evaluator, here for a simple linear regression estimator:

```
                                    begin make_mynormal_lf_m.do
clear all

mata:mata set matastrict on
mata:

void mynormal_lf_m(transmorphic scalar ML, real rowvector b,
                   real colvector lnfj) {

        real scalar    lnsigma
        real vector    depvar, xb

        depvar  = moptimize_util_depvar(ML, 1)
        xb      = moptimize_util_xb(ML, b, 1)
        lnsigma = moptimize_util_xb(ML, b, 2)

        lnfj = lnnormalden(depvar, xb, exp(lnsigma))
}
end

mata: mata mosave mynormal_lf_m(), replace
                                      end make_mynormal_lf_m.do
```

The key is the last line of the file. That will create (or replace) file **mynormal_lf_m.mo**, which contains a compiled version of our function. The next step is to run that do-file:

```
. clear all

. mata:mata set matastrict on

. mata:
                                             mata (type end to exit)
: void mynormal_lf_m(transmorphic scalar ML, real rowvector b,
>                    real colvector lnfj) {
>
>         real scalar    lnsigma
>         real vector    depvar, xb
>
>         depvar  = moptimize_util_depvar(ML, 1)
>         xb      = moptimize_util_xb(ML, b, 1)
>         lnsigma = moptimize_util_xb(ML, b, 2)
>
>         lnfj = lnnormalden(depvar, xb, exp(lnsigma))
> }
: end

. mata: mata mosave mynormal_lf_m(), replace
(file mynormal_lf_m.mo created)
```

Our do-file has created file **mynormal_lf_m.mo** as expected. Finally, here is the ado-file implementing our linear regression estimator that uses this log-likelihood function:

```
                                       begin mynormal_m.ado
program mynormal_m
        version 18
        syntax varlist [if] [in] [, vce(passthru) ]
        gettoken y rhs : varlist
        ml model lf mynormal_lf_m() (`y' = `rhs') () `if' `in', `vce'
        ml maximize
end
                                        end mynormal_m.ado
```

You may verify for yourself that the command works.

If you want to distribute an ML estimator for which you implemented the log-likelihood evaluator in Mata, you must provide both the estimation command's ado-file (`.ado`) and the Mata object file (`.mo`). You do not absolutely need to distribute the file like `make_mynormal_lf_m.do`, which creates the `.mo` file, though in the spirit of open-source code and reproducibility, we strongly encourage you to do so.

16 Mata's moptimize() function

In chapter 15, we showed how to write likelihood evaluators in Mata, and we used Stata's `ml` commands to set up our problem and obtain estimates. That division of labor is perfectly adequate for most estimators and is how many Stata estimation commands are implemented.

Perhaps you are instead writing a command for which obtaining maximum likelihood estimates is only one task among many that the command performs to obtain its results. Perhaps you are doing a Monte Carlo simulation and find that doing all of the data creation, estimation, and result-gathering is easier in Mata. Perhaps you are implementing some other type of estimator that requires numerically optimizing an objective function, and you want to use something akin to `ml`'s model notation. For all of these situations, you can use the `moptimize()` function. In fact, `ml` itself uses `moptimize()`.

`moptimize()` is not just for maximum likelihood estimation. It can be used for both function maximization (the default) or function minimization. The leading "m" in `moptimize` stands for "model", meaning that it uses the same parameter notation and equations as `ml` that we described in chapter 4. Here we cover just enough of `moptimize()` for you to begin using it on the types of examples we have illustrated throughout this book. For complete details, see appendix C and [M-5] **moptimize( )**.

This chapter assumes that you have read chapter 15 and are comfortable working with Mata. We reuse many of the examples, but the difference is that here we show how to define the problem and obtain estimates within Mata instead of using `ml`. Our examples here are written as do-files that load in a sample dataset, switch to Mata, and then execute Mata code. When doing the initial versions of your estimator, you will find that is the quickest way to get results.

In the penultimate section of this chapter, we discuss some of the issues you must consider when implementing an estimator using `moptimize()` rather than `ml`. In chapter 13, we showed how to write an ado-file to implement an estimation command using `ml`. We could write all the same estimators using `moptimize()` rather than `ml`. However, there is additional work required to specify the problem, control the estimation sample, account for weights and options, and post results, all of which are arguably easier to do in the `ml` world rather than in the `moptimize()` world. In the final section, we revisit the linear regression command we developed in chapter 13.

For straightforward estimators like those we have implemented in this book, writing the likelihood evaluator in Mata as covered in chapter 15 and using `ml` to control the maximization process represents the sweet spot among those who want the flexibility of Mata and the ease of use of `ml`. Hardcore Mata aficionados, this chapter is for you.

16.1   Introductory examples

16.1.1   The probit model

In section 5.2.1 and section 15.1.1, we fit a probit model of `died` on `i.drug` and `age`. To do the same using `moptimize()`, we create the following do-file:

```
————————————————————————————————————— begin myprobit_lf_mopt.do ———————
version 18
mata: mata clear
sysuse cancer, clear

mata:
void myprobit_lf(transmorphic scalar ML, real rowvector b,
                 real colvector lnfj)
{
        real vector depvar, xb

        depvar = moptimize_util_depvar(ML, 1)
        xb     = moptimize_util_xb(ML, b, 1)

        lnfj = depvar:*ln(normal(xb)) + (1:-depvar):*ln(normal(-xb))
}
ML = moptimize_init()
moptimize_init_evaluator(ML, &myprobit_lf())
moptimize_init_depvar(ML, 1, "died")
moptimize_init_eq_indepvars(ML, 1, "i.drug age")

moptimize(ML)

moptimize_result_display(ML)
end
————————————————————————————————————— end myprobit_lf_mopt.do ———————
```

The likelihood evaluator function `myprobit_lf()` is unchanged from what is in section 15.1.1. Writing the likelihood evaluator function when working with `moptimize()` is no different from writing a Mata-based likelihood evaluator function for use with `ml`. All the utilities we described in the previous chapter continue to work.

To begin our model definition, we coded

```
    ML = moptimize_init()
```

All model optimization problems begin by calling `moptimize_init()`, which takes no options. That function returns an object called a program handle that we assign to variable ML, though of course you can use a different name. All subsequent calls to `moptimize()` utilities will take that variable as their first argument. If you are declaring types for your variables, you will want to declare the variable returned by `moptimize_init()` as a `transmorphic scalar`. We cannot do so here because Mata does not allow us to declare types when executing statements interactively or directly from a do-file. We did, however, declare types when defining our likelihood evaluator function.

You may have noticed that once we set `ML`, none of the following statements actually assign results to new variables; there are no equal signs to be seen. Instead, as each function is called, it accesses and updates information stored in the program handle. The program handle stores all the information known about the problem at hand, and various functions are used to either add information to it or else retrieve information from it. To learn more about that style of programming, see chapter 10 of Gould (2018).

Next we called `moptimize_init_evaluator()`. This function takes two arguments, our program handle `ML` as well as the name of the function that defines our likelihood evaluator. Actually, the second argument to `moptimize_init_evaluator()` is a pointer to a function, which is why we preceded the name of the function with an ampersand (`&`) and included left and right parentheses at the end. If the previous sentence confused you, do not worry. You do not need to know anything about pointers to use `moptimize()`. Just know that with `moptimize_init_evaluator()`, you precede your evaluator function's name with an ampersand and include the left and right parentheses after the name.

`moptimize_init_depvar()` is how you declare dependent variables in your model. This function takes three arguments: the problem handle, the parameter number for which you are declaring the dependent variable, and the variable's name. Recall that with our `ml` terminology, shared by `moptimize()`, a "parameter" can be either a scalar or a linear combination of variables. Our model has just one dependent variable. If you have a more complicated model with multiple dependent variables, then you would call this function multiple times, as we do in the next example.

`moptimize_init_eq_indepvars()` is how you declare the independent variables for each parameter in your model. It again takes three arguments: the problem handle, the parameter to which you are referring, and the list of variables that make up the equation. As with `ml`, a constant term is included by default. Suppose for whatever reason that we did not want a constant term in this equation. Then we would use the function `moptimize_init_eq_cons()` to turn it off:

```
moptimize_init_eq_cons(ML, 1, "off")
```

For a model with multiple parameters, you would call this function multiple times as well.

With our model fully specified, we are ready to maximize the log-likelihood function. We do that by calling `moptimize()` with our problem handle. We have not specified initial values for any coefficients, so they are set to zero. `moptimize()` then performs a search for better initial values, much like `ml maximize` and `ml search`; recall the discussion in chapter 9. Once the search for starting values is done, `moptimize()` begins the iterative process to obtain the maximum of our likelihood evaluator function.

Executing our do-file, we obtain

```
. version 18

. mata: mata clear

. sysuse cancer, clear
(Patient survival in drug trial)

. mata:
───────────────────────────────────────────────── mata (type end to exit) ─────
: void myprobit_lf(transmorphic scalar ML, real rowvector b,
>                   real colvector lnfj)
> {
>         real vector depvar, xb
>         depvar = moptimize_util_depvar(ML, 1)
>         xb     = moptimize_util_xb(ML, b, 1)
>
>         lnfj = depvar:*ln(normal(xb)) + (1:-depvar):*ln(normal(-xb))
> }

: ML = moptimize_init()

: moptimize_init_evaluator(ML, &myprobit_lf())

: moptimize_init_depvar(ML, 1, "died")

: moptimize_init_eq_indepvars(ML, 1, "i.drug age")

: moptimize(ML)
Initial:      f(p) = -33.271065
Alternative:  f(p) = -31.427839
Rescale:      f(p) = -31.424556
Iteration 0:  f(p) = -31.424556
Iteration 1:  f(p) = -21.883453
Iteration 2:  f(p) = -21.710901
Iteration 3:  f(p) = -21.710799
Iteration 4:  f(p) = -21.710799

: moptimize_result_display(ML)
```

Number of obs = 48

died	Coefficient	Std. err.	z	P>|z|	[95% conf. interval]	
drug						
Placebo	0	(empty)				
Other	-1.941275	.597057	-3.25	0.001	-3.111485	-.7710644
NA	-1.792924	.5845829	-3.07	0.002	-2.938686	-.6471626
age	.0666733	.0410666	1.62	0.104	-.0138159	.1471624
_cons	-2.044876	2.284283	-0.90	0.371	-6.521988	2.432235

```
: end
```

The coefficient estimates and standard errors match those we obtained in sections 5.2.1, 15.1.1, and elsewhere. We used the utility function `moptimize_result_display()` to display the coefficient table. Unlike `ml display`, this function displays all the coefficients, including the zero coefficients on the base levels of factor variables. Later we will see how to transfer the results back to Stata so that you can use `ml display` or `ereturn display`.

❏ Technical note

The second line of our do-file clears all objects from Mata's memory, including any previously defined functions. If we tried to run our do-file multiple times and we did not include `mata: mata clear`, then the second time we ran our do-file, we would obtain an error message:

```
. version 18

. sysuse cancer, clear
(Patient survival in drug trial)

. mata:
──────────────────────────────────── mata (type end to exit) ────────
: void myprobit_lf(transmorphic scalar ML, real rowvector b,
myprobit_lf() already exists
(8 lines skipped)
──────────────────────────────────────────────────────────────────────
```

❏

16.1.2   The Weibull model

In section 15.1.2, we wrote a method-`lf2` likelihood evaluator in Mata for the Weibull model. Here we write a do-file that uses `moptimize()` to define the problem and obtain coefficient estimates:

```
───────────────────────── begin myweibull_lf2_mopt.do ─────────────
version 18
mata: mata clear
sysuse cancer, clear

mata:
void myweibull_lf2(transmorphic scalar ML, real scalar todo,
                   real rowvector b, real colvector lnfj,
                   real matrix S, real matrix H)
{
        real scalar lgam, p
        real vector leta, t, d, M, R
        real matrix H11, H12, H22

        leta = moptimize_util_xb(ML, b, 1)
        lgam = moptimize_util_xb(ML, b, 2)
        t    = moptimize_util_depvar(ML, 1)
        d    = moptimize_util_depvar(ML, 2)

        p = exp(lgam)
        M = (t:*exp(-1*leta)):^p
        R = ln(t) - leta
        lnfj = -M + d:*(lgam:-leta + (p-1):*R)
        if (todo == 0) return

        S = J(rows(leta), 2, .)
        S[.,1] = p*(M - d)
        S[.,2] = d - p*R:*(M - d)
        if (todo == 1) return

        H11 = moptimize_util_matsum(ML, 1, 1, p^2 :*M, lnfj[1])
        H12 = moptimize_util_matsum(ML, 1, 2, -p*(M - d + p*R:*M), lnfj[1])
        H22 = moptimize_util_matsum(ML, 2, 2, p*R:*(p*R:*M + M - d), lnfj[1])
        H = -1*(H11, H12 \ H12', H22)
}
```

```
ML = moptimize_init()
moptimize_init_evaluator(ML, &myweibull_lf2())
moptimize_init_evaluatortype(ML, "lf2")
moptimize_init_depvar(ML, 1, "studytime")
moptimize_init_depvar(ML, 2, "died")
moptimize_init_eq_indepvars(ML, 1, "i.drug age")
moptimize_init_eq_indepvars(ML, 2, "")
moptimize_init_eq_freeparm(ML, 2, "on")

moptimize(ML)

moptimize_result_display(ML)
end
```
———————————————————————————— end `myweibull_lf2_mopt.do` ————————

The likelihood evaluator function is unchanged from section 15.1.2. We created our problem handle named ML and then set our evaluator function. After that, we used `moptimize_init_evaluatortype()` to declare that our evaluator function is a method-lf2 one. We did not have to do this in the previous example because we used a method-lf evaluator, which is the default evaluator type with `moptimize()`.

As discussed in section 5.2.3, the Weibull model has two dependent variables: t_j, the exposure time for observation j, and d_j, a binary variable indicating whether observation j died. We therefore used `moptimize_init_depvar()` twice. Also recall that the Weibull model has both a scale parameter $\ln \eta_j$ and a shape parameter $\ln \gamma_j$ that here we are calling equations 1 and 2, respectively. In the first call to `moptimize_init_eq_indepvars()`, we parameterize $\ln \eta_j$ to be a function of drug type and patient age.

In this example, as is typical with Weibull regression, we are treating the shape parameter $\ln \gamma_j$ as a constant value across observations (for example, a scalar parameter). In our second call to `moptimize_init_eq_indepvars()`, we specified an empty string as our variable list, which is how we indicate to `moptimize()` that an equation contains only a constant term. In section 4.7, we discussed the concept of free parameters, parameters that are not parts of linear combinations. Declaring a parameter to be a free parameter affects the column labels of the estimated coefficient vector, and free parameters are displayed more compactly than constant-only equations in coefficient tables. The function `moptimize_init_eq_freeparm()` is how we tell `moptimize()` to treat our second equation as a free parameter: the third argument is either "on" or "off". By default, this setting is off for every equation. Setting it to on for an equation declares the parameter to be a free parameter.

Executing our do-file, we obtain

```
. version 18
. mata: mata clear
. sysuse cancer, clear
(Patient survival in drug trial)
. mata:
——————————————————————————————————————— mata (type end to exit) ———
: void myweibull_lf2(transmorphic scalar ML, real scalar todo,
>                    real rowvector b, real colvector lnfj,
>                    real matrix S, real matrix H)
> {
>        real scalar lgam, p
>        real vector leta, t, d, M, R
>        real matrix H11, H12, H22
>
>        leta = moptimize_util_xb(ML, b, 1)
>        lgam = moptimize_util_xb(ML, b, 2)
>        t    = moptimize_util_depvar(ML, 1)
>        d    = moptimize_util_depvar(ML, 2)
>
>        p = exp(lgam)
>        M = (t:*exp(-1*leta)):^p
>        R = ln(t) - leta
>        lnfj = -M + d:*(lgam:-leta + (p-1):*R)
>        if (todo == 0) return
>
>        S = J(rows(leta), 2, .)
>        S[.,1] = p*(M - d)
>        S[.,2] = d - p*R:*(M - d)
>        if (todo == 1) return
>
>        H11 = moptimize_util_matsum(ML, 1, 1, p^2 :*M, lnfj[1])
>        H12 = moptimize_util_matsum(ML, 1, 2, -p*(M - d + p*R:*M), lnfj[1])
>        H22 = moptimize_util_matsum(ML, 2, 2, p*R:*(p*R:*M + M - d), lnfj[1])
>        H = -1*(H11, H12 \ H12', H22)
> }
: ML = moptimize_init()
: moptimize_init_evaluator(ML, &myweibull_lf2())
: moptimize_init_evaluatortype(ML, "lf2")
: moptimize_init_depvar(ML, 1, "studytime")
: moptimize_init_depvar(ML, 2, "died")
: moptimize_init_eq_indepvars(ML, 1, "i.drug age")
: moptimize_init_eq_indepvars(ML, 2, "")
: moptimize_init_eq_freeparm(ML, 2, "on")
: moptimize(ML)
Initial:      f(p) =        -744
Alternative:  f(p) = -356.14276
Rescale:      f(p) = -200.80201
Rescale eq:   f(p) = -136.69232
Iteration 0:  f(p) = -136.69232  (not concave)
Iteration 1:  f(p) = -124.13044
Iteration 2:  f(p) = -113.99047
Iteration 3:  f(p) = -110.30732
Iteration 4:  f(p) = -110.26748
```

```
Iteration 5:   f(p) = -110.26736
Iteration 6:   f(p) = -110.26736

: moptimize_result_display(ML)
```

Number of obs = 48

	Coefficient	Std. err.	z	P>|z|	[95% conf.	interval]
drug						
Placebo	0	(empty)				
Other	1.012966	.2903917	3.49	0.000	.4438086	1.582123
NA	1.45917	.2821195	5.17	0.000	.9062261	2.012114
age	-.0671728	.0205688	-3.27	0.001	-.1074868	-.0268587
_cons	6.060723	1.152845	5.26	0.000	3.801188	8.320259
/eq2	.5573333	.1402154	3.97	0.000	.2825163	.8321504

```
: end
```

These results match those we obtained in section 5.2.3, section 6.4.3, section 15.1.2, and elsewhere.

16.2   Restricting the estimation sample

In section 4.5.1, we showed that if you wish to fit your model on a subsample of the data, you specify the restriction in the `ml model` statement using `if` or `in`. There are a couple of different ways to accomplish the same thing using `moptimize()`, an easy way that is appropriate for most applications and a slightly more involved way for more elaborate cases.

16.2.1   Using moptimize_init_touse()

We discussed writing ado-files in chapter 13, and we mentioned in section 13.5.2 that one of the first things you typically do in your code is parse the syntax and mark the estimation sample. For example, in your program that carries out estimation, you might have

```
syntax varlist(fv) [fweight pweight] [if] [in] [, ...]
marksample touse
```

The `marksample` command creates temporary variable 'touse', which contains 1 if an observation is to be used and 0 otherwise. Then when you use a command that works on Stata data, you append `if 'touse'` at the end to restrict attention to the estimation subsample. You also specify `if 'touse'` when you define the model using `ml model`.

If in your Stata code you have created 'touse', then you can specify that variable using the function `moptimize_init_touse()`. The only trick is that 'touse' is a temporary variable whose actual name may change each time you use your estimator, so you cannot hardcode 'touse' as your variable name. Instead, you can pass 'touse' as a string to your Mata function. An outline of your estimation command would be

```
program Estimate, eclass sortpreserve
        ...
        // Parse syntax
        syntax varlist(fv) [if] [in] ...

        // Mark the estimation sample
        marksample touse
        ...
        mata: myMataEstimator(..., "'touse'", ...)
        ...
end

mata:
void myMataEstimator(..., string scalar tousevar, ...)
{
        ...
        ML = moptimize_init()
        moptimize_init_touse(ML, tousevar)
        ...
}
```

When Stata executes the line `marksample touse`, it creates a temporary variable that we are to refer to as 'touse'. The single quotes should remind you of local macros: 'touse' might resolve to a variable with a name such as __000005, but that name may change each time you use your estimator. When we pass 'touse' as a string in our code, what actually gets passed is the name of that temporary variable.

The actual syntax of `moptimize_init_touse()` is quite simple. It takes as inputs our problem handle here called ML as well as the name of a Stata variable that marks the estimation sample. The name of that Stata variable is stored in the Mata string scalar `tousevar`.

16.2.2   Not using moptimize_init_touse()

When we used `moptimize_init_depvar()` and `moptimize_init_indepvar()` in the two examples at the start of this chapter, we specified strings containing one or more variables in our dataset; for variable **drug**, we even adorned it with factor-variable notation. When you use utilities like `moptimize_util_xb()` to get parameter values, `moptimize()` takes care of creating Mata views of your data and calculating linear combinations, accounting for the estimation subsample you may have specified using `moptimize_init_touse()`. When you specify Stata variable names with these functions for specifying data, `moptimize()` takes care of much tedious work for you automatically.

In some cases, you may want to work directly with matrices or view matrices containing your data. Perhaps you are performing a simulation and are creating data as matrices to which you will then apply your estimator. Both `moptimize_init_depvar()` and `moptimize_init_indepvar()` allow you to specify as the third argument the name of a vector or matrix in place of a string containing one or more variable names.

Our example of probit regression included the line

```
moptimize_init_eq_indepvars(ML, 1, "i.drug age")
```

You could replace that line with the following two,

```
X = st_data(., "i.drug age")
moptimize_init_eq_indepvars(ML, 1, X)
```

or, using a view matrix,

```
st_view(X=., ., "i.drug age")
moptimize_init_eq_indepvars(ML, 1, X)
```

and obtain the same coefficient estimates. If you were writing an estimator in Mata and your function accepted a string scalar `tousevar` as we described above, then you can control the estimation sample by specifying `tousevar` as the final argument to `st_data()` or `st_view()`:

```
st_view(X=., ., "i.drug age", tousevar)
moptimize_init_eq_indepvars(ML, 1, X)
```

If you specify some or all of your model's variables as vectors or matrices, you must ensure that they all have the same number of rows; otherwise, Mata will complain with a conformability error.

❏ **Technical note**

If you pass matrices to `moptimize_init_depvar()` and `moptimize_init_indepvar()` and intend for your estimator to display a standard coefficient table, then you must do more work. When you specify Stata variable names to those functions, `moptimize()` is able to create the proper row and column names for the estimated coefficient vector and covariance matrix. Then when you use `moptimize_result_display()` or, as we will show how later, `ml display`, the coefficient table will have the proper variable names, factor-variable labels, etc.

When you specify your data as matrices, `moptimize()` has no idea what the names of your variables are, so it uses generic names like `x1`, `x2`, and so on. Function `moptimize_init_eq_colnames()` allows you to set the names with which to label each parameter's coefficients, but this is an additional step you must take if you intend to display results in the standard Stata format.

❏

16.3   Estimation preliminaries

In this section, we describe some features of `moptimize()` that we have previously discussed in the context of `ml`. You call these functions before `moptimize()` to maximize your log-likelihood function. Later, we describe functions you use to obtain the results after maximization, including obtaining robust variance estimates.

16.3.1   Weights

Performing weighted estimation is straightforward as long as you use the utility functions provided as part of the `moptimize()` command. Before calling `moptimize()` to perform the actual maximization of your log-likelihood function, you first declare your weight variable and specify what type of weight it represents. You would code

```
moptimize_init_weight(ML,w)
moptimize_init_weighttype(ML,weighttype)
```

where here we assume you have called your problem handle `ML`. w represents the weight variable and can be either a string scalar containing a Stata variable name or a Mata column vector. If you specify a Stata variable name here, `moptimize()` will take care to select the subsample specified in `moptimize_init_touse()` (even if you specify the `tousevar` variable after declaring weights). If you specify a Mata column vector, you must ensure it conforms with the rest of your data. The default for w is "", yielding unweighted estimation. *weighttype* can be `"fweight"`, `"aweight"`, `"pweight"`, or `"iweight"`. If you specify a weight variable but do not declare what type of weight it represents, it is treated as an `fweight`.

In section 15.3, we introduced various Mata functions for calculating the types of sums you will encounter when writing your log-likelihood evaluator function. In particular, we mentioned four functions:

```
moptimize_util_sum()
moptimize_util_vecsum()
moptimize_util_matsum()
moptimize_util_matbysum()
```

As long as you use these functions to calculate sums in your log-likelihood function and in your gradient and Hessian calculations, `moptimize()` will calculate the appropriate weighted sums and perform weighted estimation. While a pertinacious programmer could make `moptimize()` perform weighted estimation without using these functions, we prefer to let these functions handle weights for us.

16.3.2   Panel data and clusters

To write panel-data maximum-likelihood estimators, you must tell `moptimize()` about your panel variable. You do that using `moptimize_init_by()`, which takes two arguments: the problem handle and a variable that identifies panel. That variable can be either a string scalar containing the name of a Stata variable (as you would specify with `xtset`) or a Mata column vector. Setting the panel variable enables you to use the function `moptimize_init_matbysum()` in your log-likelihood and derivative calculations. Also recall from section 7.4 that panel-data models require method-d0, method-d1, or method-d2 evaluators.

You can also specify a cluster variable so that you can compute the cluster–robust variance estimator by using the function `moptimize_init_cluster()`, which takes as arguments the problem handle and either a string scalar containing a Stata variable name or a Mata column vector. When you specify a cluster variable, the default is to compute the cluster–robust variance estimator.

Statistically, log-likelihood functions for panel-data models require that panels be nested within clusters. If you specify both a panel variable and a cluster variable, you must ensure that is the case.

16.3.3 Survey data

If you are writing an estimator for survey data, `moptimize()` will take care of setting the weights and obtaining the appropriate robust variance estimator for you. To have `moptimize()` inherit the full set of survey characteristics declared using `svyset`, include the following line in your code,

```
moptimize_init_svy(ML, "on")
```

assuming ML is again the name of your problem handle. If you do this, do not use the functions we described earlier in this section: avoid using `moptimize_init_weight()`, `moptimize_init_weighttype()`, or `moptimize_init_cluster()`. Do, however, use the utility functions we called out for computing sums in your log-likelihood function, gradient, and Hessian.

16.3.4 Initial values

In chapter 9, we described how `ml` searches for initial values, and we described the `ml search` command that does a more thorough search for initial values than `ml maximize` does on its own. That chapter also described `ml init` when you want to specify your own initial values. We emphasized that `ml maximize` and `ml search` usually do a good job of finding initial values, so unless you are willing to do extra work to obtain good initial values, `ml init` should be considered a last resort.

`moptimize()` provides similar functionality. By default, `moptimize()` initializes all parameters set to zero. It then performs a search to improve upon the initial values and follows up with rescaling them to obtain a further improvement. You can override the default values parameter by parameter using the function

```
moptimize_init_eq_coefs(ML, i, b0)
```

where we again use ML for the problem handle, i represents the parameter whose initial values of the coefficients you are providing, and $b0$ is a $1 \times (k_i + c_i)$ row vector. Here k_i is the number of independent variables in parameter i (which may be zero) and c_i equals one unless the parameter does not contain a constant, in which case c_i is zero.

We will have more to say about specifying initial values in the context of writing estimation commands in section 16.6.2.

Regarding `ml search`, we will make a couple of remarks and refer you to appendix C for details. The function to control whether `moptimize()` performs a search to improve on initial values is

```
moptimize_init_search(ML, {"on"|"off"})
```

By default a search is performed, just like with `ml maximize`.

As we mention in chapter 10, after `ml maximize` evaluates the log-likelihood function at the initial parameter vector, it next performs a random search for alternative initial values. You can request `moptimize()` do the same thing:

```
moptimize_init_search_random(ML, {"on"|"off"})
```

The default for `moptimize()` is to not perform this random search. There are additional functions that allow you to specify bounds for parameters and the number of times random variables are tried; see appendix C.

The final step `ml maximize` does before actually performing the iterative optimization technique is to rescale the coefficient vector. By default, `moptimize()` does the same thing. To override the default, you can use

```
moptimize_init_search_rescale(ML, {"on"|"off"})
```

16.4   Estimation

After setting up your problem by specifying the log-likelihood function, dependent and independent variables, and other features about your data and the problem at hand, you are now ready to perform the estimation. To do so, include the line

```
moptimize(ML)
```

assuming your problem handle is named `ML`. By default, `moptimize()` will display an iteration log much like `ml maximize`. The only difference is that each line will read, for example,

```
Iteration 0:  f(p) = ........
```

rather than

```
Iteration 0:  Log likelihood = ........
```

To make `moptimize()` display `Log likelihood` rather than `f(p)`, include the following statement, where `ML` is the problem handle, before calling `moptimize()`:

```
moptimize_init_valueid(ML, "Log likelihood")
```

16.5   Results

Once you have called `moptimize()` and it has succeeded, the next step is to get results. We will cover just a few Mata functions here, because in the next section, we will show an elegant way to pass results back to Stata that makes writing estimation commands much easier. If you are working on a problem in Mata and want to do your analysis after estimation in Mata as well, do not despair, because there are many `moptimize()` functions available to work with results. See appendix C and [M-5] **moptimize( )**.

16.5.1   Displaying results

In the introductory examples, we used one simple method of obtaining results. We used the following line to display them on screen:

```
moptimize_result_display(ML)
```

In section 1.2, we derived several ways in which we can estimate the variance–covariance matrix of the maximum-likelihood coefficient vector. Instead of the conventional variance estimator, we could use the outer product of gradients estimator. We could also compute a robust variance estimator that can provide correct inference even when we misspecify the likelihood function. If we have clustered data, we could compute a cluster–robust variance estimator to perform inference when observations are not independent within clusters.

If you want to display standard errors based on a different *vcetype*, then you can specify that *vcetype* when you call `moptimize_result_display()`. The full syntax is

`moptimize_result_display(ML, *vcetype*)`

where *vcetype* is the covariance matrix estimator you desire:

vcetype	Description
`""`	use default
`"oim"`	observed information matrix
`"opg"`	outer product of gradients
`"robust"`	Huber/White/sandwich estimator
`"svy"`	survey estimator; equivalent to `robust`

The default *vcetype* is `oim` unless you requested optimization technique `bhhh`, in which case the default is `opg`. If you specified `pweight`s or clusters as we illustrated earlier, then the default becomes `robust`. If you turned on survey settings using `moptimize_init_svy()`, then the default is `svy`.

Say we want to display our results using the robust covariance matrix estimator. The solution requiring the least amount of typing is

```
moptimize_result_display(ML, "robust")
```

though if you intend to work with the robust variance–covariance estimate later in Mata or return the robust variance–covariance estimate to Stata, see the next subsection.

Recall that you can obtain the `opg` or `robust` covariance matrix estimator only if you used a method-`lf`, method-`lf*`, or method-`gf` likelihood evaluator. Otherwise, the observation-level scores are not available, and they are needed to compute `opg` and `robust` covariance matrix estimators.

16.5.2   Retrieving results

In most cases, you want to store the estimated coefficient vector and its covariance matrix. Retrieving them is very easy. If you code

```
b = moptimize_result_coefs(ML)
V = moptimize_result_V(ML)
```

then row vector `b` will contain the vector of estimated coefficients, and `V` will contain the estimated covariance matrix based on the default *vcetype* we explained above.

If you want something other than the default *vcetype*, there are two ways to proceed. First, you can use `moptimize_init_vcetype()` before calling `moptimize()` to change the default to the type you want:

```
moptimize_init_vcetype(ML, vcetype)
...
moptimize(ML)
```

Alternatively, if you do not want to change to default or want to obtain another type in addition to the default, there are separate accessor functions for each type of covariance matrix available after estimation,

```
Vrobust = moptimize_result_V_robust(ML)
Voim = moptimize_result_V_oim(ML)
Vopg = moptimize_result_V_opg(ML)
```

where we have again assumed the problem handle is named `ML`.

`moptimize()` also has other accessor functions to retrieve coefficients and covariance matrices for individual parameters, to obtain scores, and to obtain the column stripe needed to produce output tables. In addition, you can check the iteration log, the final value of the log-likelihood function, and error codes if problems occurred. We do not describe them further here, but their usage is straightforward. See appendix C and [M-5] **moptimize( )** for details.

16.5.3   Storing results in e()

When you write an estimation command using `ml`, after `ml model` has successfully maximized the likelihood function, it automatically stores standard results in `e()` for you. In none of the commands we have written so far in this book have we had to use `ereturn post` to store the estimated coefficient vector, variance matrix, or estimation sample. `ml` made sure to do that for us.

If you are writing an estimation command using `moptimize()`, then you can have it store the standard results for you in `e()`, just as if you had used `ml` instead. After you have coded

```
moptimize(ML)
```

to carry out the estimation, include the line

```
moptimize_result_post(ML)
```

Doing that transfers the results into `e()` just as `ml` does. Once you have done that, you can then use Stata command `ml display` or `ereturn display` to display results just like any other command that stores results in `e()`.

When you call `moptimize_result_post()` with just the problem handle, it posts the default *vcetype*, which as we explained above depends on the optimization technique, the presence of `pweights`, and whether you turned on survey settings. How do you post, say, a robust covariance matrix instead? Again you have two alternatives. First, you could change the default *vcetype* for the current program before calling `moptimize()` using the function `moptimize_init_vcetype()` as we described above. Alternatively, `moptimize_result_post()` takes as an optional second argument the *vcetype* you wish to post, so you could instead code

```
moptimize(ML)
moptimize_result_post(ML,vcetype)
```

16.6   Estimation commands

In chapter 13, we showed how to write a complete estimation command in ado-code using `ml`, and in chapter 14, we showed how to amend the command to work with survey data. The final version of the command we wrote, `mynormal13`, has many features and options, some of which we did not bother to mention but work thanks to `ml`. All the features documented in [R] **Maximize** are available with `mynormal13`.

All the same features can also be implemented if you write your estimator using `moptimize()` instead of `ml`. Common options like `technique()` and `trace` require just a couple of lines of code. Handling the `from()` option and linear constraints specified using the `constraints` command is more tedious but straightforward because Stata has several utilities to aid you. Essentially, implementing those two options requires you to know exactly how the coefficient vector `e(b)` will be labeled after estimation—before you estimate it. We recommend implementing all other features of your estimation command and testing it before writing code for these two options. That way you can check to make sure `e(b)` is labeled the way you expect.

16.6.1 Common maximization options

Several maximization options allow you to turn a feature on that is by default turned off. For example, with an `ml` estimator you can specify options `trace`, `gradient`, and `hessian` to control the level of detail displayed in the iteration log. `moptimize()` has a set of functions you can call to turn on the same details for the iteration log:

```
moptimize_init_trace_coefs()
moptimize_init_trace_gradient()
moptimize_init_trace_Hessian()
```

Other maximization options like `iterate()` and `technique()` require you to specify a number or a string. There are also functions for specifying these options with `moptimize()`:

```
moptimize_init_conv_maxiter()
moptimize_init_technique()
```

You could parse these options yourself, pass them to Mata, and then use functions like those we just named to implement them.

There is an easier way. In writing our ado-code, we used command `mlopts` to check syntax and parse options related to maximization; see section 13.5.3. That command produced a cleaned list of options that we could pass on to `ml model`. With these simple steps, our estimation command supports all the options documented in [R] **Maximize**.

We can again use `mlopts` to produce a cleaned list of maximization options, pass the resulting list of options to Mata, and call function `moptimize_init_mlopts()` to take care of setting each individual option the user specified. In particular, that Mata function will set the following maximization options for you:

trace	gradient	hessian
showstep	nonrtolerance	showtolerance
difficult	technique()	iterate()
tolerance()	ltolerance()	nrtolerance()

`moptimize_init_mlopts()` will quietly ignore options not in that list. See `help mlopts` and `help mata moptimize_init_mlopts()`.

Expanding the outline we began in section 16.2.1, here is how we can handle all the options in that list in one fell swoop:

```
program Estimate, eclass sortpreserve
        ...
        // Parse syntax
        syntax varlist(fv) [if] [in] [, options_to_parse * ]

        // Mark the estimation sample
        marksample touse
        ...
        // Parse maximization options
        mlopts mlopts, 'options'
```

```
        mata: myMataEstimator(...,           ///
                              "'touse'",       ///
                              "'mlopts'",      ///
                              ...)
        ...
end

mata:
void myMataEstimator(...,
                     string scalar tousevar,
                     string scalar mlopts,
                     ...)
{
        ...
        ML = moptimize_init()
        moptimize_init_touse(ML, tousevar)
        moptimize_init_mlopts(ML, mlopts)
        ...
}
```

Program **Estimate** as written will accept the **constraints()** option, but it will do nothing with it. **Estimate** will not accept the **from()** option as written, and it will produce an error message if you attempt to do so. That is because we wrote

```
mlopts mlopts, 'options'
```

If we had instead written

```
mlopts mlopts otheropts, 'options'
```

then local macro **'otheropts'** would have contained the **from()** option if the user had specified it along with any other options not specified in *options_to_parse* or recognized by **mlopts**.

Later when we implement yet another estimator for linear regression, we will specify our **syntax** command to recognize **from()** and **constraints()** explicitly rather than rely on **mlopts** to capture them.

16.6.2   Initial values

We mentioned how to specify initial values in section 16.3.4, and we made it sound easy thanks to function **moptimize_init_eq_coefs()**.

There are a couple of complications that arise if you want your estimation command to accept the **from()** option. First, function **moptimize_init_mlopts()** does not handle option **from()**. That function could be called before you specify major aspects of your model such as the variables included or the number of equations. Without those details, it cannot parse what is specified in **from()**.

Second, with the **from()** option, a user can specify individual coefficients along with the values to which they should be initialized. Say we fit a simple probit model:

```
. sysuse auto
. probit foreign mpg gear, from(gear=2)
```

The `from()` option lets us specify an individual coefficient, an equal sign, and the value to which we want to initialize it. With `moptimize()`, we cannot specify initial values for individual coefficients; we can specify only initial values for parameters. Knowing that this particular specification has two independent variables and a constant term, we would have to specify a 1×3 vector if we were to use `moptimize_init_eq_coefs()`. In Mata, you could code

```
b = J(1, 3, 0)
b[1,2] = 2
moptimize_init_eq_coefs(ML, 1, b)
...
```

but as soon as the user specified different variables or different initial values, your program might not work anymore. A robust solution requires us to know how big the full coefficient vector will be, and we need to know how the columns will be labeled.

We added a third complication when we specified the `from()` option with `probit`. We specified `from(gear=2)`, but `gear` is not a variable in our dataset! The dataset does have a variable named `gear_ratio`, though. We also need to allow for variable name abbreviations.

The solution is to parse the `from()` option in Stata using the `_mkvec` command. That command processes the elements specified in `from()` and returns a coefficient vector with our initial values. The syntax of `_mkvec` is

_mkvec *matname* $\big[$, from(*init_specs* $\big[$, copy skip$\big]$) <u>up</u>date

    <u>col</u>names(*list_of_colfullnames*) first <u>err</u>or(*string*) $\big]$

where *init_specs* are the initial values supplied by the user and described in [R] **Maximize**. *matname* is the name of the initial value vector that `_mkvec` is to create and populate.

Options `copy` and `skip` are actually options one specifies in the `from()` option of the command being used. We do not specify these ourselves as developers. Option `copy` is specified when the user specifies a list of initial values like `from(1 2 3 4, copy)`, meaning that the coefficient vector is to be initialized to $(1, 2, 3, 4)$. Option `skip` specifies that the user may specify coefficients in *init_specs* that might not actually appear in the model; without option `skip`, the user would receive an error message if a coefficient was specified that did not appear in the model.

Option `first` controls what happens if a user specifies a coefficient name but does not include the equation name. With option `first`, if the user specifies `from(mpg=2)`, `_mkvec` will assume the user means `from(eq1:mpg=2)` or whatever the first equation is named.

Option `update` is for use when you have already created *matname* and want _mkvec
to update it rather than to create a new one. We will not use this option in our example
in the next section. You would use this option if your estimator first fit a simple model
(like a constant-only model) and then wanted to merge any initial values provided by
the user with the values obtained from your simple model.

Option `error()` allows you to customize the error message printed when the user
specifies a parameter not in the model or makes a syntactic error. We will just use the
default error message in our examples.

Calling _mkvec when the `from()` option is empty produces an initial value vector of
all zeros. That makes our code slightly simpler because then we do not have to test
whether the user specified the `from()` option, and we can write our Mata code so it
always receives initial values for all the parameters.

To use _mkvec, we need to know the full column names that our coefficient vector will
have; that is what we are to specify in _mkvec's `colnames()` option. In the next section,
we will write another command to implement linear regression, and to keep this version
relatively short, we will not allow the user to model conditional heteroskedasticity. Our
coefficient vector will have two equations, one for the parameter $\mathbf{x}'_j\boldsymbol{\beta}$ and one for $\ln\sigma$.
Following what `mynormal13` does, we are going to call the first equation `mu:`. The second
equation will be named `/:` because it contains just the free parameter `lnsigma`.

We need to handle the case where the user specifies one or more independent variables
that are adorned with factor-variable notation. For example, the user might specify
`i.rep78`, which expands to a set of four indicator variables as well as a base level. We
can use `fvexpand` to expand our list of independent variables before building our column
stripe; see `help fvexpand` for the details. The following code fragment will allow the
user to specify `from()`:

```
program Estimate, eclass sortpreserve

    syntax varlist(fv) [if] [in]                    ///
                [fweight pweight iweight] [,         ///
                noCONStant                           ///
                vce(passthru)                        ///
                FROM(string)                         ///  <- NEW
                *                                    ///
        ]
    ...
    // Build coefficient column stripe
    local mueq
    fvexpand `rhs'
    local rhs_list = r(varlist)
    foreach v in `rhs_list' {
            local mueq `mueq' mu:`v'
    }
    if "`constant'" == "" {
            local mueq `mueq' mu:_cons
    }
    local lseq /:lnsigma
    local alleq `mueq' `lseq'
```

```
                // _mkvec wants the full coefficient vector, but moptimize()
                // requires us to initialize each param. separately.
                tempname b0 b0_1 b0_2
                _mkvec 'b0', from('from') colnames('alleq')
                matrix 'b0_1' = 'b0'[1, "mu:"]
                matrix 'b0_2' = 'b0'[1, "/:"]
                ...
```

Say the user specifies a regression model with three independent variables: turn, gear_ratio, and weight. Here are just a few valid ways to specify the from() option:

```
            from(gear=2)
            from(gear=2 _cons=3 /lnsigma=1)
            from(1 2 3 4 5, copy)
            from(gear=2 rep78=5, skip)
```

Say the user left off the skip option in the last example:

```
            from(gear=2 rep78=5)
            extra parameter rep78 found
            specify skip option if necessary
            r(111);
```

You can use option error() to customize the error message if you wish.

16.6.3 Constraints

In section 4.5.6, we illustrated how to specify constraints with ml model. Our command mynormal13 from chapter 14 also accepts constraints:

```
    . use http://www.stata-press.com/data/r11/nhanes2a
    . svyset [pweight=leadwt], strata(stratid) psu(psuid)
  Sampling weights: leadwt
               VCE: linearized
       Single unit: missing
          Strata 1: stratid
  Sampling unit 1: psuid
             FPC 1: <zero>
    . constraint define 1 zinc = copper
```

```
. svy: mynormal13 loglead age female zinc copper, constraint(1)
(running mynormal13 on estimation sample)
Survey data analysis
Number of strata = 31                     Number of obs     =       4,345
Number of PSUs   = 62                     Population size = 99,646,837
                                          Design df         =          31
                                          F(3, 29)          =      310.60
                                          Prob > F          =      0.0000
```

(1) [mu]zinc - [mu]copper = 0

loglead	Coefficient	Linearized std. err.	t	P>|t|	[95% conf. interval]	
age	.0026388	.0004472	5.90	0.000	.0017268	.0035509
female	-.3919745	.0129038	-30.38	0.000	-.4182919	-.3656571
zinc	.000999	.0001974	5.06	0.000	.0005964	.0014015
copper	.000999	.0001974	5.06	0.000	.0005964	.0014015
_cons	2.453603	.0605497	40.52	0.000	2.330111	2.577095
/lnsigma	-.9926975	.0284634	-34.88	0.000	-1.050749	-.934646
sigma	.3705757	.0105479			.3496757	.3927249

We did no extra work to make constraints work. `mlopt` and `ml model` did all the work for us. With `moptimize()`, we have to do some of the work ourselves.

Before we dive into details of making `moptimize()` work with constraints specified using Stata's `constraints` command, we start with the mathematics of linear constraints. Suppose we have a linear regression model,

$$y_j = \mathbf{x}_j \boldsymbol{\beta} + \epsilon_j$$

where $\mathbf{x}_j$ is $1 \times k$ and $\boldsymbol{\beta}$ is $k \times 1$. For concreteness, we specify

$$\mathbf{x}_j = \begin{bmatrix} \mathtt{mpg}_j & \mathtt{gear}_j & \mathtt{weight}_j & 1 \end{bmatrix}$$

so that $\boldsymbol{\beta}$ is a 4×1 vector in this example.

Linear constraints can be written in the form $\mathbf{R}\boldsymbol{\beta} = \mathbf{r}$, where $\mathbf{R}$ is a $q \times k$ matrix, $\mathbf{r}$ is a $q \times 1$ vector, and q is the number of constraints. For example, say we want to specify the constraint that β_1, the coefficient on $\mathtt{mpg}_j$, is equal to 2. Then we have $q = 1$ constraint with

$$\mathbf{R} = \begin{bmatrix} 1 & 0 & 0 & 0 \end{bmatrix} \quad \text{and} \quad \mathbf{r} = \begin{bmatrix} 2 \end{bmatrix}$$

Now suppose we want to constrain the coefficients on $\mathtt{mpg}_j$, $\mathtt{gear}_j$, and $\mathtt{weight}_j$ to be equal. Now we have $q = 2$ constraints with

$$\mathbf{R} = \begin{bmatrix} 1 & -1 & 0 & 0 \\ 0 & 1 & -1 & 0 \end{bmatrix} \quad \text{and} \quad \mathbf{r} = \begin{bmatrix} 0 \\ 0 \end{bmatrix}$$

The first row of $\mathbf{R}$ and $\mathbf{r}$ implies that $\beta_1 = \beta_2$, and the second row implies that $\beta_2 = \beta_3$. Thus, we have constrained $\beta_1 = \beta_2 = \beta_3$ as desired.

If you can write your constraints using $\mathbf{R}$ and $\mathbf{r}$, then imposing those constraints during estimation is easy with `moptimize()`. Define new $q \times (k+1)$ matrix $\mathbf{C}$ as

$$\mathbf{C} \equiv \begin{bmatrix} \mathbf{R} & \mathbf{r} \end{bmatrix}$$

Then call function `moptimize_init_constraints()` before you call `moptimize()`,

```
moptimize_init_constraints(ML, C)
...
moptimize(ML)
```

where `ML` is the problem handle and `C` is the constraint matrix.

Your users will employ the command `constraint define` to define their linear constraints, and they will want to be able to specify the `constraints()` option of your command to impose the constraints during estimation. How do we go from a list of numbers the user specified in the `constraints()` option to the matrix $\mathbf{C}$ we need to provide in `moptimize_init_constraints()`? Fortunately, Stata has a utility that does all the hard work for us.

Command `makecns` takes a list of constraints and returns in `e(Cns)` the matrix $\mathbf{C}$ we need. When we processed the `from()` option, we needed to specify the column names we would use to label the coefficient vector in our call to `_mkvec`; without the column names, `_mkvec` could not determine whether an initial value made sense in the context of the model being fit. Similarly, `makecns` needs to know the coefficients appearing in the model so that it can determine whether a constraint can be imposed in the present model.

`makecns` looks in `e(b)` to obtain the column names before we have even estimated the coefficient vector. The actual values stored in `e(b)` do not matter, so we can use `ereturn post` to post a coefficient vector and variance matrix that contain all zeros or other initial values. Moreover, our code to handle the `from()` option above produces a properly labeled initial value vector, so we can just use that. The additional ado-code needed to process the `constraint()` option is

```
program Estimate, eclass sortpreserve

    syntax varlist(fv) [if] [in]                  ///
                  [fweight pweight iweight] [,     ///
                  noCONStant                       ///
                  vce(passthru)                    ///
                  FROM(string)                     ///
                  CONSTraints(string)              /// <- NEW
                  *                                ///
          ]
    ...
    // _mkvec wants the full coefficient vector, but moptimize()
    // requires us to initialize each param. separately.
    tempname b0 b0_1 b0_2
    _mkvec `b0', from(`from') colnames(`alleq')
    matrix `b0_1' = `b0'[1, "mu:"]
    matrix `b0_2' = `b0'[1, "/:"]
```

```
// Now that we have a coefficient vector with a column stripe,
// we can parse constraints if any.
if "`constraints'" != "" {
        tempname C V
                // To post we need V too
        matrix `V' = `b0''*`b0'
                // A dummy solution
        ereturn post `b0' `V'
        makecns `constraints'
        matrix `C' = e(Cns)
}
// If there are no constraints, "`C'" will be ""
...
```

In the next section, you will see the corresponding Mata code that receives `C` and acts accordingly.

16.7   Regression redux

Here we implement another program for classical normal linear regression, emphasizing the issues we addressed in the previous section. To keep the program short, in this version we do not implement the ability to model conditional heteroskedasticity, nor do we implement the survey features described in chapter 14. Finally, we do not bother with fitting a constant-only model to obtain starting values, because with normal linear regression, `moptimize()` is almost surely going to find feasible initial values on its own. As an exercise, you can add features to `mynormal14` so that it matches (or exceeds) the capabilities of `mynormal13`.

The following listing shows the code implementing our command `mynormal14`:

```
                                              begin mynormal14.ado
program mynormal14, properties(swml)
        version 18
        if replay() {
                if (`"`e(cmd)'"' != "mynormal14") error 301
                Replay `0'
        }
        else    Estimate `0'
end
program Estimate, eclass sortpreserve
        syntax varlist(fv) [if] [in]                ///
                        [fweight pweight iweight] [, ///
                        noLOG                        ///
                        noCONStant                   ///
                        vce(passthru)                ///
                        FROM(string)                 ///
                        CONSTraints(string)          ///
                        *                            ///
                ]
        // Mark estimation sample
        marksample touse
        // check syntax
```

```
        mlopts mlopts other, 'options'
        // parse Replay options
        _get_diopts diopts, 'other'
        gettoken lhs rhs : varlist
        _fv_check_depvar 'lhs'

        _vce_parse 'touse', opt(Robust oim opg) argopt(CLuster): 'wgt', 'vce'
        local vcespec  'r(vce)'
        local vceclust 'r(cluster)'

        // Build coefficient column stripe, accounting for
        local mueq
        fvexpand 'rhs'
        local rhs_list = r(varlist)
        foreach v in 'rhs_list' {
                local mueq 'mueq' mu:'v'
        }
        if "'constant'" == "" {
                local mueq 'mueq' mu:_cons
        }
        local lseq /:lnsigma
        local alleq 'mueq' 'lseq'

        // _mkvec wants the full coefficient vector, but moptimize()
        // requires us to initialize each param. separately.
        tempname b0 b0_1 b0_2
        _mkvec 'b0', from('from') colnames('alleq')
        matrix 'b0_1' = 'b0'[1, "mu:"]
        matrix 'b0_2' = 'b0'[1, "/:"]

        // Now that we have a coefficient vector with a column stripe,
        // we can parse constraints if any.
        if "'constraints'" != "" {
                tempname C V
                        // To post we need V too
                matrix 'V' = 'b0''*'b0'
                        // A dummy solution
                ereturn post 'b0' 'V'
                makecns 'constraints'
                matrix 'C' = e(Cns)
        }

        // This will hold the maximized log-likelihood value
        tempname ll
        mata: myMataRegress("'lhs'",      "'rhs'",      "'touse'",   ///
                        "'constant'", "'weight'",  "'wtvar'",   ///
                        "'vcespec'",  "'vceclust'", "'mlopts'",  ///
                        "'b0_1'",     "'b0_2'",     "'C'",       ///
                        "'log'",      "'ll'")
        ereturn scalar ll = 'll'
        ereturn local predict mynormal_p
        ereturn local cmd mynormal14

        Replay , 'diopts'
end

program Replay

        syntax [, *]
        ml display , 'options' diparm(/lnsigma, exp label("sigma"))

end
```

```
        // This allows us to type 'SS' rather than string scalar
        // when we declare string scalar variables.
local SS string scalar
mata:
mata set matastrict on
void mynormal_lf(transmorphic scalar ML, real rowvector b,
                 real colvector lnfj)
{
    real scalar lnsigma
    real vector depvar, xb

    depvar  = moptimize_util_depvar(ML, 1)
    xb      = moptimize_util_xb(ML, b, 1)
    lnsigma = moptimize_util_xb(ML, b, 2)

    lnfj    = ln(normalden(depvar, xb, exp(lnsigma)))
}
void myMataRegress('SS' lhs,       'SS' rhs,      'SS' touse,
                   'SS' noconst,   'SS' weight,   'SS' wtvar,
                   'SS' vcespec,   'SS' vceclust, 'SS' mlopts,
                   'SS' b0_1,      'SS' b0_2,     'SS' Cmat,
                   'SS' logopt,    'SS' lltemp) {

        transmorphic scalar ML

        ML = moptimize_init()
        moptimize_init_which(ML, "max")
        moptimize_init_evaluator(ML, &mynormal_lf())
        moptimize_init_evaluatortype(ML, "lf")
        moptimize_init_mlopts(ML, mlopts)
        moptimize_init_valueid(ML, "Log likelihood")

        moptimize_init_touse(ML, touse)
        moptimize_init_depvar(ML, 1, lhs)
        moptimize_init_eq_indepvars(ML, 1, rhs)
        if (noconst != "") {
                moptimize_init_eq_cons(ML, 1, "off")
        }
        moptimize_init_eq_indepvars(ML, 2, "")
        moptimize_init_eq_freeparm(ML, 2, "on")
        moptimize_init_eq_coefs(ML, 1, st_matrix(b0_1))
        moptimize_init_eq_coefs(ML, 2, st_matrix(b0_2))

        moptimize_init_eq_name(ML, 1, "mu")
        moptimize_init_eq_name(ML, 2, "lnsigma")

        if (wtvar != "") {
                moptimize_init_weighttype(ML, weight)
                moptimize_init_weight(ML, wtvar)
        }

        if (vceclust != "") {
                moptimize_init_cluster(ML, vceclust)
        }
        else if (vcespec != "") {
                moptimize_init_vcetype(ML, vcespec)
        }
```

```
                  if (Cmat != "") {
                          moptimize_init_constraints(ML, st_matrix(Cmat))
                  }

                  if (logopt != "") {
                          moptimize_init_tracelevel(ML, "none")
                  }

                  moptimize(ML)
                          // We changed the vcetype above, so we
                          // don't need to specify vcetype here.
                  moptimize_result_post(ML)
                          // Store the value of the log-likelihood function
                  st_numscalar(lltemp, moptimize_result_value(ML))
          }
   end
   exit
```
──────────────────────────────────── end mynormal14.ado ────────

Here we demonstrate `mynormal14`, specifying an initial value in the `from()` option and controlling the coefficients' number formatting:

```
. sysuse auto
(1978 automobile data)
. mynormal14 mpg gear weight i.rep78, from(/lnsigma=5) cformat(%9.3f)
Initial:     Log likelihood = -409.16982
Rescale:     Log likelihood =  -349.1548
Rescale eq:  Log likelihood =  -349.1548
Iteration 0: Log likelihood =  -349.1548   (not concave)
Iteration 1: Log likelihood = -245.02116   (not concave)
Iteration 2: Log likelihood = -216.14524   (not concave)
Iteration 3: Log likelihood = -209.50808   (not concave)
Iteration 4: Log likelihood = -202.15254
Iteration 5: Log likelihood = -192.12685
Iteration 6: Log likelihood = -180.95985
Iteration 7: Log likelihood = -180.87104
Iteration 8: Log likelihood = -180.87091
Iteration 9: Log likelihood = -180.87091
```

| Log likelihood = -180.87091 | | | | | Number of obs = 69 | |
```
```

mpg	Coefficient	Std. err.	z	P>|z|	[95% conf. interval]	
gear_ratio	0.901	1.484	0.61	0.544	-2.007	3.810
weight	-0.005	0.001	-5.85	0.000	-0.007	-0.003
rep78						
2	-0.383	2.640	-0.15	0.885	-5.557	4.791
3	-0.520	2.434	-0.21	0.831	-5.292	4.251
4	-0.751	2.497	-0.30	0.763	-5.645	4.142
5	2.037	2.598	0.78	0.433	-3.055	7.129
_cons	34.201	7.003	4.88	0.000	20.476	47.926
/lnsigma	1.202	0.085	14.12	0.000	1.036	1.369
sigma	3.328	0.283			2.817	3.932

We can tell that the `from()` option was parsed correctly by looking at the iteration log. When we run `mynormal14` (or `mynormal13`) without specifying the `from()` option, the first line of the iteration log reads

```
Initial:  Log likelihood =     -<inf>  (could not be evaluated)
```

However, when we initialize `/lnsigma` to be anything much over zero, the log-likelihood function can be computed. We also see that our request for three digits after the decimal point for coefficients, standard errors, and confidence intervals was respected.

Now let's test the `constraints()` option:

```
. sysuse auto
(1978 automobile data)

. constraint define 1 gear=5.rep78

. mynormal14 mpg gear weight i.rep78, constraints(1)
Initial:     Log likelihood =     -<inf>  (could not be evaluated)
Feasible:    Log likelihood =  -6014.006
Rescale:     Log likelihood = -343.25904
Rescale eq:  Log likelihood = -240.51981
Iteration 0: Log likelihood = -240.51981  (not concave)
Iteration 1: Log likelihood = -216.12167  (not concave)
Iteration 2: Log likelihood = -204.34083
Iteration 3: Log likelihood = -183.92934  (backed up)
Iteration 4: Log likelihood = -180.96125
Iteration 5: Log likelihood = -180.94097
Iteration 6: Log likelihood = -180.94096
```

```
Log likelihood = -180.94096                       Number of obs = 69

 ( 1)  [mu]1b.rep78 = 0
 ( 2)  [mu]gear_ratio - [mu]5.rep78 = 0
```

mpg	Coefficient	Std. err.	z	P>|z|	[95% conf. interval]	
gear_ratio	1.188382	1.272266	0.93	0.350	-1.305212	3.681977
weight	-.0050073	.0008351	-6.00	0.000	-.0066442	-.0033705
rep78						
2	-1.028437	2.001232	-0.51	0.607	-4.950779	2.893905
3	-1.213809	1.582459	-0.77	0.443	-4.315372	1.887754
4	-1.491213	1.52846	-0.98	0.329	-4.486941	1.504514
5	1.188382	1.272266	0.93	0.350	-1.305212	3.681977
_cons	33.75451	6.907482	4.89	0.000	20.2161	47.29293
/lnsigma	1.203394	.0851256	14.14	0.000	1.036551	1.370237
sigma	3.331405	.283588			2.819476	3.936285

The coefficients on `gear_ratio` and `5.rep78` are identical as we constrained them to be. The list of constraints shown above the coefficient table includes the base level of `rep78` to be constrained to be zero (because one level must be dropped to avoid perfect collinearity). You will not see that constraint listed there unless you do specify the `constraints()` option to impose additional constraints. With some more tedious but straightforward coding, we could make Stata hide that constraint from the output even when the user specifies `constraints()`.

❑ Technical note

To pass information between Stata and Mata, we wrote our Mata function to accept string arguments for each feature or option we need to process. We then call the Mata function from ado-code, passing each item in the corresponding order. We like this method because it makes clear what our Mata function requires of our ado-code.

For complicated estimators, the amount of information your Mata function requires can become quite lengthy. An alternative is to make the Mata function you call from Stata accept no arguments. Instead, your Mata function can use Stata access utilities like `st_local()` to obtain information stored in local macros you have set in your ado-code. The upside is that this does shorten your code, removing what can be a long Mata function call from within your ado-code and removing a long function definition in Mata. Moreover, adding features is somewhat easier because you do not need to change your function call or function definition to accommodate the additional information shared between Stata and Mata. The downside is that maintaining your code can be more difficult. When you are writing your code, you know the names of the local macros and other Stata objects your Mata code references, and you know how you pass information from Mata back to Stata. What happens six months or a year later? What if someone reports an obscure bug, and you must dive deep into your code to find the problem? Will you remember everything your Mata code accesses in Stata? What if you share your code with another user who wants to modify (after renaming) your command or make a bug fix?

By explicitly passing arguments between Stata and Mata, you know right away what information Stata expects your Mata code to use and vice versa. Convenience is often at odds with long-term maintainability.

❑

17 Other examples

This chapter contains example estimation commands. In each example, we derive the log-likelihood function, show the derivatives that make up the gradient and Hessian functions (where applicable), list one or more likelihood evaluators, provide a fully functional estimation command, and use the estimation command to fit the model to a dataset. Appendix E contains the ado-files for each estimation command discussed in this chapter.

17.1 The logit model

In the logit model, the dependent variable Y is an indicator variable, meaning that it takes on only the values 0 or 1. The probability mass function from the Bernoulli distribution (basically, the only distribution for random indicator variables) is

$$f(y_j; \pi_j) = \begin{cases} \pi_j & \text{if } y_j = 1 \\ 1 - \pi_j & \text{if } y_j = 0 \end{cases}$$

where $0 \leq \pi_j \leq 1$ and we identify π_j as the probability for a success (we arbitrarily call $y_j = 1$ a success). The logit model then uses the log of the odds ratio (the logit function) to describe the relationship between the independent variables $\mathbf{x}_j$ and π_j by

$$h^{-1}(\pi_j) = \ln \frac{\pi_j}{1 - \pi_j} = \mathbf{x}_j \boldsymbol{\beta}$$

which can be rewritten as

$$\pi_j = h(\mathbf{x}_j \boldsymbol{\beta}) = \frac{e^{\mathbf{x}_j \boldsymbol{\beta}}}{1 + e^{\mathbf{x}_j \boldsymbol{\beta}}} = \frac{1}{1 + e^{-\mathbf{x}_j \boldsymbol{\beta}}}$$

where $h(\cdot)$ is generically called the *link function*. This particular function is commonly referred to as the *logistic function* or *inverse logit function*. The log-likelihood function for the jth observation is

$$\ln \ell_j = \begin{cases} \ln h(\mathbf{x}_j \boldsymbol{\beta}) & \text{if } y_j = 1 \\ \ln h(-\mathbf{x}_j \boldsymbol{\beta}) & \text{if } y_j = 0 \end{cases}$$

Thus the logit model meets the linear-form restrictions. Our method-`lf` likelihood evaluator is

```
─────────────────────────────────────────── begin mylogit_lf.ado ───────────
program mylogit_lf
        version 18
        args lnf xb
        quietly replace 'lnf' = ln(invlogit( 'xb')) if $ML_y1 == 1
        quietly replace 'lnf' = ln(invlogit(-'xb')) if $ML_y1 == 0
end
                                            ──────────── end mylogit_lf.ado ───────────
```

Note that

$$\frac{\partial h(\mathbf{x}_j\boldsymbol{\beta})}{\partial \mathbf{x}_j\boldsymbol{\beta}} = h(\mathbf{x}_j\boldsymbol{\beta})h(-\mathbf{x}_j\boldsymbol{\beta})$$

Thus the derivatives of the log-likelihood function are

$$g_j = \frac{\partial \ln \ell_j}{\partial \mathbf{x}_j\boldsymbol{\beta}} = \begin{cases} h(-\mathbf{x}_j\boldsymbol{\beta}) & \text{if } y_j = 1 \\ -h(\mathbf{x}_j\boldsymbol{\beta}) & \text{if } y_j = 0 \end{cases}$$

$$H_j = \frac{\partial^2 \ln \ell_j}{\partial^2 \mathbf{x}_j\boldsymbol{\beta}} = -h(\mathbf{x}_j\boldsymbol{\beta})h(-\mathbf{x}_j\boldsymbol{\beta})$$

and our method-lf2 likelihood evaluator is

```
─────────────────────────────────────────── begin mylogit_lf2.ado ───────────
program mylogit_lf2
        version 18
        args todo b lnfj g1 H
        tempvar xb lj
        mleval 'xb' = 'b'
        quietly {
                generate double 'lj' = invlogit( 'xb')  if $ML_y1 == 1
                replace     'lj' = invlogit(-'xb')  if $ML_y1 == 0
                replace 'lnfj' = ln('lj')
                if ('todo'==0) exit

                replace 'g1' =  invlogit(-'xb')  if $ML_y1 == 1
                replace 'g1' = -invlogit( 'xb')  if $ML_y1 == 0
                if ('todo'==1) exit

                mlmatsum 'lnfj' 'H' = -1*abs('g1')*'lj', eq(1,1)
        }
end
                                            ──────────── end mylogit_lf2.ado ───────────
```

Our estimation command for this model is called `mylogit`; it is listed in appendix E.1. `mylogit` uses `mylogit_lf2` as its likelihood evaluator. Because `mylogit_lf2` computes the equation-level scores, `mylogit` allows all optimization algorithms, robust and outer product of the gradients (OPG) variance estimates, and the `svy` option. The maximum likelihood estimate (MLE) for π in the constant-only model is $\bar{p}$, the average number of positive outcomes observed in the estimation sample. Thus the MLE for β_0 is $\widehat{\beta}_0 = h^{-1}(\bar{p})$.

Here is `mylogit` in use:

```
. sysuse cancer
(Patient survival in drug trial)

. mylogit died i.drug age

Fitting full model:

Iteration 0:   Log likelihood = -31.199418
Iteration 1:   Log likelihood = -22.259892
Iteration 2:   Log likelihood = -21.539075
Iteration 3:   Log likelihood = -21.525583
Iteration 4:   Log likelihood = -21.525537
Iteration 5:   Log likelihood = -21.525537

My logit estimates                                  Number of obs =       48
                                                    LR chi2(3)    =    19.35
Log likelihood = -21.525537                         Prob > chi2   =   0.0002
```

died	Coefficient	Std. err.	z	P>|z|	[95% conf. interval]	
drug						
Other	-3.550783	1.222919	-2.90	0.004	-5.94766	-1.153905
NA	-3.249504	1.180797	-2.75	0.006	-5.563824	-.935184
age	.1203223	.0718246	1.68	0.094	-.0204513	.2610959
_cons	-3.625338	3.966999	-0.91	0.361	-11.40051	4.149838

```
. mylogit, or

My logit estimates                                  Number of obs =       48
                                                    LR chi2(3)    =    19.35
Log likelihood = -21.525537                         Prob > chi2   =   0.0002
```

died	Odds ratio	Std. err.	z	P>|z|	[95% conf. interval]	
drug						
Other	.0287022	.0351004	-2.90	0.004	.0026119	.3154027
NA	.0387934	.0458072	-2.75	0.006	.0038341	.3925136
age	1.12786	.0810081	1.68	0.094	.9797564	1.298352
_cons	.0266401	.1056813	-0.91	0.361	.0000112	63.42372

Note: _cons estimates baseline odds.

17.2   The probit model

Similarly to the logit model in the previous section, the dependent variable Y is an indicator variable; however, the probit model uses the standard normal cumulative distribution function $\Phi(\cdot)$ to link the independent variables $\mathbf{x}_j$ and the probability of success, π_j, by

$$\pi_j = \Phi(\mathbf{x}_j \boldsymbol{\beta})$$

The log-likelihood function for the jth observation is

$$\ln \ell_j = \begin{cases} \ln \Phi(\mathbf{x}_j \boldsymbol{\beta}) & \text{if } y_j = 1 \\ \ln \Phi(-\mathbf{x}_j \boldsymbol{\beta}) & \text{if } y_j = 0 \end{cases}$$

Thus the probit model meets the linear-form restrictions. Our method-`1f` likelihood evaluator is

```
 ──────────────────────────────────────── begin myprobit_1f.ado ────────
program myprobit_1f
        version 18
        args lnfj xb
        quietly replace `lnfj' = ln(normal( `xb')) if $ML_y1 == 1
        quietly replace `lnfj' = ln(normal(-`xb')) if $ML_y1 == 0
end
 ──────────────────────────────────────────── end myprobit_1f.ado ────────
```

Taking the derivatives of the log-likelihood function yields

$$
g_j = \frac{\partial \ln \ell_j}{\partial \mathbf{x}_j \boldsymbol{\beta}} = \left\{ \begin{array}{ll} \phi(\mathbf{x}_j\boldsymbol{\beta})/\Phi(\mathbf{x}_j\boldsymbol{\beta}) & \text{if } y_j = 1 \\ -\phi(\mathbf{x}_j\boldsymbol{\beta})/\Phi(-\mathbf{x}_j\boldsymbol{\beta}) & \text{if } y_j = 0 \end{array} \right.
$$

$$
H_j = \frac{\partial^2 \ln \ell_j}{\partial^2 \mathbf{x}_j \boldsymbol{\beta}} = -g_j(g_j + \mathbf{x}_j\boldsymbol{\beta})
$$

and our method-`1f2` likelihood evaluator is

```
 ──────────────────────────────────────── begin myprobit_1f2.ado ────────
program myprobit_1f2
        version 18
        args todo b lnfj g1 H
        tempvar xb lj
        mleval `xb' = `b'
        quietly {
                // Create temporary variable used in both likelihood
                // and scores
                generate double `lj'  = normal( `xb')  if $ML_y1 == 1
                replace     `lj'  = normal(-`xb')  if $ML_y1 == 0
                replace     `lnfj' = log(`lj')
                if (`todo'==0) exit

                replace `g1' =  normalden(`xb')/`lj'  if $ML_y1 == 1
                replace `g1' = -normalden(`xb')/`lj'  if $ML_y1 == 0
                if (`todo'==1) exit

                mlmatsum `lnfj' `H' = -`g1'*(`g1'+`xb'), eq(1,1)
        }
end
 ──────────────────────────────────────────── end myprobit_1f2.ado ────────
```

Our estimation command for this model is called `myprobit` and is listed in appendix E.2. `myprobit` uses `myprobit_1f2` as its likelihood evaluator. Because `myprobit_1f2` computes the equation-level scores, `myprobit` allows all optimization algorithms, robust and OPG variance estimates, and the `svy` option. As with the logit model, the MLE for π in the constant-only model is $\bar{p}$, so the MLE for β_0 is $\widehat{\beta}_0 = \Phi^{-1}(\bar{p})$.

Here is `myprobit` in use:

```
. use http://www.stata-press.com/data/r11/nhanes2a

. svyset [pweight=leadwt], strata(stratid) psu(psuid)

Sampling weights: leadwt
             VCE: linearized
     Single unit: missing
        Strata 1: stratid
   Sampling unit 1: psuid
           FPC 1: <zero>

. svy: myprobit heartatk diabetes female age
(running myprobit on estimation sample)

Survey: My probit estimates
```

```
Number of strata = 31                    Number of obs   =      4,946
Number of PSUs   = 62                     Population size = 112,861,252
                                          Design df       =         31
                                          F(3, 29)        =      57.62
                                          Prob > F        =     0.0000
```

heartatk	Coefficient	Linearized std. err.	t	P>|t|	[95% conf. interval]	
diabetes	.1375559	.1223278	1.12	0.269	-.1119333	.3870451
female	-.2875907	.0928203	-3.10	0.004	-.476899	-.0982825
age	.0383615	.0030946	12.40	0.000	.0320501	.044673
_cons	-3.741134	.1901302	-19.68	0.000	-4.128907	-3.353361

```
. estat effects, deft
```

heartatk	Coefficient	Linearized std. err.	DEFT
diabetes	.1375559	.1223278	.737652
female	-.2875907	.0928203	1.09954
age	.0383615	.0030946	1.16025
_cons	-3.741134	.1901302	1.28455

17.3 Normal linear regression

In classical normal linear regression, the dependent variable Y is assumed to be normally distributed. We denote the mean and standard deviation of the jth observation with $\mu_j = \mathbf{x}_j\boldsymbol{\beta}$ and $\sigma_j > 0$, respectively. The probability density function is

$$f(y; \mu_j, \sigma_j) = \frac{1}{\sigma_j\sqrt{2\pi}} \exp\left\{ -\frac{1}{2}\left(\frac{y - \mu_j}{\sigma_j}\right)^2 \right\} = \frac{1}{\sigma_j}\phi\left(\frac{y - \mu_j}{\sigma_j}\right)$$

where $\phi(\cdot)$ is the probability density function of the standard normal distribution. The log likelihood for the jth observation is

$$\ln \ell_j = \ln \phi \left(\frac{y_j - \mu_j}{\sigma_j} \right) - \ln \sigma_j$$

Thus the linear regression model meets the linear-form restrictions. We will model the standard deviation in the log space,

$$\theta_{2j} = \ln \sigma_j$$

so our method-1f likelihood evaluator is

```
                                              begin mynormal_lf.ado
program mynormal_lf
        version 18
        args lnf mu lnsigma
        quietly replace 'lnf' = ln(normalden($ML_y1,'mu',exp('lnsigma')))
end
                                                end mynormal_lf.ado
```

The derivative of $\phi(\cdot)$ is

$$\phi'(z) = -z\phi(z)$$

so taking derivatives of the log-likelihood function yields

$$g_{1j} = \frac{\partial \ln \ell_j}{\partial \mu_j} = \frac{y_j - \mu_j}{\sigma_j^2} \qquad\qquad H_{11j} = \frac{\partial^2 \ell_j}{\partial^2 \mu_j} = -\frac{1}{\sigma_j^2}$$

$$g_{2j} = \frac{\partial \ln \ell_j}{\partial \ln \sigma_j} = \left(\frac{y_j - \mu_j}{\sigma_j} \right)^2 - 1 \qquad\qquad H_{22j} = \frac{\partial^2 \ell_j}{\partial^2 \ln \sigma_j} = -2 \left(\frac{y_j - \mu_j}{\sigma_j} \right)^2$$

$$H_{12j} = \frac{\partial^2 \ell_j}{\partial \mu_j \, \partial \ln \sigma_j} = -2 \frac{y_j - \mu_j}{\sigma_j^2}$$

and our method-1f2 likelihood evaluator is

```
                                              begin mynormal_lf2.ado
program mynormal_lf2
        version 18
        args todo b lnfj g1 g2 H
        tempvar mu lnsigma sigma
        mleval 'mu' = 'b', eq(1)
        mleval 'lnsigma' = 'b', eq(2)
        quietly {
                generate double 'sigma' = exp('lnsigma')
                replace 'lnfj' = ln( normalden($ML_y1,'mu','sigma') )
                if ('todo'==0) exit

                tempvar z
                generate double 'z' = ($ML_y1-'mu')/'sigma'
                replace 'g1' = 'z'/'sigma'
                replace 'g2' = 'z'*'z'-1
                if ('todo'==1) exit

                tempname d11 d12 d22
```

```
                    mlmatsum 'lnfj' 'd11' = -1/'sigma'^2       , eq(1)
                    mlmatsum 'lnfj' 'd12' = -2*'z'/'sigma'     , eq(1,2)
                    mlmatsum 'lnfj' 'd22' = -2*'z'*'z'         , eq(2)
                    matrix 'H' = ('d11', 'd12' \ 'd12'', 'd22')
            }
    end
```
———————————————————————————————— end mynormal_lf2.ado ————————

Our estimation command for this model is called **mynormal** and is listed in appendix E.3. **mynormal** uses **mynormal_lf2** as its likelihood evaluator. Because **mynormal_lf2** computes the equation-level scores, **mynormal** allows all optimization algorithms, robust and OPG variance estimates, and the **svy** option. The MLEs for the constant-only model are

$$\widehat{\mu} = \overline{y} = \frac{1}{N} \sum_{j=1}^{N} y_j \quad \text{and} \quad \widehat{\sigma} = \left\{ \frac{1}{N} \sum_{j=1}^{N} (y_j - \overline{y})^2 \right\}^{1/2}$$

Here is **mynormal** in use:

```
. use http://www.stata-press.com/data/r11/nhanes2a

. svyset [pweight=leadwt], strata(stratid) psu(psuid)

Sampling weights: leadwt
            VCE: linearized
    Single unit: missing
        Strata 1: stratid
 Sampling unit 1: psuid
           FPC 1: <zero>

. svy: mynormal loglead age female
(running mynormal on estimation sample)

Survey: My normal estimates

Number of strata = 31            Number of obs     =        4,948
Number of PSUs   = 62            Population size = 112,915,016
                                 Design df         =           31
                                 F(2, 30)          =       625.61
                                 Prob > F          =       0.0000
```

loglead	Coefficient	Linearized std. err.	t	P>|t|	[95% conf. interval]	
age	.0024985	.0004326	5.78	0.000	.0016162	.0033807
female	-.3648611	.0101531	-35.94	0.000	-.3855684	-.3441537
_cons	2.663709	.0295542	90.13	0.000	2.603432	2.723985
/lnsigma	-.9892149	.0294101	-33.64	0.000	-1.049197	-.9292325

```
. estat effects, deff meff
```

loglead	Coefficient	Linearized std. err.	DEFF	MEFF
age	.0024985	.0004326	1.63378	1.96588
female	-.3648611	.0101531	.921835	.920059
_cons	2.663709	.0295542	3.52743	3.25028
/lnsigma	-.9892149	.0294101	6.40399	8.5596

17.4   The Weibull model

In the Weibull model, the dependent variable measures time, $T > 0$, until failure (such as death or some other specific event). The Weibull density function is

$$f(t_j; \eta_j, \gamma_j) = \frac{\gamma_j}{\eta_j} \left(\frac{t_j}{\eta_j}\right)^{\gamma_j - 1} \exp\left\{-\left(\frac{t_j}{\eta_j}\right)^{\gamma_j}\right\}$$

with $\eta_j > 0$ and $\gamma_j > 0$. Because the Weibull model is often used in survival analysis, we should also account for censoring, for example, the last time the subject came under observation when failure was not observed during the study.

We need a variable to distinguish between observed failures and censored observations, so let $d_j = 1$ indicate that t_j is the time of failure and $d_j = 0$ indicate that t_j is the time of censoring. If an observation is censored, we know that the time of failure is greater than the time of censoring.

$$\Pr(T > t_j) = \int_{t_j}^{\infty} \frac{\gamma_j}{\eta_j} \left(\frac{s}{\eta_j}\right)^{\gamma_j - 1} \exp\left\{-\left(\frac{s}{\eta_j}\right)^{\gamma_j}\right\} ds$$

$$= \exp\left\{-\left(\frac{t_j}{\eta_j}\right)^{\gamma_j}\right\}$$

Thus the log likelihood for the jth observation is

$$\ln \ell_j = \begin{cases} -(t_j/\eta_j)^{\gamma_j} + \ln \gamma_j - \ln \eta_j + (\gamma_j - 1)(\ln t_j - \ln \eta_j) & \text{if } d_j = 1 \\ -(t_j/\eta_j)^{\gamma_j} & \text{if } d_j = 0 \end{cases}$$

with $\eta_j > 0$ and $\gamma_j > 0$, and this Weibull model meets the linear-form restrictions.

We will model both η_j and γ_j in the log space. To emphasize this, we rewrite the log likelihood as

$$\ln \ell_j = \begin{cases} -(t_j e^{-\theta_{1j}})^{e^{\theta_{2j}}} + \theta_{2j} - \theta_{1j} + (e^{\theta_{2j}} - 1)(\ln t_j - \theta_{1j}) & \text{if } d_j = 1 \\ -(t_j e^{-\theta_{1j}})^{e^{\theta_{2j}}} & \text{if } d_j = 0 \end{cases} \quad (17.1)$$

$$\theta_{1j} = \ln \eta_j = \mathbf{x}_{1j}\boldsymbol{\beta}_1$$
$$\theta_{2j} = \ln \gamma$$

and our method-`lf` likelihood evaluator is

```
————————————————————————————————— begin myweibull_lf.ado —————————
program myweibull_lf
        version 18
        args lnfj leta lgam
        tempvar p M R
        quietly {
                generate double 'p' = exp('lgam')
                generate double 'M' = ($ML_y1*exp(-'leta'))^'p'
                generate double 'R' = ln($ML_y1)-'leta'
                replace 'lnfj' = -'M' + $ML_y2*('lgam'-'leta'+('p'-1)*'R')
        }
end
—————————————————————————————————— end myweibull_lf.ado —————————
```

where `$ML_y1` is the time variable t_j and `$ML_y2` is the censor variable d_j. We use temporary variables to evaluate some of the terms in the log likelihood. This is not necessary, but it makes the program slightly more readable and keeps us from having to evaluate `exp('lgam')` more than once. Remember, those temporary variables were generated as `doubles`.

Before we take derivatives of the log likelihood, let's get better acquainted with the above intermediate results to help us with notation. Using the notation in (17.1), we have

$$p_j = e^{\theta_{2j}}$$
$$R_j = \ln t_j - \theta_{1j}$$
$$M_j = (t_j e^{-\theta_{1j}})^{e^{\theta_{2j}}}$$

so our log likelihood is

$$\ln \ell_j = d_j\{\theta_{2j} - \theta_{1j} + (p_j - 1)R_j\} - M_j$$

The derivatives of p_j, R_j, and M_j are

$$\frac{\partial p_j}{\partial \theta_{1j}} = 0 \qquad\qquad \frac{\partial p_j}{\partial \theta_{2j}} = p_j$$

$$\frac{\partial R_j}{\partial \theta_{1j}} = -1 \qquad\qquad \frac{\partial R_j}{\partial \theta_{2j}} = 0$$

$$\frac{\partial M_j}{\partial \theta_{1j}} = -p_j M_j \qquad\qquad \frac{\partial M_j}{\partial \theta_{2j}} = R_j p_j M_j$$

Thus the derivatives of the log-likelihood function are

$$g_{1j} = \frac{\partial \ln \ell_j}{\partial \theta_{1j}} = p_j(M_j - d_j) \qquad\qquad H_{11j} = \frac{\partial^2 \ln \ell_j}{\partial^2 \theta_{1j}} = -p_j^2 M_j$$

$$g_{2j} = \frac{\partial \ln \ell_j}{\partial \theta_{2j}} = d_j - R_j p_j(M_j - d_j) \quad H_{22j} = \frac{\partial^2 \ln \ell_j}{\partial^2 \theta_{2j}} = -p_j R_j(R_j p_j M_j + M_j - d_j)$$

$$H_{12j} = \frac{\partial^2 \ln \ell_j}{\partial \theta_{1j}\partial \theta_{2j}} = p_j(M_j - d_j + R_j p_j M_j)$$

and our method-lf2 likelihood evaluator is

——————————————————————————————————— begin `myweibull_lf2.ado` ————————

```
program myweibull_lf2
        version 18
        args todo b lnf g1 g2 H
        tempvar leta lgam p M R
        mleval 'leta' = 'b', eq(1)
        mleval 'lgam' = 'b', eq(2)
        local t "$ML_y1"
        local d "$ML_y2"
        quietly {
                generate double 'p' = exp('lgam')
                generate double 'M' = ('t'*exp(-'leta'))^'p'
                generate double 'R' = ln('t')-'leta'
                replace 'lnf' = -'M' + 'd'*('lgam'-'leta' + ('p'-1)*'R')
                if ('todo'==0) exit

                replace 'g1' = 'p'*('M'-'d')
                replace 'g2' = 'd' - 'R'*'p'*('M'-'d')
                if ('todo'==1) exit

                tempname d11 d12 d22
                mlmatsum 'lnf' 'd11' = -'p'^2 * 'M'                        , eq(1)
                mlmatsum 'lnf' 'd12' = 'p'*('M'-'d'+'R'*'p'*'M')      , eq(1,2)
                mlmatsum 'lnf' 'd22' = -'p'*'R'*('R'*'p'*'M'+'M'-'d') , eq(2)
                matrix 'H' = ('d11','d12' \ 'd12'','d22')
        }
end
```

——————————————————————————————————— end `myweibull_lf2.ado` ————————

Our estimation command for this model is called `myweibull` and is listed in appendix E.4. `myweibull` uses `myweibull_lf2` as its likelihood evaluator. Because `myweibull_lf2` computes the equation-level scores, `myweibull` allows all optimization algorithms, robust and OPG variance estimates, and the `svy` option. When $\gamma_j = 1$, we have the exponential model (with censoring). The MLE for η in the constant-only exponential model is

$$\widehat{\eta} = \frac{\sum_{j=1}^{N} t_j}{\sum_{j=1}^{N} d_j}$$

Thus `myweibull` uses $\widehat{\beta}_0 = \ln \widehat{\eta}$ as the initial value.

Here is `myweibull` in use:

```
. sysuse cancer
(Patient survival in drug trial)
. constraint 1 /lngamma = 0
. myweibull studytime i.drug age, fail(died) constraint(1) hr
Fitting constant-only model:
Iteration 0:  Log likelihood = -129.51967
Iteration 1:  Log likelihood = -129.51967

Fitting full model:
Iteration 0:  Log likelihood = -129.51967
Iteration 1:  Log likelihood = -122.88547
Iteration 2:  Log likelihood = -116.36177
Iteration 3:  Log likelihood = -116.34506
Iteration 4:  Log likelihood = -116.34505
Iteration 5:  Log likelihood = -116.34505
```

```
My Weibull estimates                          Number of obs =      48
                                              Wald chi2(3)  =   25.16
Log likelihood = -116.34505                   Prob > chi2   = 0.0000

 ( 1)   [/]lngamma = 0
```

	Haz. ratio	Std. err.	z	P>|z|	[95% conf. interval]	
drug						
Other	4.313332	2.053639	3.07	0.002	1.69646	10.96686
NA	6.407428	3.000648	3.97	0.000	2.558936	16.04383
age	.9181277	.0307545	-2.55	0.011	.8597862	.9804281
_cons	1056.781	2007.241	3.67	0.000	25.53988	43727.11
/lngamma	0	(constrained)				
gamma	1	(constrained)				
1/gamma	1	(constrained)				

```
Note: Estimates are transformed only in the first equation to hazard ratios.
Note: _cons estimates baseline hazard.
```

17.5   The Cox proportional hazards model

In the Cox proportional hazards model, the dependent variable measures time, $T > 0$, until an individual fails or is censored. Given the vector of covariates $\mathbf{x}_j$ for the jth individual, the log-likelihood function for this model is

$$\ln L = \sum_{i=1}^{N} \left[\sum_{k \in D_i} \mathbf{x}_k \boldsymbol{\beta} - d_i \ln \left\{ \sum_{j \in R_i} \exp(\mathbf{x}_j \boldsymbol{\beta}) \right\} \right]$$

where i indexes the ordered failure times $t_{(i)}$ $(i = 1, \ldots, N)$, d_i is the number who fail at $t_{(i)}$, D_i is the set of observations that fail at $t_{(i)}$, and R_i (the risk pool at time $t_{(i)}$) is the set of observations that are at risk at time $t_{(i)}$. A more complete discussion of this model, including other methods for handling tied failure times, is in Cleves, Gould, and Marchenko (2016).

This model fails to meet the linear-form restrictions, so we will have to use method d0 rather than method lf to maximize the log-likelihood function. Without equation-level scores, the BHHH algorithm, robust and OPG variance estimates, and the svy option are not allowed.

We can sort the data from within our estimation command, so in accordance with the technical notes in section 7.5 (pages 169–170), our method-d0 likelihood evaluator is

```
──────────────────────────────────── begin mycox2_d0.ado ──────────
program mycox2_d0
        version 18
        args todo b lnf
        local negt "$ML_y1"               // $ML_y1 is -t
        local d "$ML_y2"                   // $ML_y2 is d
        tempvar xb B A sumd last L
        mleval `xb' = `b'
        // data assumed already sorted by `negt' and `d'
        local byby "by `negt' `d'"
        local wxb "$ML_w*`xb'"
        quietly {
                generate double `B' = sum($ML_w*exp(`xb'))
                `byby': generate double `A' = cond(_n==_N, sum(`wxb'), .) if `d'==1
                `byby': generate `sumd' = cond(_n==_N, sum($ML_w*`d'), .)
                `byby': generate byte `last' = (_n==_N & `d' == 1)
                generate double `L' = `A' - `sumd'*ln(`B') if `last'
                mlsum `lnf' = `L' if `last', noweight
        }
end
                                      ──────── end mycox2_d0.ado ──────────
```

Our estimation command for this model is called mycox; see appendix E.5. We use the default initial values for this estimation command; however, fitting a constant-only model is a little tricky because of the lack of a constant. To get around this, we use the iterate(0) option to compute the value of the likelihood ll_0 at the default initial values (vector of zeros) and then pass lf0(0 ll_0) to the ml model statement in the second step. The relevant code fragment is

```
────────────────────────────────────── fragment mycox.ado ──────────
        ...
        gettoken time rhs : varlist
        tempvar negt
        quietly generate double `negt' = -`time'
        sort `negt' `failure'
        ...
        // fit constant-only model
        if "`lrtest'" == "" {
                quietly ml model d0 mycox2_d0                    ///
                        (xb: `negt' `failure' = `rhs',           ///
                                noconstant                       ///
                        )                                        ///
                        `wgt' if `touse',                        ///
                        iterate(0)                               ///
                        nocnsnotes missing maximize
                local initopt lf0(0 `e(ll)')
        }

        // fit the full model
        `qui' di as txt _n "Fitting full model:"
        ml model d0 mycox2_d0                            ///
                (xb: `negt' `failure' = `rhs',           ///
                        noconstant                       ///
                )                                        ///
                `wgt' if `touse',                        ///
                `log' `mlopts' `vce' `initopt'           ///
                missing maximize
        ...
────────────────────────────────────── end mycox.ado ──────────
```

Here is `mycox` in use:

```
. sysuse cancer
(Patient survival in drug trial)

. mycox studytime i.drug age, fail(died)

Fitting full model:

Initial:      Log likelihood = -99.911448
Alternative:  Log likelihood = -99.606964
Rescale:      Log likelihood = -99.101334
Iteration 0:  Log likelihood = -99.101334
Iteration 1:  Log likelihood = -82.649192
Iteration 2:  Log likelihood = -81.729635
Iteration 3:  Log likelihood = -81.652727
Iteration 4:  Log likelihood = -81.652567
Iteration 5:  Log likelihood = -81.652567

My Cox estimates                                  Number of obs =      48
                                                  LR chi2(3)    =   36.52
Log likelihood = -81.652567                       Prob > chi2   =  0.0000
```

	Coefficient	Std. err.	z	P>|z|	[95% conf. interval]
drug					
Other	-1.71156	.4943639	-3.46	0.001	-2.680495 -.7426242
NA	-2.956384	.6557432	-4.51	0.000	-4.241617 -1.671151
age	.11184	.0365789	3.06	0.002	.0401467 .1835333

17.6   The random-effects regression model

The random-effects regression model is typically written

$$y_{it} = \mathbf{x}_{it}\boldsymbol{\beta} + u_i + e_{it}$$

where u_i (the random effects) are assumed independent and identically distributed $N(0, \sigma_u^2)$, e_{it} are assumed independent and identically distributed $N(0, \sigma_e^2)$, and u_i and e_{it} are assumed uncorrelated (independent). The joint likelihood function for the ith group or panel is given by

$$f(y_{i1}, \dots, y_{iT_i}) = \int_{-\infty}^{\infty} f(u_i) \prod_{t=1}^{T_i} f(y_{it}|u_i)\, du_i$$

where

$$f(u_i) = \frac{1}{\sigma_u}\phi\left(\frac{u_i}{\sigma_u}\right) \quad \text{and} \quad f(y_{it}|u_i) = \frac{1}{\sigma_e}\phi\left(\frac{y_{it} - \mathbf{x}_{it}\boldsymbol{\beta} - u_i}{\sigma_e}\right)$$

Thus the log likelihood for the ith group of observations is

$$\ln L_i = -\frac{1}{2}\left\{ \frac{\sum_t z_{it}^2 - a_i(\sum_t z_{it})^2}{\sigma_e^2} + \ln(T_i\sigma_u^2/\sigma_e^2 + 1) + T_i\ln(2\pi\sigma_e^2) \right\}$$

T_i = number of observations in the ith group

$z_{it} = y_{it} - \mathbf{x}_{it}\boldsymbol{\beta}$

$a_i = \sigma_u^2/(T_i\sigma_u^2 + \sigma_e^2)$

and this model does not meet the linear-form restrictions. Without equation-level scores, the BHHH algorithm, robust and OPG variance estimates, and the svy option are not allowed.

We wish to model the variances in the log space, so the equations for our evaluator are

$$\theta_{1it} = \mathbf{x}_{it}\boldsymbol{\beta}$$
$$\theta_{2it} = \ln\sigma_u$$
$$\theta_{3it} = \ln\sigma_e$$

Taking derivatives with respect to $\boldsymbol{\beta}$, $\ln \sigma_u$, and $\ln \sigma_e$ yields

$$\frac{\partial \ln L_i}{\partial \boldsymbol{\beta}} = \sum_{t=1}^{T_i} \frac{\partial \ln L_i}{\partial \mathbf{x}_{it}\boldsymbol{\beta}} \mathbf{x}_{it}$$

$$\frac{\partial \ln L_i}{\partial \mathbf{x}_{it}\boldsymbol{\beta}} = \frac{1}{\sigma_e^2} \left(z_{it} - a_i \sum_{s=1}^{T_i} z_{is} \right)$$

$$\frac{\partial \ln L_i}{\partial \ln \sigma_u} = \frac{a_i^2}{\sigma_u^2} \left(\sum_{t=1}^{T_i} z_{it} \right)^2 - T_i a_i$$

$$\frac{\partial \ln L_i}{\partial \ln \sigma_e} = \frac{1}{\sigma_e^2} \left\{ \sum_{t=1}^{T_i} z_{it}^2 - a_i \left(\sum_{t=1}^{T_i} z_{it} \right)^2 \right\} - (T_i - 1) - \frac{a_i^2}{\sigma_u^2} \left(\sum_{t=1}^{T_i} z_{it} \right)^2 - \frac{\sigma_e^2}{\sigma_u^2} a_i$$

The second (and cross) derivatives, with respect to $\ln \sigma_u$ and $\ln \sigma_e$, are

$$\frac{\partial^2 \ln L_i}{\partial^2 \ln \sigma_u} = - \left\{ \frac{2a_i^2}{\sigma_u^2} \left(\sum_{t=1}^{T_i} z_{it} \right)^2 - \frac{4a_i^3 \sigma_e^2}{\sigma_u^4} \left(\sum_{t=1}^{T_i} z_{it} \right)^2 + \frac{2T_i a_i^2 \sigma_e^2}{\sigma_u^2} \right\}$$

$$\frac{\partial^2 \ln L_i}{\partial^2 \ln \sigma_e} = - \left[\frac{2}{\sigma_e^2} \left\{ \sum_{t=1}^{T_i} z_{it}^2 - a_i \left(\sum_{t=1}^{T_i} z_{it} \right)^2 \right\} \right.$$
$$\left. - \left(\frac{2a_i^2}{\sigma_u^2} + \frac{4a_i^3 \sigma_e^2}{\sigma_u^4} \right) \left(\sum_{t=1}^{T_i} z_{it} \right)^2 + \frac{2a_i \sigma_e^2}{\sigma_u^2} - \frac{2a_i^2 \sigma_e^4}{\sigma_u^4} \right]$$

$$\frac{\partial^2 \ln L_i}{\partial \ln \sigma_e \, \partial \ln \sigma_u} = - \left\{ \frac{4a_i^3 \sigma_e^2}{\sigma_u^4} \left(\sum_{t=1}^{T_i} z_{it} \right)^2 - \frac{2T_i a_i^2 \sigma_e^2}{\sigma_u^2} \right\}$$

The rest of the second derivatives are

$$\frac{\partial^2 \ln L_i}{\partial \boldsymbol{\beta} \partial \boldsymbol{\beta}'} = -\frac{1}{\sigma_e^2}\left\{\sum_{t=1}^{T_i}\mathbf{x}'_{it}\mathbf{x}_{it} - a_i\left(\sum_{s=1}^{T_i}\mathbf{x}'_{is}\right)\left(\sum_{t=1}^{T_i}\mathbf{x}_{it}\right)\right\}$$

$$\frac{\partial^2 \ln L_i}{\partial \boldsymbol{\beta} \partial \ln \sigma_u} = \sum_{t=1}^{T_i}\frac{\partial^2 \ln L_i}{\partial \mathbf{x}_{it}\boldsymbol{\beta}\, \partial \ln \sigma_u}\mathbf{x}_{it}$$

$$\frac{\partial^2 \ln L_i}{\partial \mathbf{x}_{it}\boldsymbol{\beta}\, \partial \ln \sigma_u} = -\frac{2a_i^2}{\sigma_u^2}\sum_{t=1}^{T_i}z_{it}$$

$$\frac{\partial^2 \ln L_i}{\partial \boldsymbol{\beta} \partial \ln \sigma_e} = \sum_{t=1}^{T_i}\frac{\partial^2 \ln L_i}{\partial \mathbf{x}_{it}\boldsymbol{\beta}\, \partial \ln \sigma_e}\mathbf{x}_{it}$$

$$\frac{\partial^2 \ln L_i}{\partial \mathbf{x}_{it}\boldsymbol{\beta}\, \partial \ln \sigma_e} = -\left\{\frac{2}{\sigma_e^2}\left(z_{it} - a_i\sum_{t=1}^{T_i}z_{it}\right) - \frac{2a_i^2}{\sigma_u^2}\sum_{t=1}^{T_i}z_{it}\right\}$$

We cannot use `mlmatsum` to calculate $\partial^2 \ln L_i/\partial \boldsymbol{\beta} \partial \boldsymbol{\beta}'$ directly, but we can use it to calculate

$$\sum_{t=1}^{T_i}\mathbf{x}'_{it}\mathbf{x}_{it}$$

and we can use `mlmatbysum` to calculate

$$a_i\left(\sum_{s=1}^{T_i}\mathbf{x}'_{is}\right)\left(\sum_{t=1}^{T_i}\mathbf{x}_{it}\right)$$

Our method-d2 evaluator for this model is called `myrereg2_d2` and is listed in appendix E.6. If you read closely, you will notice that we took the `sort` command from the evaluator (`myrereg_d2` from section 7.4.3) and placed it in the estimation command; this prevents the evaluator from performing unnecessary tasks.

Our estimation command for this model is called `myrereg` and is listed in appendix E.6. The initial value for the intercept in the constant-only model is the sample mean of the response variable. The initial values for the variances are taken from the variance components from a one-way analysis of variance of the constant-only model with the groups treated as random factors:

$$\ln s_e = \frac{1}{2}\ln(\mathrm{MS}_e)$$

$$\ln s_u = \frac{1}{2}\ln\left(\frac{\mathrm{MS}_b - \mathrm{MS}_e}{\overline{\overline{T}}}\right)$$

where MS$_b$ is the mean squares between groups, MS$_e$ is the within-group (or error) mean squares, and $\overline{T}$ is the average number of observations per group.

Here is `myrereg` in use:

```
. use http://www.stata-press.com/data/ml5/tablef7-1
. myrereg lnC lnQ lnPF lf, panel(i) nolog
```

My rereg estimates

					Number of obs =	90
					LR chi2(3)	= 436.32
					Prob > chi2	= 0.0000

Log likelihood = 114.72904

lnC	Coefficient	Std. err.	z	P>|z|	[95% conf. interval]	
lnQ	.90531	.0253759	35.68	0.000	.8555742	.9550459
lnPF	.4233757	.013888	30.48	0.000	.3961557	.4505957
lf	-1.064456	.1962308	-5.42	0.000	-1.449061	-.6798508
_cons	9.618648	.2066219	46.55	0.000	9.213677	10.02362
/lns_u	-2.170817	.3026644	-7.17	0.000	-2.764028	-1.577605
/lns_e	-2.828404	.077319	-36.58	0.000	-2.979947	-2.676862
sigma_u	.1140844	.0345293			.0630373	.2064689
sigma_e	.0591071	.0045701			.0507955	.0687786

17.7   The seemingly unrelated regression model

Models for systems of equations can also be fit using `ml`. To illustrate, we will implement seemingly unrelated regression. An early challenge we will face is determining a strategy to ensure convergence when fitting this model; all the other models we have looked at thus far rarely have this problem.

The seemingly unrelated regression model is a system of regression equations where the errors are assumed correlated between equations. A system of p equations may be written

$$y_{1j} = \beta_{1,0} + \beta_{1,1}x_{1j} + \cdots + \beta_{1,k}x_{kj} + e_{1j}$$
$$y_{2j} = \beta_{2,0} + \beta_{2,1}x_{1j} + \cdots + \beta_{2,k}x_{kj} + e_{2j}$$
$$\vdots$$
$$y_{pj} = \beta_{p,0} + \beta_{p,1}x_{1j} + \cdots + \beta_{p,k}x_{kj} + e_{pj}$$

Using vector notation, we have

$$\mathbf{y}_j = \boldsymbol{\beta}'\mathbf{x}_j + \mathbf{e}_j = \boldsymbol{\mu}_j + \mathbf{e}_j$$

where

$$\mathbf{y}_j = (y_{1j}, \ldots, y_{pj})'$$
$$\boldsymbol{\beta} = (\boldsymbol{\beta}_1, \ldots, \boldsymbol{\beta}_p)$$
$$\mathbf{x}_j = (x_{1j}, \ldots, x_{kj})'$$
$$\mathbf{e}_j = (e_{1j}, \ldots, e_{pj})'$$

Here $\boldsymbol{\beta}$ is a matrix with p columns and k rows, and the error vectors $\mathbf{e}_j$ are assumed to be independent across observations and distributed multivariate normal with a zero-mean vector and variance matrix Σ.

Even though $\mathbf{x}_j$ is the same for all p equations, we can impose exclusion restrictions on the elements of the coefficient matrix to remove a variable from an equation. Given ml's flexible equation notation, we can just specify which independent variables go with a given dependent variable. For example, if our system was

$$y_{1j} = \beta_{1,0} + \beta_{1,1}x_{1j} + \beta_{1,2}x_{2j} + \beta_{1,3}x_{3j} + e_{1j}$$
$$y_{2j} = \beta_{2,0} + \beta_{2,1}x_{1j} + \beta_{2,2}x_{2j} + \beta_{2,3}x_{3j} + e_{2j}$$

we would type something like

```
. ml model lf1 mysureg_lf1 (y1 = x1 x2 x3) (y2 = x1 x2 x3)
```

But if our system was

$$y_{1j} = \beta_{1,0} + \beta_{1,1}x_{1j} + e_{1j}$$
$$y_{2j} = \beta_{2,0} + \beta_{2,1}x_{2j} + e_{2j}$$
$$y_{3j} = \beta_{3,0} + \beta_{3,1}x_{3j} + e_{3j}$$

we would type

```
. ml model lf1 mysureg_lf1 (y1 = x1) (y2 = x2) (y3 = x3)
```

The log likelihood for the jth observation is

$$\ln L_j = -\frac{1}{2}\left\{p\ln(2\pi) + \ln|\Sigma| + (\mathbf{y}_j - \boldsymbol{\beta}'\mathbf{x}_j)'\Sigma^{-1}(\mathbf{y}_j - \boldsymbol{\beta}'\mathbf{x}_j)\right\}$$

where $|\Sigma|$ is the determinant of the error variance matrix. Although this model meets the linear-form restrictions, we will need to write an evaluator for the *concentrated likelihood* (to be explained shortly), so we will develop a method-d1 evaluator instead (we started with a d0 evaluator and then added derivatives for the d1 evaluator). Later we will develop a method-lf0 evaluator for the unconcentrated likelihood so that we can obtain scores and hence compute robust standard errors.

As in the previous examples, we will break this rather large formula into smaller, more manageable pieces. First, we will denote the jth residual vector by

$$\mathbf{r}_j = \mathbf{y}_j - \boldsymbol{\beta}'\mathbf{x}_j = (r_{1j}, \ldots, r_{pj})'$$

and the jth quadratic form by

$$Q_j = (\mathbf{y}_j - \boldsymbol{\beta}'\mathbf{x}_j)'\Sigma^{-1}(\mathbf{y}_j - \boldsymbol{\beta}'\mathbf{x}_j) = \mathbf{r}_j'\Sigma^{-1}\mathbf{r}_j$$

By σ_{ab}, we mean the element of Σ from row a and column b and, similarly, for σ^{ab} from Σ^{-1}. I_{ab} is the matrix with one in row a and column b and zero everywhere else, and

$$\delta_{ab} = \begin{cases} 1 & \text{if } a = b \\ 0 & \text{if } a \neq b \end{cases}$$

The relevant derivatives, using Lütkepohl (1996) as a reference, are then

$$\frac{\partial Q_j}{\partial \boldsymbol{\beta}'\mathbf{x}_j} = -2\Sigma^{-1}\mathbf{r}_j$$

$$\frac{\partial \ln|\Sigma|}{\partial \sigma_{ab}} = (2 - \delta_{ab})\sigma^{ab}$$

$$\frac{\partial Q_j}{\partial \sigma_{ab}} = -(2 - \delta_{ab})\mathbf{r}_j'\Sigma^{-1}I_{ab}\Sigma^{-1}\mathbf{r}_j$$

Thus the derivatives we will code in our likelihood evaluator are

$$\frac{\partial L_j}{\partial \boldsymbol{\beta}'\mathbf{x}_j} = \Sigma^{-1}\mathbf{r}_j$$

$$\frac{\partial L_j}{\partial \sigma_{ab}} = \frac{1}{2}(2 - \delta_{ab})\left(\mathbf{r}_j'\Sigma^{-1}I_{ab}\Sigma^{-1}\mathbf{r}_j - \sigma^{ab}\right)$$

The analytical derivatives necessary for implementing a method-`d2` evaluator are not difficult; however, the investment adds little to this discussion, so we will let `ml` compute the Hessian numerically. Our `d1` evaluator for this model is

────────────────────────────────────── begin `mysureg_d1.ado` ──────────

```
program mysureg_d1
        version 18
        args todo b lnf g H
        local p : word count $ML_y
        local k = `p'*(`p'+1) / 2 + `p'

        tempname S iS sij isij isi ip
        matrix `S' = J(`p',`p',0)

quietly {
        // get residuals and build variance matrix
        local e 1
        forval i = 1/`p' {
                tempvar r`i'
                mleval `r`i'' = `b', eq(`i')
                replace `r`i'' = ${ML_y`i'} - `r`i''
                local resids `resids' `r`i''
                mleval `sij' = `b', eq(`=`p'+`e'') scalar
                matrix `S'[`i',`i'] = `sij'
                local ++e
                forval j = `=`i'+1'/`p' {
                        mleval `sij' = `b', eq(`=`p'+`e'') scalar
                        matrix `S'[`i',`j'] = `sij'
                        matrix `S'[`j',`i'] = `sij'
                        local ++e
                }
        }
        matrix `iS' = invsym(`S')
        // get score variables
        forval i = 1/`k' {
                tempvar g`i'
                quietly generate double `g`i'' = .
        }
        generate double `ip' = 0
        forval i = 1/`p' {
                matrix `isi' = `iS'[`i',1...]
                matrix colnames `isi' = `resids'
                matrix score `g`i'' = `isi', replace
                replace `ip' = `ip' + `r`i''*`g`i''
        }
        mlsum `lnf' = -0.5*(`p'*ln(2*c(pi))+ln(det(`S'))+`ip')
} // quietly
        if (`todo' == 0 | missing(`lnf')) exit

        // compute the scores and gradient
        tempname gi gs
        capture matrix drop `g'

quietly {
        local e = `p'+1
        forval i = 1/`p' {
                mlvecsum `lnf' `gi' = `g`i'', eq(`i')
                matrix `g' = nullmat(`g'), `gi'
                replace `g`e'' = 0.5*(`g`i''*`g`i''-`iS'[`i',`i'])
                mlvecsum `lnf' `gi' = `g`e'' , eq(`e')
                matrix `gs' = nullmat(`gs'), `gi'
                local ++e
                forval j = `=`i'+1'/`p' {
                        replace `g`e'' = `g`i''*`g`j'' - `iS'[`i',`j']
                        mlvecsum `lnf' `gi' = `g`e'' , eq(`e')
```

```
                              matrix 'gs' = nullmat('gs'), 'gi'
                              local ++e
                    }
           }
           matrix 'g' = 'g' , 'gs'
    } // quietly
    end
```
── end `mysureg_d1.ado` ───────

Here we try to fit this model using a subset of the variables in the dataset from Grunfeld and Griliches (1960); the data were taken from the website for Greene (2018). Unfortunately, optimizing this likelihood is difficult.

```
. set seed 98

. use http://www.stata-press.com/data/ml5/tablef13-1wide

. ml model d1 mysureg_d1
>         (iGM: iGM = fGM cGM)
>         (iCH: iCH = fCH cCH)
>         (iGE: iGE = fGE cGE)
>         /sigma1_1 /sigma1_2 /sigma1_3
>         /sigma2_2 /sigma2_3
>         /sigma3_3

. ml max, difficult iterate(300)
Initial:       Log likelihood =      -<inf>  (could not be evaluated)
Feasible:      Log likelihood = -1100983.3
Rescale:       Log likelihood = -39215.742
Rescale eq:    Log likelihood = -414.16649
Iteration 0:   Log likelihood = -414.16649  (not concave)
Iteration 1:   Log likelihood = -403.02592  (not concave)
Iteration 2:   Log likelihood = -402.17352  (not concave)
(output omitted)
Iteration 300: Log likelihood = -388.02459  (not concave)
convergence not achieved

(output omitted)
```

One clear problem with optimizing this likelihood is that the current guess for Σ must always be positive definite, something that cannot be guaranteed until `ml` is already close to the solution. We might be able to get around this problem by using transformations of the elements of Σ, but there is a less complicated solution.

The log likelihood for this model can be written as

$$\ln L = -\frac{N}{2}\left[p\ln(2\pi) + \ln|\Sigma| + \text{trace}\{\Sigma^{-1}\mathbf{W}(\beta)\}\right]$$

where

$$\mathbf{W}(\boldsymbol{\beta}) = \frac{1}{N}\sum_{j=1}^{N}(\mathbf{y}_j - \boldsymbol{\beta}'\mathbf{x}_j)(\mathbf{y}_j - \boldsymbol{\beta}'\mathbf{x}_j)'$$

is the average of the outer product of the residual vectors. It turns out that the MLE of Σ is

$$\widehat{\Sigma} = \mathbf{W}(\widehat{\boldsymbol{\beta}}) \tag{17.2}$$

If we were to substitute $\mathbf{W}(\beta)$ for Σ in our log-likelihood function, we get what is called a *concentrated* log-likelihood function:

$$\ln L_{\boldsymbol{\beta}} = -\frac{N}{2}\Big[p\{\ln(2\pi) + 1\} + \ln|\mathbf{W}(\beta)|\Big]$$

(see Greene 2018, sec. E.4.3). Let w_{ab} be the element of $\mathbf{W}(\boldsymbol{\beta})$ from row a and column b, and then

$$w_{ab} = \frac{1}{N}\sum_{j=1}^{N} r_{aj} r_{bj}$$

where r_{aj} is the ath element of $\mathbf{r}_j$. Taking derivatives, we have

$$\frac{\partial w_{ab}}{\partial \beta_a' \mathbf{x}_j} = -\frac{1 + \delta_{ab}}{N}\sum_{j=1}^{N} r_{bj}$$

$$\frac{\partial \ln L_{\boldsymbol{\beta}}}{\partial \beta_a' \mathbf{x}_j} = -\frac{N}{2}\sum_{b=1}^{p} \frac{\partial \ln|\mathbf{W}|}{\partial w_{ab}}\frac{\partial w_{ab}}{\partial \beta_a' \mathbf{x}_j} = \sum_{b=1}^{p} w^{ab}\sum_{j=1}^{N} r_{bj}$$

and

$$\frac{\partial^2 \ln L_{\boldsymbol{\beta}}}{\partial \beta_a' \mathbf{x}_j \partial \beta_b' \mathbf{x}_j} = -w^{ab} + \frac{\partial w^{ab}}{\partial \beta_b' \mathbf{x}_j}\sum_{j=1}^{N} r_{bj}$$

$$\approx -w^{ab} \tag{17.3}$$

which we will use to approximate the Hessian. In fact, the second term in (17.3) will go to zero at the same rate as the elements of the gradient vector, so this approximation becomes the correct calculation close to the solution. Our d2 evaluator for the concentrated model is

─────────────────────────── begin `mysuregc_d2.ado` ───────────────

```
program mysuregc_d2
        // concentrated likelihood for the SUR model
        version 18
        args todo b lnf g H
        local p : word count $ML_y
        tempname W sumw

        // get residuals and build variance matrix
        forval i = 1/`p' {
                tempvar r`i'
                mleval `r`i'' = `b', eq(`i')
                quietly replace `r`i'' = ${ML_y`i'} - `r`i''
                local resids `resids' `r`i''
        }
        quietly matrix accum `W' = `resids' [iw=$ML_w] if $ML_samp, noconstant
        scalar `sumw' = r(N)
        matrix `W' = `W'/`sumw'
        scalar `lnf' = -0.5*`sumw'*(`p'*(ln(2*c(pi))+1)+ln(det(`W')))

        if (`todo' == 0 | missing(`lnf')) exit

        // compute gradient
        tempname gi iW iWi
        tempvar  scorei
        quietly generate double `scorei' = .
        capture matrix drop `g'
        matrix `iW' = invsym(`W')
        forval i = 1/`p' {
                matrix `iWi' = `iW'[`i',1...]
                matrix colnames `iWi' = `resids'
                matrix score `scorei' = `iWi', replace
                mlvecsum `lnf' `gi' = `scorei', eq(`i')
                matrix `g' = nullmat(`g'), `gi'
        }

        if (`todo' == 1 | missing(`lnf')) exit

        // compute the Hessian, as if we were near the solution
        tempname hij
        local k = colsof(`b')
        matrix `H' = J(`k',`k',0)
        local r 1
        forval i = 1/`p' {
                local c `r'
                mlmatsum `lnf' `hij' = -1*`iW'[`i',`i'], eq(`i')
                matrix `H'[`r',`c'] = `hij'
                local c = `c' + colsof(`hij')
                forval j = `=`i'+1'/`p' {
                        mlmatsum `lnf' `hij' = -1*`iW'[`i',`j'], eq(`i',`j')
                        matrix `H'[`r',`c'] = `hij'
                        matrix `H'[`c',`r'] = `hij''
                        local c = `c' + colsof(`hij')
                }
                local r = `r' + rowsof(`hij')
        }
end
```

─────────────────────────── end `mysuregc_d2.ado` ───────────────

We can now interactively compute the maximum likelihood estimates of β for the concentrated log likelihood using the data from Grunfeld and Griliches (1960).

```
. use http://www.stata-press.com/data/ml5/tablef13-1wide, clear
. ml model d2 mysuregc_d2
>    (iGM: iGM = fGM cGM)
>    (iCH: iCH = fCH cCH)
>    (iGE: iGE = fGE cGE)
. ml max
Initial:       Log likelihood = -341.95727
Alternative:   Log likelihood = -341.86494
Rescale:       Log likelihood = -340.01308
Rescale eq:    Log likelihood = -334.19438
Iteration 0:   Log likelihood = -334.19438
Iteration 1:   Log likelihood =  -294.5609
Iteration 2:   Log likelihood = -287.39098
Iteration 3:   Log likelihood = -287.24995
Iteration 4:   Log likelihood = -287.24908
Iteration 5:   Log likelihood = -287.24907
```

	Number of obs	= 20
	Wald chi2(2)	= 234.83
Log likelihood = -287.24907	Prob > chi2	= 0.0000

	Coefficient	Std. err.	z	P>|z|	[95% conf. interval]	
iGM						
fGM	.1149511	.0229982	5.00	0.000	.0698755	.1600268
cGM	.3756614	.0336697	11.16	0.000	.3096699	.4416528
_cons	-133.7523	94.7783	-1.41	0.158	-319.5144	52.00973
iCH						
fCH	.0720636	.0181543	3.97	0.000	.0364818	.1076454
cCH	.3211742	.0264877	12.13	0.000	.2692593	.3730891
_cons	-2.772468	12.32262	-0.22	0.822	-26.92436	21.37942
iGE						
fGE	.0279556	.0140716	1.99	0.047	.0003758	.0555353
cGE	.1518845	.0236063	6.43	0.000	.105617	.1981521
_cons	-12.75894	28.52304	-0.45	0.655	-68.66307	43.14519

Although we can compute $\widehat{\Sigma}$ using (17.2), doing so would not provide us with its variance estimates. This obstacle is easily remedied by using the estimated parameters from the concentrated model as initial values for maximizing the unconcentrated likelihood function. Moreover, as we previously indicated, the unconcentrated model satisfies the linear-form restrictions, so we can obtain scores. Because we have excellent initial values based on the concentrated model, there is little advantage to providing analytic derivatives for the unconcentrated likelihood. Thus a method-lf0 evaluator suffices. In fact, we could write a method-lf evaluator, but it is slightly easier to convert a d-family evaluator into an lf-family one because both d- and lf-family evaluators receive the coefficient vector 'b'. Our method-lf0 evaluator is

────────────────────────────────── begin `mysureg_lf0.ado` ──────────

```
program mysureg_lf0
        version 18
        args todo b lnfj
        local p : word count $ML_y
        local k = 'p'*('p'+1) / 2 + 'p'

        tempname S iS sij isij isi
        tempvar ip
        matrix 'S' = J('p','p',0)
quietly {
        // get residuals and build variance matrix
        local e 1
        forval i = 1/'p' {
                tempvar r'i'
                mleval 'r'i'' = 'b', eq('i')
                replace 'r'i'' = ${ML_y'i'} - 'r'i''
                local resids 'resids' 'r'i''
                mleval 'sij' = 'b', eq('='p'+'e'') scalar
                matrix 'S'['i','i'] = 'sij'
                local ++e
                forval j = '='i'+1'/'p' {
                        mleval 'sij' = 'b', eq('='p'+'e'') scalar
                        matrix 'S'['i','j'] = 'sij'
                        matrix 'S'['j','i'] = 'sij'
                        local ++e
                }
        }
        matrix 'iS' = invsym('S')
        // get score variables
        tempvar scorei
        generate double 'scorei' = .
        generate double 'ip' = 0
        forval i = 1/'p' {
                matrix 'isi' = 'iS'['i',1...]
                matrix colnames 'isi' = 'resids'
                matrix score 'scorei' = 'isi', replace
                replace 'ip' = 'ip' + 'r'i''*'scorei'
        }
        replace 'lnfj' = -0.5*('p'*ln(2*c(pi))+ln(det('S'))+'ip')
} // quietly
end
```

────────────────────────────────── end `mysureg_lf0.ado` ──────────

In the following example, we compute the residuals from each equation, use them and `matrix accum` to compute $\widehat{\Sigma}$, and then supply $\widehat{\beta}$ and the upper-triangle elements of $\widehat{\Sigma}$ (with the appropriate equation and column names) as initial values using `ml init`.

```
. // get the residuals
. predict r1, eq(#1)

. replace r1 = iGM - r1
(20 real changes made)

. predict r2, eq(#2)

. replace r2 = iCH - r2
(20 real changes made)

. predict r3, eq(#3)

. replace r3 = iGE - r3
(20 real changes made)

. // compute the MLE of Sigma from the residuals
. matrix accum Sigma = r1 r2 r3, noconstant
(obs=20)

. matrix Sigma = Sigma/r(N)

. // initial values of the coefficients
. matrix b = e(b)

. // initial values of the upper triangle elements of Sigma
. matrix s = Sigma[1,1..3]

. matrix s = s, Sigma[2,2..3]

. matrix s = s, Sigma[3,3]

. matrix colnames s = /sigma1_1 /sigma1_2 /sigma1_3 /sigma2_2 /sigma2_3 /sigma3_3

. ml model lf0 mysureg_lf0
>         (iGM: iGM = fGM cGM)
>         (iCH: iCH = fCH cCH)
>         (iGE: iGE = fGE cGE)
>         /sigma1_1 /sigma1_2 /sigma1_3
>                   /sigma2_2 /sigma2_3
>                             /sigma3_3

. ml init b

. ml init s
```

```
. ml max
Initial:      Log likelihood = -287.24907
Rescale:      Log likelihood = -287.24907
Rescale eq:   Log likelihood = -287.24907
Iteration 0:  Log likelihood = -287.24907
Iteration 1:  Log likelihood = -287.24907
```

```
                                        Number of obs =        20
                                        Wald chi2(2)  =  234.75
Log likelihood = -287.24907             Prob > chi2   =  0.0000
```

	Coefficient	Std. err.	z	P>|z|	[95% conf. interval]	
iGM						
fGM	.1149513	.0231193	4.97	0.000	.0696382	.1602643
cGM	.3756612	.0337378	11.13	0.000	.3095364	.441786
_cons	-133.7528	95.19374	-1.41	0.160	-320.3291	52.8235
iCH						
fCH	.0720611	.0190324	3.79	0.000	.0347583	.1093639
cCH	.3211765	.0269839	11.90	0.000	.2682891	.3740639
_cons	-2.770999	12.76672	-0.22	0.828	-27.79332	22.25132
iGE						
fGE	.0279566	.0146924	1.90	0.057	-.00084	.0567533
cGE	.1518841	.0236933	6.41	0.000	.105446	.1983221
_cons	-12.76079	29.48049	-0.43	0.665	-70.54148	45.0199
/sigma1_1	7173.009	2272.075	3.16	0.002	2719.824	11626.19
/sigma1_2	-307.9035	248.6466	-1.24	0.216	-795.2419	179.4348
/sigma1_3	615.6388	507.0188	1.21	0.225	-378.0998	1609.377
/sigma2_2	150.7304	47.95484	3.14	0.002	56.74066	244.7202
/sigma2_3	-16.29927	73.92042	-0.22	0.825	-161.1806	128.5821
/sigma3_3	661.1597	209.1859	3.16	0.002	251.1629	1071.156

The value of the log likelihood does not change between the concentrated and unconcentrated model fits; in fact, `ml` converges in one iteration (as it should). The variance
estimates change because `ml` can compute the covariances between the MLEs of $\boldsymbol{\beta}$ and
Σ. Note that

$$\text{Var}\{\text{vec}(\widehat{\boldsymbol{\beta}})\} = \widehat{\Sigma} \otimes (\mathbf{X}'\mathbf{X})^{-1}$$

(with the implied exclusion restrictions) is true only asymptotically and for some special cases of the seemingly unrelated regression model. One special case is multivariate normal regression where there are no exclusion restrictions; that is, the same independent variables are present in all the equations.

Our estimation command for this model is called `mysureg` and is listed in appendix E.7. In the constant-only model, the MLE for $\boldsymbol{\beta}_0$ is

$$\widehat{\boldsymbol{\beta}}_0 = \overline{\mathbf{y}} = \frac{1}{N}\sum_{j=1}^{N}\mathbf{y}_j$$

the sample mean of the dependent variables. `mysureg` uses `matrix vecaccum` to compute $\widehat{\boldsymbol{\beta}}_0$. The initial values for the full model fit are $\widehat{\boldsymbol{\beta}}$ and $\widehat{\Sigma}$ from the concentrated likelihood fit.

Here is `mysureg` in use:

```
. use http://www.stata-press.com/data/ml5/tablef13-1wide, clear
. mysureg (iGM fGM cGM)
>         (iCH fCH cCH)
>         (iGE fGE cGE), vce(robust)
Fitting constant-only model:
Iteration 0:  Log likelihood = -323.88722
Iteration 1:  Log likelihood = -323.88722
Fitting concentrated model:
Iteration 0:  Log likelihood = -323.88722
Iteration 1:  Log likelihood = -291.22787
Iteration 2:  Log likelihood = -287.27972
Iteration 3:  Log likelihood = -287.24921
Iteration 4:  Log likelihood = -287.24908
Iteration 5:  Log likelihood = -287.24907
Fitting full model:
Iteration 0:  Log pseudolikelihood = -287.24907
Iteration 1:  Log pseudolikelihood = -287.24907
```

My sureg estimates

Number of obs =   20
Wald chi2(6) = 1396.39

Log pseudolikelihood = -287.24907

Prob > chi2   = 0.0000

	Coefficient	Robust std. err.	z	P>|z|	[95% conf. interval]	
iGM						
fGM	.1149513	.0228856	5.02	0.000	.0700963	.1598062
cGM	.3756612	.0429121	8.75	0.000	.291555	.4597674
_cons	-133.7528	88.83564	-1.51	0.132	-307.8675	40.36185
iCH						
fCH	.0720611	.0144682	4.98	0.000	.0437039	.1004182
cCH	.3211765	.0244714	13.12	0.000	.2732135	.3691395
_cons	-2.770999	8.789279	-0.32	0.753	-19.99767	14.45567
iGE						
fGE	.0279566	.0122976	2.27	0.023	.0038537	.0520595
cGE	.1518841	.0161167	9.42	0.000	.120296	.1834722
_cons	-12.76079	23.70002	-0.54	0.590	-59.21198	33.6904
/sigma1_1	7173.009	1449.301	4.95	0.000	4332.43	10013.59
/sigma1_2	-307.9035	299.5417	-1.03	0.304	-894.9944	279.1873
/sigma1_3	615.6388	449.4178	1.37	0.171	-265.2039	1496.482
/sigma2_2	150.7304	77.33421	1.95	0.051	-.8418416	302.3027
/sigma2_3	-16.29927	59.50743	-0.27	0.784	-132.9317	100.3332
/sigma3_3	661.1597	196.7761	3.36	0.001	275.4856	1046.834

```
. mysureg, notable corr
Correlation matrix of residuals:
        iGM     iCH     iGE
iGM       1
iCH   -.2961       1
iGE    .2827   -.0516       1
```

17.8   A bivariate Poisson regression model

Say we have a count variable, that is, a variable that takes on values 0, 1, 2, ... that represents the number of times some event happens, and we want to model the expected count for each observation as a function of covariates. We could use linear regression, but there is no guarantee that the predictions from linear regression will be positive as a count must be. In many cases, the appropriate course of action is to assume the counts follow a Poisson distribution and use Poisson regression analysis. Stata already has a command for Poisson regression; we refer you to [R] **poisson** and the sources cited therein.

In section 17.7, we discussed the seemingly unrelated regression model used when we have two or more continuous unbounded dependent variables we want to model jointly, either to obtain more efficient coefficient estimates than those we would obtain from fitting each equation individually or else to impose cross-equation coefficient restrictions. Now we want to consider the case of modeling two count variables jointly. We focus on the case of two variables, rather than an arbitrary number, to keep both the math and the computer code tractable. The command we develop is based on the estimator described in King (1989), Ozuna and Gomez (1994), AlMuhayfith et al. (2016), and elsewhere.

This estimator is unlike all others we have developed in this chapter. All of those are widely used statistical models, and they are already available in Stata. In contrast, bivariate Poisson regression is relatively rare, Stata does not have a built-in command for it, and it is not to our knowledge discussed in popular regression modeling or econometrics textbooks.

We take you through the steps necessary to go from an estimator you may have read about in a journal article or a book to having a reasonably well-written estimation command that implements it. Follow along on our journey as we do some math, do some coding, revisit some of our decisions, do some more math, do some more coding, and finally arrive at our destination of a working estimation command. Whether you have a need to fit a bivariate Poisson regression, we hope this example motivates you to write an estimation command for an interesting model you may come across in the literature or derive on your own.

We will write the log-likelihood evaluator function and some other functions using Mata, so you should be familiar with Mata and the content in chapter 15. We will use `ml`, however, because as we stated in chapter 16, `ml` takes care of many details for us that we would have to deal with ourselves if we were to use `moptimize()`. Using Mata is not strictly necessary here, but it both simplifies the code and makes our estimator execute much faster.

17.8.1   A bivariate Poisson distribution

Underlying bivariate Poisson regression is the bivariate Poisson distribution, the joint distribution of outcomes from two possibly correlated Poisson random variables. There are actually multiple distributions that characterize the joint distribution of two Poisson random variables that have been proposed in the literature. Here we pursue one of the simplest and most common ways to obtain such a distribution; an early reference is Campbell (1934). We may at times be sloppy in our writing and say the bivariate Poisson distribution, just as one would say the bivariate normal distribution, but you should understand that to mean the version we are deriving here.

We will use the notation $X \sim \text{Pois}(\mu)$ to indicate that random variable X follows a Poisson distribution with mean μ. Recall that such a random variable has the following probability mass function (PMF):

$$\Pr(X = x; \mu) = e^{-\mu} \frac{\mu^x}{x!}$$

We will also stick to the convention that uppercase letters like X represent random variables, while lowercase letters like x represent a realization of a random variable. Greek letters represent parameters. We will adopt the shorthand notation $\Pr(x; \mu)$ to represent $\Pr(X = x; \mu)$ given that $X \sim \text{Pois}(\mu)$.

To obtain the bivariate PMF, we consider three Poisson random variables. This general technique is therefore known as *trivariate reduction*. Let

$$U_0 \sim \text{Pois}(\xi_0) \qquad U_1 \sim \text{Pois}(\xi_1) \qquad U_2 \sim \text{Pois}(\xi_2)$$

be independently distributed; and for our bivariate Poisson distribution to be defined, we must have $\xi_1 > 0$, $\xi_2 > 0$, and $\xi_0 \geq 0$. Now let

$$Y_1 = U_1 + U_0 \qquad Y_2 = U_2 + U_0 \tag{17.4}$$

Then by using the moment-generating function for the univariate Poisson distribution, we can show that

$$Y_1 \sim \text{Pois}(\xi_1 + \xi_0) \qquad Y_2 \sim \text{Pois}(\xi_2 + \xi_0)$$

Moreover, because U_0, U_1, and U_2 are independently distributed, (17.4) implies that $\text{cov}(Y_1, Y_2) = \xi_0$. By construction, the covariance between Y_1 and Y_2 can therefore be no less than zero.

If you have reason to believe that the two count variables you are modeling may be negatively correlated, the estimator we develop here will not work for you. We discuss alternatives at the end of the section.

The joint PMF follows directly from the axioms of probability. Consider the joint outcome $Y_1 = y_1$ and $Y_2 = y_2$. There are several ways this could occur. We could have $U_1 = y_1$, $U_2 = y_2$, and $U_0 = 0$. Alternatively, we could have $U_1 = y_1 - 1$, $U_2 = y_2 - 1$, and $U_0 = 1$, we could have $U_1 = y_1 - 2$, $U_2 = y_2 - 2$, and $U_0 = 2$, and so on. Each of these ways is distinct and mutually exclusive: one, and only one, can occur to determine our joint outcome. Therefore, the axioms of probability say that the probability of observing $Y_1 = y_1$ and $Y_2 = y_2$ is equal to the sum of the probabilities of these outcomes for U_0, U_1, and U_2

$$\begin{aligned}
\Pr(Y_1 = y_1, Y_2 = y_2; \boldsymbol{\xi}) = {}& \Pr(U_1 = y_1, U_2 = y_2, U_0 = 0; \boldsymbol{\xi}) + \\
& \Pr(U_1 = y_1 - 1, U_2 = y_2 - 1, U_0 = 1; \boldsymbol{\xi}) + \\
& \Pr(U_1 = y_1 - 2, U_2 = y_2 - 2, U_0 = 2; \boldsymbol{\xi}) + \\
& \cdots
\end{aligned}$$

where $\boldsymbol{\xi} = (\xi_1, \xi_2, \xi_0)$. Because Poisson distributions are defined only for nonnegative counts, we can quit the sum as soon as $U_0 = \min\{y_1, y_2\}$, and we have

$$\Pr(Y_1 = y_1, Y_2 = y_2; \boldsymbol{\xi}) = \sum_{r=0}^{\min\{y_1, y_2\}} \Pr(U_1 = y_1 - r, U_2 = y_2 - r, U_0 = r; \boldsymbol{\xi})$$

U_1, U_2, and U_0 are independent, so we can write

$$\Pr(y_1, y_2; \boldsymbol{\xi}) = \sum_{r=0}^{\min\{y_1, y_2\}} \Pr(y_1 - r; \xi_1) \cdot \Pr(y_2 - r; \xi_2) \cdot \Pr(r; \xi_0) \tag{17.5}$$

If you substitute the PMF for the univariate Poisson random variable we gave earlier into (17.5), you will arrive at the bivariate PMF shown in King (1989). Upon rearranging terms, we obtain the probability of the joint outcome $(Y_1 = y_1, Y_2 = y_2)$ in a form that will be easier to work with,

$$\Pr(y_1, y_2; \boldsymbol{\xi}) = e^{(-\xi_0 - \xi_1 - \xi_2)} \frac{\xi_1^{y_1}}{y_1!} \frac{\xi_2^{y_2}}{y_2!} \sum_{r=0}^{\min\{y_1, y_2\}} r! \binom{y_1}{r} \binom{y_2}{r} \left(\frac{\xi_0}{\xi_1 \xi_2}\right)^r \tag{17.6}$$

where $\binom{n}{k} = \frac{n!}{k!(n-k)!}$.

The bivariate Poisson distribution possesses an interesting property when the covariance between Y_1 and Y_2 is 0. To see this, suppose $\xi_0 = 0$. Then we have

$$\begin{aligned}
\Pr(y_1, y_2; \xi_1, \xi_2, 0) &= e^{(-\xi_1 - \xi_2)} \frac{\xi_1^{y_1}}{y_1!} \frac{\xi_2^{y_2}}{y_2!} \\
&= \left\{ e^{-\xi_1} \frac{\xi_i^{y_1}}{y_1!} \right\} \times \left\{ e^{-\xi_2} \frac{\xi_i^{y_2}}{y_2!} \right\} \\
&= \Pr(y_1; \xi_1) \times \Pr(y_2; \xi_2)
\end{aligned}$$

implying that Y_1 and Y_2 are independent. In general, it is not true that if two random variables have zero covariance that they must be independent. Students often learn that for the bivariate normal distribution, as a special case, zero covariance does imply independence. This bivariate Poisson distribution is another such special case.

As a first step toward implementing bivariate Poisson regression, we decided to implement the bivariate Poisson PMF in Mata because such a function is not currently built in and we will need it to implement our log-likelihood evaluator function. You may be tempted to implement the PMF as written in (17.6). After all, we went through the algebra to derive this function and write it in a relatively compact form so that must be the right one to use. However, as statistical developers, we know that writing accurate density and distribution functions is challenging. Implementing a function like (17.6) may be accurate when called with "reasonable" outcomes and parameter values but may produce entirely wrong answers when called with less probable events.

We are going to show two different ways to implement our bivariate Poisson PMF. In the first way, we show how to go from (17.6) to a representation that will have much better numerical properties, and we take the opportunity to discuss how you might go about verifying that a function you have written produces accurate results. The second way is trivially easy to program, and we will use that way when implementing our bivariate Poisson regression estimator.

A brute-force way

In the case of (17.6), what concerns us are the factorials and the fact that the parameters ξ_1, ξ_2, and ξ_0 may be raised to very high powers. Mata has function `factorial()` to compute factorials, but it does not accept arguments greater than 167. In fact, 168 is larger than the largest double precision floating-point number a computer can handle. While we often think of Poisson models as having relatively small counts, we cannot guarantee a user will not try to use our bivariate Poisson regression command with counts that exceed 167.

We want a numerically stable way of computing our bivariate PMF. We first take the natural logarithm of (17.6) and rewrite it as

$$\ln \Pr(y_1, y_2; \boldsymbol{\xi}) \tag{17.7}$$
$$= (-\xi_0 - \xi_1 - \xi_2) + (y_1 \ln \xi_1 - \ln y_1!) + (y_2 \ln \xi_2 - \ln y_2!) + \ln A$$

$$A = \sum_{r=0}^{\min\{y_1, y_2\}} r! \binom{y_1}{r} \binom{y_2}{r} \left(\frac{\xi_0}{\xi_1 \xi_2} \right)^r$$

Because A is a sum, taking logarithms did not help us there, but we have made progress. We have gone from having to compute some potentially large powers to simple multiplications, which will help with accuracy. For a moment, we will leave A alone and focus on refining the first part of the right-hand side of (17.7).

We want to see whether we can eliminate the factorials. The solution is to use the gamma function defined as $\Gamma(z) = \int_0^\infty t^{z-1} e^{-t}\, dt$. Introducing improper integrals into our function might not strike you as an improvement, but it is. For any positive integer x, we have $\Gamma(x) = (x-1)!$, so rather than compute, say, $y_1!$ in our function, we can instead compute $\Gamma(y_1 + 1)$. The gamma function is widely used in many disciplines, and algorithms to compute it very accurately exist. Moreover, the natural logarithm of gamma is itself a widely used function with very accurate algorithms available, and that is what appears in (17.7). In Mata, function `gamma(z)` computes $\Gamma(z)$, and function `lngamma`(z) computes $\ln \Gamma(z)$. We will not need them here, but Mata also has the so-called digamma and trigamma functions that are the first and second derivatives of the log gamma function. Replacing the factorials in (17.7), we have

$$\ln \Pr(y_1, y_2; \boldsymbol{\xi}) = \ln A - \xi_0 - \xi_1 - \xi_2 + y_1 \ln \xi_1 - \ln \Gamma(y_1 + 1) \tag{17.8}$$
$$+ y_2 \ln \xi_2 - \ln \Gamma(y_2 + 1)$$

Now we return to calculating $A = A_0 + A_1 + A_2 + \cdots$. If we have $\xi_0 > 0$, we can take the logarithm of element A_r of that sum and make use of the gamma function to write

$$
\begin{aligned}
\ln A_r = {} & \ln \Gamma(r+1) + \ln \binom{y_1}{r} + \ln \binom{y_2}{r} + r \cdot (\ln \xi_0 - \ln \xi_1 - \ln \xi_2) \\
= {} & \ln \Gamma(r+1) + r \cdot (\ln \xi_0 - \ln \xi_1 - \ln \xi_2) \\
& + \ln \Gamma(y_1 + 1) - \ln \Gamma(r+1) - \ln \Gamma(y_1 - r + 1) \\
& + \ln \Gamma(y_2 + 1) - \ln \Gamma(r+1) - \ln \Gamma(y_2 - r + 1) \\
= {} & r \cdot (\ln \xi_0 - \ln \xi_1 - \ln \xi_2) - \ln \Gamma(r+1) \\
& + \ln \Gamma(y_1 + 1) - \ln \Gamma(y_1 - r + 1) + \ln \Gamma(y_2 + 1) - \ln \Gamma(y_2 - r + 1)
\end{aligned} \tag{17.9}
$$

Assuming that $\xi_1 > 0$ and $\xi_2 > 0$ are perfectly reasonable for our purposes, because if, say, $\xi_1 = 0$, then Y_1 is a point mass at zero. In our regression estimator, we will exit with an error if a dependent variable contains all zeros, because the model is inappropriate in that case. However, in arriving at (17.9), we assumed $\xi_0 > 0$, which we want to avoid: we want our function to return the bivariate probability even when Y_1 and Y_2 are uncorrelated.

So suppose $\xi_0 = 0$. Then as we saw earlier, $\Pr(y_1, y_2; \xi_1, \xi_2, 0) = \Pr(y_1; \xi_1) \cdot \Pr(y_2; \xi_2)$. If we abuse notation slightly and define $\ln A$ to be zero when $\xi_0 = 0$, then (17.8) and (17.9) along with this treatment for the case where $\xi_0 = 0$ provide us with a numerically good way to calculate our bivariate Poisson PMF.

In our derivation, we have obtained an expression for $\Pr(Y_1 = y_1, Y_2 = y_2; \boldsymbol{\xi})$, but the mean of Y_1 is $\mu_1 = \xi_1 + \xi_0$ and the mean of Y_2 is $\mu_2 = \xi_2 + \xi_0$. For the sake of convenience, we wrote our function to take the means μ_1 and μ_2 rather than the parameters ξ_1 and ξ_2. Our function thus takes five arguments in the order y_1, y_2, μ_1, μ_2, and ξ_0 and returns $\Pr(Y_1 = y_1, Y_2 = y_2; \mu_1, \mu_2, \xi_0)$. Later, we will want to calculate this probability for every observation in our sample, so we wrote our function to accept column vectors of all arguments and return a column vector of results. To keep the code easy to follow and because the upper limit of the sum that defines $\ln A$ is observation specific, we did not attempt to "vectorize" the function. Instead, we loop over the vectors of inputs and calculate probabilities one by one. Looping over observations in Stata is slow, but it is very fast in Mata. We also did not bother with sanity checks like making sure that all parameter vectors have the same length and that neither count variable contains negative numbers. Our code reads

```
──────────────────────────────────────── fragment ex_bipois1.do ────────────
clear all
set seed 12345
mata: mata clear
mata: mata set matastrict on

mata:
real colvector bipoissonp_1(real colvector y1,
                            real colvector y2,
                            real colvector mu1,
                            real colvector mu2,
                            real colvector xi0) {
        real scalar    i, r, lim, A, lnc, lnAr, lnpr
        real colvector pr, xi1, xi2

        pr = J(rows(y1), 1, .)
        // Go from (mu1, mu2, xi0)-space to (xi1, xi2, xi0)-space
        xi1 = mu1 - xi0
        xi2 = mu2 - xi0
        for(i=1; i<=rows(y1); ++i) {
                lnpr = -xi0[i] - xi1[i] - xi2[i] +
                        y1[i]*ln(xi1[i]) - lngamma(y1[i]+1) +
                        y2[i]*ln(xi2[i]) - lngamma(y2[i]+1)
                if (xi0[i] > 0) {
                        lim = min( (y1[i], y2[i]) )
                        A   = 0
                        lnc = ln(xi0[i]) - ln(xi1[i]) - ln(xi2[i])
                        for(r=0; r<=lim; ++r) {
                                lnAr = r*lnc - lngamma(r+1) +
                                        lngamma(y1[i]+1) - lngamma(y1[i]-r+1) +
                                        lngamma(y2[i]+1) - lngamma(y2[i]-r+1)
                                A = A + exp(lnAr)
                        }
                }
                else {
                        A = 1              // ln(1) = 0
                }
                pr[i] = exp(lnpr + ln(A))
        }
        return(pr)
}

...
──────────────────────────────────────── end ex_bipois1.do ────────────
```

We named this function `bipoissonp_1()`, with the _1 at the end, because we want
to reserve the name `bipoissonp()` for a better implementation we will describe shortly.
A quick test of our code shows that it "works", at least for scalars:

```
: bipoissonp_1(3, 4, 4.5, 3.5, 0.5)
  .027398122
```

Whenever you write a new function or estimator, one of the first things to consider is
how you will prove that your program returns correct results. As we have previously
remarked, if the covariance term $\xi_0 = 0$ for our bivariate Poisson distribution, then
$\Pr(y_1, y_2; \boldsymbol{\xi}) = \Pr(y_1; \xi_1) \cdot \Pr(y_2; \xi_2)$, and we made sure our function could handle that.
Let us test that case first. We have

```
: // Test case where covariance is 0
: bivp = bipoissonp_1(2, 3, 2.5, 4.1, 0) // bipoisson_1() takes outcomes first
: p1   = poissonp(2.5, 2)                // poissonp() takes mean first
: p2   = poissonp(4.1, 3)
: // Pr(Y1=2 | Y1 ~ Pois(2.5))
: p1
  .2565156207
: // Pr(Y2=3 | Y2 ~ Pois(4.1))
: p2
  .1903675602
: // Pr(Y1=2, Y2=3) given independence
: p1*p2
  .0488322529
: // Pr(Y1=2, Y2=3) using bipoissonp_1()
: bivp
  .0488322529
: reldif(bivp, p1*p2)
  1.98475e-17
```

In this example, the probability from `bipoissonp_1()` matches what we calculated using the independence of the two random variables to a high tolerance.

To check whether `bipoissonp_1()` produces accurate results in the general case, we return to the derivation of our bivariate Poisson distribution. If we draw u_1, u_2, and u_0 from independent Poisson variates U_1, U_2, and U_0 with means ξ_1, ξ_2, and ξ_0, respectively, then the random variables $Y_1 = U_1 + U_0$ and $Y_2 = U_2 + U_0$ with realized values $y_1 = u_1 + u_0$ and $y_2 = u_2 + u_0$ follow our bivariate Poisson distribution with means $\xi_1 + \xi_0$ and $\xi_2 + \xi_0$, respectively, and covariance ξ_0. Moreover, we obtained (17.5), which expresses our bivariate probability as a sum of products of three independent univariate Poisson probabilities. Therefore, we can use `poissonp()` to verify `bipoissonp_1()`.

Here is a simple test we wrote in **ex_bipois1.do** to see whether our method works. In fact, it took us a couple of tries to get all our logic straight. In our first try, we got different results from `bipoissonp_1()` and our method using `poissonp()`. Was our derivation of a bivariate Poisson distribution wrong? Did we make a mistake in writing our function to implement it? Was our code to test the function wrong? In our case, it turned out the test code was wrong: in one case, we had mistyped and called function `poisson()` rather than `poissonp()`.

```
: // Test a case where covariance > 0
: xi0  = 1
: xi1  = 5
: xi2  = 12
: u0   = rpoisson(1, 1, xi0)
: u1   = rpoisson(1, 1, xi1)
: u2   = rpoisson(1, 1, xi2)
: y1   = u1 + u0
: y2   = u2 + u0
: mu1  = xi1 + xi0
: mu2  = xi2 + xi0
: bivp = bipoissonp_1(y1, y2, mu1, mu2, xi0)
: lim  = min( (y1,y2) )
: p    = 0
: for(r=0; r<=lim; ++r) {
>         p = p + poissonp(xi0, r) * poissonp(xi1, y1-r) * poissonp(xi2, y2-r)
> }
: bivp
  .0083719062
: p
  .0083719062
: reldif(bivp, p)
  8.60161e-18
```

Our function value and our manual computation based on one random experiment agree to an acceptable level of tolerance. To put our function to a more rigorous test, we wrote a few lines of code to consider values of ξ_1 and ξ_2 from near 0 to as high as 500. For each (ξ_1, ξ_2), we randomly chose a feasible value of ξ_0 using the fact that a correlation coefficient cannot be greater than one. This portion of our code reads

───────────────────────────────── fragment ex_bipois1.do ───────

```
// Harder test
rdifs = J(500, 500, .)
for(i=0; i<500; ++i) {
        for(j=0; j<500; ++j) {
                xi1 = i + runiform(1,1)          // So we don't just test
                xi2 = j + runiform(1,1)          // integer means.
                lim = sqrt(xi1*xi2)              // Limit for xi0 based
                xi0 = runiform(1,1,0,lim)        // on rho <= 1

                u0  = rpoisson(1, 1, xi0)
                u1  = rpoisson(1, 1, xi1)
                u2  = rpoisson(1, 1, xi2)

                y1  = u1 + u0
                y2  = u2 + u0
                mu1 = xi1 + xi0
                mu2 = xi2 + xi0

                bivp = bipoissonp_1(y1, y2, mu1, mu2, xi0)

                lim = min( (y1,y2) )
                p   = 0
                for(r=0; r<=lim; ++r) {
                        p = p + poissonp(xi0, r) * poissonp(xi1, y1-r) *
                                                   poissonp(xi2, y2-r)
                }
                rdifs[i+1, j+1] = reldif(bivp, p)
        }
}
max(rdifs)
```

─────────────────────────────────────── end ex_bipois1.do ───────

Running this code, we obtain

```
: max(rdifs)
  4.24847e-15
```

Our function produces accurate results for means as high as 500. If we increase the limits to 1,000, the maximum relative difference is still only 8.76071×10^{-15}, which is sufficiently accurate.

The easy, accurate way

We derived the bivariate Poisson PMF as shown in (17.6), and we used that equation to show an interesting property regarding zero correlation and bivariate Poisson random variables. Campbell (1934), King (1989), Ozuna and Gomez (1994), and AlMuhayfith et al. (2016) all show either that equation or something similar to it. However, when verifying our function bipoissonp_1(), we hinted at an easier way to implement our PMF. We can just use (17.5).

As a general rule, if a probability distribution, density, or mass function is already available, then you should use it. It has already been carefully developed and thoroughly tested to produce accurate results across a broad range of inputs. Equation (17.5) allows us to do that. Our preferred implementation of the bivariate Poisson PMF is

```
————————————————————————————— fragment ex_bipois2.do —————————
real colvector bipoissonp(real colvector y1,
                          real colvector y2,
                          real colvector mu1,
                          real colvector mu2,
                          real colvector xi0) {
        real scalar      i, lim, p, r
        real colvector   pr, xi1, xi2

        pr = J(rows(y1), 1, .)
        // Go from (mu1, mu2, xi0)-space to (xi1, xi2, xi0)-space
        xi1 = mu1 - xi0
        xi2 = mu2 - xi0
        for(i=1; i<=rows(y1); ++i) {
                if (xi0[i] == 0) {
                        pr[i] = poissonp(xi1[i], y1[i]) *
                                poissonp(xi2[i], y2[i])
                }
                else {
                        lim = min( (y1[i], y2[i]) )
                        p = 0
                        for(r=0; r<=lim; ++r) {
                                p = p + poissonp(xi0[i], r) *
                                        poissonp(xi1[i], y1[i]-r) *
                                        poissonp(xi2[i], y2[i]-r)
                        }
                        pr[i] = p
                }
        }
        return(pr)
}
...
————————————————————————————————————— end ex_bipois2.do —————————
```

Testing function `bipoissonp()` is easy. We have written tests for `bipoissonp_1()` to verify that it works correctly. We can therefore use `bipoissonp_1()` to verify `bipoisson()`. In the do-file `ex_bipois2.do`, we include code to do that.

17.8.2   Bivariate Poisson regression

Mathematics

Now that we are familiar with the bivariate Poisson distribution, we can develop our regression estimator for bivariate counts. We have two dependent variables y_{1j} and y_{2j} for $j = 1, \ldots, N$. We want to model those counts as a function of regressors $\mathbf{x}_{1j}$ and $\mathbf{x}_{2j}$. We are agnostic as to whether the two sets of regressors coincide, are completely disjoint, or have some common elements and some unique elements.

Summarizing the discussion in [R] **poisson**, in the univariate case we have incidence rate r_j and exposure time E_j. r_j is parameterized as a function of covariates:

$$r_j = e^{\beta_0 + \beta_1 x_{1,j} + \cdots + \beta_k x_{k,j}}$$

The expected number of events C_j over the exposure window E_j is

$$C_j = E_j \cdot e^{\beta_0 + \beta_1 x_{1,j} + \cdots + \beta_k x_{k,j}}$$
$$= e^{\ln E_j + \beta_0 + \beta_1 x_{1,j} + \cdots + \beta_k x_{k,j}}$$

If you have variable E_j in your dataset, then you use option `exposure()` to include it in your model. If you happen to have variable $\ln E_j$ instead, then you use option `offset()`. In either case, the coefficient is constrained to be one. The probability model is, for realized count c_j,

$$\Pr(C_j = c_j; \mathbf{x}_j, \mathtt{offset}_j, \boldsymbol{\beta}) = e^{-\mu_j} \frac{\mu_j^{c_j}}{c_j!}$$
$$\ln \mu_j = \mathbf{x}_j'\boldsymbol{\beta} + \mathtt{offset}_j$$

The conditional mean of the jth outcome is μ_j. Modeling $\ln \mu_j$ as a linear combination of covariates ensures that μ_j is positive regardless of $\mathbf{x}_j$, $\boldsymbol{\beta}$, or $\mathtt{offset}_j$.

The bivariate Poisson PMF can be written in the (ξ_1, ξ_2, ξ_0) space as we derived in (17.6) and implied in (17.5). Alternatively, we made our implementations `bipoissonp()` and `bipoissonp_1()` work in the (μ_1, μ_2, ξ_0) space because the marginal mean of Y_i is $\xi_i + \xi_0$ for $i \in \{1, 2\}$ and we felt it would be more natural to specify the two means rather than ξ_1 and ξ_2.

The question at hand is whether we can follow `poisson` and have the two regression equations represent μ_{1j} and μ_{2j} or whether we must resign ourselves to modeling ξ_{1j} and ξ_{2j}. For the bivariate Poisson distribution, we must have $\xi_1 > 0$, $\xi_2 > 0$, and $\xi_0 \geq 0$. Imposing the strict inequalities is easy, but the constraint $\xi_0 \geq 0$ poses too many theoretical and practical difficulties we want to avoid. We are therefore going to instead require $\xi_0 > 0$.

For simplicity, our estimator will treat ξ_0 as a scalar parameter and not allow the user to make it a linear combination of variables. In section 17.7, we developed a command for seemingly unrelated linear regression, and even in the simple case of two equations, the user cannot parameterize the error correlation as a function of regressors. The same is true for Stata's `sureg` command; the bivariate probit command `biprobit` does not allow the user to do so either. We strongly suspect that allowing the correlation to depend on covariates would quite frequently lead to convergence problems.

If we model in the (ξ_1, ξ_2, ξ_0) space, defining our equations for `ml` is straightforward using the notation introduced in section 4.2:

$$\theta_{1j} = \ln \xi_{1j} = \mathbf{x}_{1j}'\boldsymbol{\beta}_1$$
$$\theta_{2j} = \ln \xi_{2j} = \mathbf{x}_{2j}'\boldsymbol{\beta}_2$$
$$\theta_{3j} = \ln \xi_{0j} = \beta_{3,1}$$

This will ensure that $\xi_1 > 0$, $\xi_2 > 0$, and $\xi_0 > 0$. If we instead model in the μ_1, μ_2, ξ_0 space and define our equations as

$$\theta_{1j} = \ln \mu_{1j} = \mathbf{x}'_{1j}\boldsymbol{\beta}_1$$
$$\theta_{2j} = \ln \mu_{2j} = \mathbf{x}'_{2j}\boldsymbol{\beta}_2$$
$$\theta_{3j} = \ln \xi_{0j} = \beta_{3,1}$$

then we will have $\mu_{1j} > 0$, $\mu_{2j} > 0$, and ξ_{0j}. However, this definition does not guarantee that $\xi_{1j} = \mu_{1j} - \xi_{0j} > 0$ or likewise for ξ_{2j}, as must be true of our bivariate Poisson distribution. We will therefore model in the (ξ_1, ξ_2, ξ_0) space. We will make sure that Stata's `margins` command works after our bivariate Poisson regression estimator so that the user can examine how the means of Y_1 and Y_2 change in response to changes in the regressors.

The log-likelihood function for the jth observation is simply a restatement of the joint probability of observing $Y_1 = y_{1j}$ and $Y_2 = y_{2j}$, where we focus on the parameters rather than the data. Letting $\boldsymbol{\theta}_j = (\theta_{1j}, \theta_{2j}, \theta_{3j})$ and $R_j = \min\{y_{1j}, y_{2j}\}$, the log likelihood for the jth observation is

$$\ln \ell_j(\boldsymbol{\theta}_j; y_{1j}, y_{2j}, \mathbf{x}_{1j}, \mathbf{x}_{2j}) = \qquad (17.10)$$

$$\ln \left\{ \sum_{r=0}^{R_j} \Pr(y_1 - r; e^{\theta_{1j}}) \cdot \Pr(y_2 - r; e^{\theta_{2j}}) \cdot \Pr(r; e^{\theta_{3j}}) \right\}$$

The log likelihood for the full sample, assuming independence, is $\ln L = \sum_{j=1}^{N} \ln \ell_j$. The first and second derivatives of this log likelihood are easy to derive; thus, our final command will use a method-`lf2` evaluator.

To obtain the necessary derivatives, first define S_{rj} as

$$S_{rj} = \Pr(y_{1j} - r; e^{\theta_{1j}}) \cdot \Pr(y_{2j} - r; e^{\theta_{2j}}) \cdot \Pr(r; e^{\theta_{3j}}) \qquad (17.11)$$

and let $S_j = \sum_{r=0}^{R_j} S_{rj}$. Differentiate S_{rj} with respect to θ_{kj} for $k \in \{1, 2, 3\}$, recalling the PMF of a Poisson random variable, to obtain

$$\frac{\partial S_{rj}}{\partial \theta_{kj}} = \begin{cases} S_{rj} \cdot (y_{kj} - r - e^{\theta_{kj}}) & k \in \{1, 2\} \\ S_{rj} \cdot (r - e^{\theta_{kj}}) & k = 3 \end{cases} \qquad (17.12)$$

Now just apply the rule for the derivative of the natural logarithm:

$$\frac{\partial \ln \ell_j(\boldsymbol{\theta}_j)}{\partial \theta_{kj}} = \frac{1}{S_j} \sum_{r=0}^{R_j} \frac{\partial S_{rj}}{\partial \theta_{kj}} \qquad (17.13)$$

The second derivatives are not much harder. Use the quotient rule on the first derivatives to obtain

$$\frac{\partial^2 \ln \ell_j(\boldsymbol{\theta}_j)}{\partial \theta_{kj} \partial \theta_{mj}} = \frac{1}{S_j^2} \left\{ S_j \cdot \sum_{r=0}^{R_j} \frac{\partial^2 S_{rj}}{\partial \theta_{kj} \partial \theta_{mj}} - \left(\sum_{r=0}^{R_j} \frac{\partial S_{rj}}{\partial \theta_{kj}} \right) \cdot \left(\sum_{r=0}^{R_j} \frac{\partial S_{rj}}{\partial \theta_{mj}} \right) \right\} \qquad (17.14)$$

for $k \in \{1, 2, 3\}$ and $m \in \{1, 2, 3\}$. Finally, use the product rule to differentiate (17.12) with respect to θ_m, substitute the equation for $\partial S_{rj}/\partial \theta_{mj}$ into its own derivative, rearrange, and get the final term we need:

$$
\frac{\partial^2 S_{rj}}{\partial \theta_{kj} \partial \theta_{mj}} = \begin{cases} S_{rj} \cdot \left\{ (y_{kj} - r - e^{\theta_{kj}})^2 - e^{\theta_{kj}} \right\} & k = m \in \{1, 2\} \\ S_{rj} \cdot (y_{kj} - r - e^{\theta_{kj}}) \cdot (y_{mj} - r - e^{\theta_{mj}}) & k \neq m, k \in \{1, 2\}, m \in \{1, 2\} \\ S_{rj} \cdot (y_{kj} - r - e^{\theta_{kj}}) \cdot (r - e^{\theta_{mj}}) & k \in \{1, 2\}, m = 3 \\ S_{rj} \cdot \left\{ (r - e^{\theta_{kj}})^2 - e^{\theta_{kj}} \right\} & k = m = 3 \end{cases}
$$

Code

In section 15.5, we wrote a log-likelihood evaluator function in Mata, and we saved the compiled function object as a .mo file using `mata mosave`. There we had just a single Mata function, so `mata mosave` was the perfect solution. Here we are going to write several functions in Mata, including our probability mass function, our log-likelihood evaluator, and some helper functions that will make evaluating the log likelihood and its derivatives easier. Because we have multiple functions that will work together to implement our command, a Mata library is a better alternative than creating a separate .mo file for each function. This way we will have just one file that ends in .mlib that we must distribute with our ado-files.

We wrote three files as we developed the estimation command; they are listed in appendix E.8.

1. `mybipoisson.ado`—This will be the main ado-file that implements our command `mybipoisson`. As we develop the log-likelihood evaluator, add first derivatives, and then add second derivatives, we will change the `ml` method in our `ml model` statement.

2. `make_lmybipoisson_mlib.do`—This file will contain the Mata functions needed to implement `mybipoisson`, including methods `lf0`, `lf1`, and `lf2` log-likelihood evaluators, our function to compute bivariate Poisson probabilities, and various helper functions. Running the `do` command on this do-file will compile all of those functions and produce Mata library `lmybipoisson.mlib`.

3. `mybipoisson_p.ado`—This file will allow us to predict conditional means, scores, and other statistics after estimation. This file will also enable us to use `margins` after we fit our model.

The ado-file `mybipoisson.ado` has a similar structure to the other ado-files we have developed in this chapter as well as in chapter 13, so we do not examine it closely here. We borrowed some of the parsing logic from `mysureg.ado` to allow the user to specify two equations bound in parentheses. Because there is nothing fundamentally different from other commands' ado-files, we will not discuss it further. Likewise, file `mybipoisson_p.ado` is rather straightforward to follow on your own.

We do want to discuss some highlights of `make_lmybipoisson_mlib.do` because we have not encountered a file like it before in this book. We first focus on the top and bottom portions of the file because the bottom portion is key to making our estimator work.

```
────────────────────────── fragment make_lmybipoisson_mlib.do ──────────
// Likelihood evaluator and other functions for -mybipoisson-.

clear all

mata:
mata set matastrict on

// This is the method-lf0 evaluator() -- no derivatives.
void mybipoisson_lf0(transmorphic scalar ML, real scalar    todo,
                     real rowvector    b , real colvector lnfj,
                     real matrix       S , real matrix    H)
{
        real colvector   y1, y2, xi1, xi2, xi0

        y1  = moptimize_util_depvar(ML, 1)
        y2  = moptimize_util_depvar(ML, 2)
        xi1 = exp(moptimize_util_xb(ML, b, 1))
        xi2 = exp(moptimize_util_xb(ML, b, 2))
        xi0 = J(rows(y1), 1, exp(moptimize_util_xb(ML, b, 3)))

        lnfj = ln( bipoissonp_xi(y1, y2, xi1, xi2, xi0) )
}
...
mata mlib create lmybipoisson, replace
mata mlib add lmybipoisson *()
mata mlib index

end
──────────────────────────── end make_lmybipoisson_mlib.do ──────────
```

The first part of this do-file is relatively straightforward. We do a `clear all`, mainly so that we can clear out previous versions of the Mata functions we are about to define as well as any others that may be lurking. We then enter the Mata environment and turn on strict enforcement of variable declarations.

Our method-lf0 evaluator is named `mybipoisson_lf0()`, and we wrote it following the prototype in section 15.2. In the first few lines of code, we obtain our dependent variables as well as our model parameters, remembering that we are modeling the natural logarithms of the bivariate Poisson distribution's parameters. Finally, we call our implementation of the distribution's PMF. One difference from section 17.8.1 is that here we wrote a version of the PMF that takes as arguments ξ_1 and ξ_2 rather than μ_1 and μ_2.

Notice the three lines that begin `mata mlib`. This is how we create our Mata `mlib` library. In the first line, we instruct Mata to create a new, empty library called `lmybipoisson`. Doing so creates the file `lmybipoisson.mlib`, and the `replace` option tells Mata to overwrite the existing version if it exists. The second line adds to our new library all functions currently in Mata's memory—the wildcard `*()` matches all functions. That is also why we typed `clear all` at the top of our do-file: we want to add to `lmybipoisson` only those functions that we have written to implement `mybipoisson`.

The third line, `mata mlib index`, will allow Mata to find the functions we just defined. When you first launch Stata, it searches your ado-path for any Mata libraries—files that end in `.mlib`—and records what functions are defined in them. Many features of Stata and Mata, in fact, are implemented using functions stored in Mata libraries; this searching and indexing is what allows Stata to operate correctly. `mata mlib index` causes Stata to rebuild that index of functions, so that if you happen to `clear` Mata, it will still know where the functions we just defined are located.

Incidentally, say you wrote your own estimator similar to `mybipoisson` using a Mata library to hold your functions and you want to share it with a colleague. Then you would need to distribute your `.mlib` file as well as your ado-files. The person who receives your files and stores them in an ado-path would then either need to exit and restart Stata or type `mata: mata mlib index` at the Stata prompt. For more information about where Stata searches for ado-files and `.mlib` files, see [U] **17.5 Where does Stata look for ado-files?**.

Later in `make_lmybipoisson_mlib.do`, we define our method-lf2 evaluator function, `mybipoisson_lf2()`. The part that computes the log likelihood is identical to our method-lf0 version, though we do want to highlight a few points that made coding the derivatives easier.

```
———————————————————————— fragment make_lmybipoisson_mlib.do ————————

...
// This is the method-lf2() evaluator. You can also use it with
// methods lf0 and lf1.
void mybipoisson_lf2(transmorphic scalar ML, real scalar      todo,
                     real rowvector      b , real colvector lnfj,
                     real matrix         S , real matrix       H)
{
        real colvector y1, y2, xi1, xi2, xi0
        real matrix       G, Hraw, H11, H12, H13, H22, H23, H33

        y1  = moptimize_util_depvar(ML, 1)
        y2  = moptimize_util_depvar(ML, 2)
        xi1 = exp(moptimize_util_xb(ML, b, 1))
        xi2 = exp(moptimize_util_xb(ML, b, 2))
        xi0 = J(rows(y1), 1, exp(moptimize_util_xb(ML, b, 3)))

        // The following lines are necessary to handle constant-
        // only models. All of our functions assume xi1 and xi2
        // are colvectors; -ml- turns constant-only equations into
        // scalars.
        if (rows(xi1) == 1) {
                xi1 = J(rows(y1), 1, xi1)
                xi2 = J(rows(y1), 1, xi2)
        }

        lnfj = ln( bipoissonp_xi(y1, y2, xi1, xi2, xi0) )
        if (todo==0) return              // No derivatives

        G = ln_bip_xi_grad(y1, y2, xi1, xi2, xi0)
        S = J(rows(G), 3, 0)
        S[., 1] = G[.,1] :/ G[.,4]
        S[., 2] = G[.,2] :/ G[.,4]
        S[., 3] = G[.,3] :/ G[.,4]
        if (todo==1) return              // Just gradient
```

```
            Hraw = ln_bip_xi_hessian(y1, y2, xi1, xi2, xi0, G)
            H11  = moptimize_util_matsum(ML, 1, 1, Hraw[.,1], lnfj[1])
            H12  = moptimize_util_matsum(ML, 1, 2, Hraw[.,2], lnfj[1])
            H13  = moptimize_util_matsum(ML, 1, 3, Hraw[.,3], lnfj[1])
            H22  = moptimize_util_matsum(ML, 2, 2, Hraw[.,4], lnfj[1])
            H23  = moptimize_util_matsum(ML, 2, 3, Hraw[.,5], lnfj[1])
            H33  = moptimize_util_matsum(ML, 3, 3, Hraw[.,6], lnfj[1])
            H    = ( H11 , H12 , H13 \
                     H12', H22 , H23 \
                     H13', H23', H33 )
    }
    ...
```

────────────────────────────────────── end `make_lmybipoisson_mlib.do` ──────────

Instead of implementing the log likelihood as well as all the calculations for the derivatives in one function, we wrote a couple of helper functions to take care of components that make up the gradient and Hessian. In general, shorter functions that do one thing are easier to write and debug. Here we were clever in how we wrote those helper functions.

Equation (17.13) shows that three parameters' first derivatives are equal to a summation that varies by parameter divided by a term S_j that is not parameter specific. Rather than return the three gradient vectors directly as an $N \times 3$ matrix, our utility for the first derivatives, `ln_bip_xi_grad()`, returns an $N \times 4$ matrix. The first three columns contain the three summation terms, and the fourth column contains the S_j term. Why did we do that?

Looking at (17.14) for the second derivatives, we see the same summations that appear in (17.13) appear here as well. Rather than compute these sums twice, the gradient function returns them to our likelihood evaluator so that we can pass them to and reuse them in our Hessian computations.

The S_{rj} term at first appears as if we could also compute it just once and use it in both the gradient and Hessian computations. The problem is that $R_j = \min\{y_{1j}, y_{2j}\}$ is observation specific, so the number of S_{rj} terms varies by observation. Simply recalculating the terms as we need them is certainly easier to understand and perhaps just as fast as any complicated solution using pointers we could dream up.

Finally, method-`lf2` requires us to return the 3×3 Hessian matrix for our three-parameter model. We wrote a utility, `ln_bip_xi_hessian()`, to return an $N \times 6$ matrix, and we then used `moptimize_util_matsum` to get the overall Hessian matrix. By using that function, we do not need to worry about accounting for any weights the user may have specified. We could have written our Hessian utility to return the 3×3 matrix for us, but it would have been more hassle: we would need to pass in our problem handle `ML` as well as the log-likelihood vector. At least by our reckoning, returning the $N \times 6$ matrix made more sense.

❏ **Technical note**

Rather than using `mata mlib` commands to create and populate our library, we can alternatively use the `lmbuild` command. See [M-3] **lmbuild**. That command is arguably easier to use, but it would require us to create a do-file to build our library as well as another file to hold all our functions' code. Using `mata mlib` allows us to put everything in one do-file, and, at least for our purposes, it is not much more difficult to use.

❏

Example

`donuts.dta` contains a fictitious survey of 1,753 commuters who stopped at a popular convenience store on their way to work. For each customer, the store recorded the number of cups of coffee and the number of donuts purchased. The store also asked commuters how many miles the commute was, whether they worked in a cubicle, the number of stop signs or traffic lights they encountered, and whether they went to a gym at least once per week. Most commuters were driving cars, but some were riding bicycles. We expect bicyclists to purchase fewer cups of coffee on average because it is difficult to carry multiple cups of coffee while riding a bicycle. We also surmise that those who work out more frequently are less likely to purchase donuts because these small bundles of joy pack much sugar and many carbohydrates.

```
                                          ─── begin ex_mybipoisson_1 ───
. use donuts

. mybipoisson (coffee miles i.cubicle i.bike n_stops)
>             (donuts miles i.cubicle i.bike i.gym)
Initial:      Log likelihood = -6362.5474
Rescale:      Log likelihood = -6065.8849
Rescale eq:   Log likelihood = -6065.8849
Iteration 0:  Log likelihood = -6065.8849
Iteration 1:  Log likelihood = -5572.9767
Iteration 2:  Log likelihood = -5403.6169
Iteration 3:  Log likelihood =  -5369.867
Iteration 4:  Log likelihood = -5369.7299
Iteration 5:  Log likelihood = -5369.7299
                                          Number of obs =  1,753
                                          Wald chi2(4)  = 360.98
Log likelihood = -5369.7299               Prob > chi2   = 0.0000
```

	Coefficient	Std. err.	z	P>|z|	[95% conf. interval]	
coffee						
miles	.0358512	.0068093	5.27	0.000	.0225052	.0491973
1.cubicle	.2129047	.060994	3.49	0.000	.0933586	.3324508
1.bike	-2.211014	.125469	-17.62	0.000	-2.456929	-1.965099
n_stops	-.0481949	.0108183	-4.45	0.000	-.0693983	-.0269915
_cons	-.0474623	.0852912	-0.56	0.578	-.21463	.1197053
donuts						
miles	.0301776	.0055329	5.45	0.000	.0193333	.041022
1.cubicle	.851857	.0595431	14.31	0.000	.7351546	.9685593
1.bike	-.0051329	.0816582	-0.06	0.950	-.1651799	.1549142
1.gym	-2.861991	.1143904	-25.02	0.000	-3.086192	-2.63779
_cons	-.1742461	.0980474	-1.78	0.076	-.3664154	.0179232
/ln_xi0	.3314428	.0228638	14.50	0.000	.2866307	.376255

```
. margins, predict(mu1) dydx(bike)
Average marginal effects                  Number of obs = 1,753
Model VCE: OIM

Expression: Expected count for coffee, predict(mu1)
dy/dx wrt:  1.bike
```

	dy/dx	Delta-method std. err.	z	P>|z|	[95% conf. interval]	
1.bike	-1.199135	.0575391	-20.84	0.000	-1.311909	-1.08636

Note: dy/dx for factor levels is the discrete change from the base level.

```
. margins, predict(mu2) dydx(gym)
```

Average marginal effects   Number of obs = 1,753
Model VCE: OIM

Expression: Expected count for donuts, predict(mu2)
dy/dx wrt: 1.gym

	dy/dx	Delta-method std. err.	z	P>|z|	[95% conf. interval]
1.gym	-2.067405	.0696793	-29.67	0.000	-2.203974 -1.930836

Note: dy/dx for factor levels is the discrete change from the base level.

——————————————————————————————————— end ex_mybipoisson_1 ———————

17.8.3   Discussion

We started out this section asking how we could extend the seemingly unrelated regression model of the previous section to handle two count variables. We then derived a popular version of the bivariate Poisson distribution and wrote Mata functions to implement it. Doing that allowed us to introduce just a few of the subtleties that arise when implementing probability distributions, and we discussed a couple of strategies for checking whether our function produced accurate results. Once we were familiar with the bivariate Poisson distribution, writing a regression estimator was straightforward. We expressed the PMF's parameters using ml's θ-notation and applied the methods we have learned in previous chapters of this book to write an estimation command.

Our command mybipoisson implements the estimator we derived, but it is certainly not state of the art and has several shortcomings. We have already noted that the bivariate Poisson distribution we used requires that the two variables be positively correlated. The community-contributed command bivcnto described in Xu and Hardin (2016) fits versions of bivariate Poisson and bivariate negative binomial regression models that allow for the two outcomes to be negatively correlated. Also see Ghosh et al. (2021) and the many articles cited therein for additional developments on correlated count models that allow for negative correlation.

Even if our model is appropriate for a particular exercise, there are still shortcomings that could be addressed. We have written our estimator based on standard Newton—Raphson or quasi-Newton optimization techniques as implemented in ml, but this model can be difficult to estimate that way. Apparently the expectation-maximization algorithm is a better choice for this model. Because a correlation coefficient is bounded from above at one, we must have $\xi_0 \leq (\xi_1\xi_2)^{1/2}$, but our command does not impose that restriction. Some sources (for example, Loukas and Kemp [1986]) give even tighter bounds on the covariance parameter ξ_0. In our estimator, the covariance term is observation specific. What if it exceeds one for some observations?

17.9 Epilogue

We hope that having read this book, you will agree that once you have a log-likelihood function like (17.10), writing an estimation command to take a user's input, maximize the log-likelihood function, and display the output is not as intimidating as you may have imagined. We have taken the covers off some of Stata's most popular estimators and shown how you can write your own commands to implement them, even allowing for weights, robust variance matrix estimators, and complex survey designs. In the last example, we took you along all the way from a research question—how can we model a pair of count variables?—to an estimation command.

Stata's `ml` command handles many details for us so we can focus on writing code to implement the log-likelihood function. Most programming languages include functionality to do numerical optimization. `ml` contributes so much more, including a consistent θ-notation for defining equations and tools to search for good initial parameters, debug your log-likelihood function and derivatives, and obtain robust variance matrices without needing to use additional packages or libraries. For the simplest models, Stata's `mlexp` command lets you specify your log-likelihood function just as you would write it and obtain estimates with no programming. For the most complicated models, you can use `moptimize()` and take advantage of `ml`'s consistent notation and features while working entirely in the Mata environment.

A Syntax of mlexp

Syntax

mlexp (*lexp*) [*if*] [*in*] [*weight*] [*options*]

where *lexp* is a substitutable expression representing the log-likelihood function.

options	Description
Model	
<u>variab</u>les(*varlist*)	specify variables in model
from(*initial_values*)	specify initial values for parameters
<u>constr</u>aints(*numlist*)	apply specified linear constraints
Derivatives	
<u>deriv</u>ative(/*name* = *dexp*)	specify derivative of *lexp* with respect to parameter *name*; can be specified more than once
SE/Robust	
vce(*vcetype*)	*vcetype* may be oim, opg, <u>r</u>obust, <u>cl</u>uster *clustvar*, <u>boot</u>strap, or <u>jack</u>knife
Reporting	
<u>l</u>evel(*#*)	set confidence level; default is level(95)
title(*string*)	display *string* as title above the table of parameter estimates
title2(*string*)	display *string* as subtitle
display_options	control columns and column formats, row spacing, line width, display of omitted variables and base and empty cells, and factor-variable labeling
Maximization	
maximize_options	control the maximization process; seldom used
debug	display debug output
<u>coll</u>inear	keep collinear variables
<u>coefl</u>egend	display legend instead of statistics

lexp and *dexp* may contain factor variables and time-series operators; see [U] **11.4.3 Factor variables** and [U] **11.4.4 Time-series varlists**.

bootstrap, by, collect, jackknife, rolling, statsby, and svy are allowed; see [U] **11.1.10 Prefix commands**.

Weights are not allowed with the bootstrap prefix; see [R] **bootstrap**.

aweights are not allowed with the jackknife prefix; see [R] **jackknife**.

vce() and weights are not allowed with the svy prefix; see [SVY] **svy**.

aweights, fweights, iweights, and pweights are allowed; see [U] **11.1.6 weight**.

debug, collinear, and coeflegend do not appear in the dialog box.

See [U] **20 Estimation and postestimation commands** for more capabilities of estimation commands.

lexp and *dexp* are substitutable expressions, Stata expressions that also contain parameters to be estimated. The parameters are enclosed in curly braces and must satisfy the naming requirements for variables; {beta} is an example of a parameter. The notation {*lc: varlist*} is allowed for linear combinations of multiple covariates and their parameters. For example, {xb: mpg price turn _cons} defines a linear combination of the variables mpg, price, turn, and _cons (the constant term). See *Substitutable expressions* under *Remarks and examples* in [R] **mlexp**.

Description

mlexp performs maximum likelihood estimation of models that satisfy the linear-form restrictions, that is, models for which you can write the log likelihood for an individual observation and for which the overall log likelihood is the sum of the individual observations' log likelihoods.

You express the observation-level log-likelihood function by using a substitutable expression. Unlike models fit using ml, you do not need to do any programming. However, ml can fit classes of models that cannot be fit by mlexp.

Options

    ⌐ Model ⌐──

variables(*varlist*) specifies the variables in the model. mlexp excludes observations for which any of these variables has missing values. If you do not specify variables(), then mlexp assumes all observations are valid. mlexp will exit with an error message if the log likelihood cannot be calculated at the initial values for any observation.

from(*initial_values*) specifies the initial values to begin the estimation. You can specify parameter names and values, or you can specify a $1 \times k$ matrix, where k is the number of parameters in the model. For example, to initialize alpha to 1.23 and delta to 4.57, you would type

```
mlexp ..., from(alpha=1.23 delta=4.57) ...
```

or equivalently

```
matrix define initval = (1.23, 4.57)
mlexp ..., from(initval) ...
```

Initial values declared in the `from()` option override any that are declared within substitutable expressions. If you specify a parameter that does not appear in your model, `mlexp` exits with an error. If you specify a matrix, the values must be in the same order in which the parameters are declared in your model.

constraints(*numlist*); see [R] **Estimation options**.

    ⌐ Derivatives ⌐――――――――――――――――――――――――――――――――――――

derivative(/*name* = *dexp*) specifies the derivative *dexp* of the observation-level log-likelihood function with respect to parameter *name*. If you wish to specify analytic derivatives, you must specify `derivative()` for each parameter in your model.

    dexp uses the same substitutable expression syntax as is used to specify the log-likelihood function. If you declare a linear combination in the log-likelihood function, you provide the derivative for the linear combination; `mlexp` then applies the chain rule for you. See *Specifying derivatives* under *Remarks and examples* in [R] **mlexp** for examples.

    If you do not specify the `derivative()` option, `mlexp` calculates derivatives numerically.

    ⌐ SE/Robust ⌐――――――――――――――――――――――――――――――――――――

vce(*vcetype*) specifies the type of standard error reported, which includes types that are derived from asymptotic theory (`oim`, `opg`), that are robust to some kinds of misspecification (`robust`), that allow for intragroup correlation (`cluster` *clustvar*), and that use bootstrap or jackknife methods (`bootstrap`, `jackknife`); see [R] ***vce_option***.

    ⌐ Reporting ⌐――――――――――――――――――――――――――――――――――――

level(*#*); see [R] **Estimation options**.

title(*string*) specifies an optional title that will be displayed just above the table of parameter estimates.

title2(*string*) specifies an optional subtitle that will be displayed between the title specified in `title()` and the table of parameter estimates. If `title2()` is specified but `title()` is not, then `title2()` has the same effect as `title()`.

display_options are the following: `noci`, `nopvalues`, `noomitted`, `vsquish`, `noemptycells`, `baselevels`, `allbaselevels`, `nofvlabel`, `fvwrap(#)`, `fvwrapon(`*style*`)`, `cformat(%`*fmt*`)`, `pformat(%`*fmt*`)`, `sformat(%`*fmt*`)`, and `nolstretch`; see [R] **Estimation options**.

| Maximization |

maximize_options are the following: <u>diff</u>icult, <u>tech</u>nique(*algorithm_spec*), <u>iter</u>ate(*#*), [no]log, <u>trace</u>, <u>gradient</u>, showstep, <u>hess</u>ian, <u>showtol</u>erance, <u>tol</u>erance(*#*), <u>ltol</u>erance(*#*), <u>nrtol</u>erance(*#*), and <u>nonrtol</u>erance; see [R] **Maximize**. These options are seldom used.

The following options are available with mlexp but are not shown in the dialog box:

debug specifies that differences between the numerically computed gradient and the gradient computed from your derivative expression are reported at each iteration. This option is only allowed with the derivative() option.

collinear, coeflegend; see [R] **Estimation options**.

Stored results

mlexp stores the following in e():

Scalars
e(N)	number of observations
e(k)	number of parameters
e(k_aux)	number of ancillary parameters
e(k_eq)	number of equations in e(b)
e(k_eq_model)	number of equations in overall model test
e(df_m)	model degrees of freedom
e(ll)	log likelihood
e(N_clust)	number of clusters
e(rank)	rank of e(V)
e(ic)	number of iterations
e(rc)	return code
e(converged)	1 if converged, 0 otherwise

Macros
e(cmd)	mlexp
e(cmdline)	command as typed
e(lexp)	likelihood expression
e(wtype)	weight type
e(wexp)	weight expression
e(usrtitle)	user-specified title
e(usrtitle2)	user-specified secondary title
e(clustvar)	name of cluster variable
e(vce)	*vcetype* specified in vce()
e(vcetype)	title used to label Std. err.
e(params)	names of parameters
e(hasderiv)	yes, if derivative() is specified
e(d_j)	derivative expression for parameter j
e(rhs)	contents of variables()
e(opt)	type of optimization
e(ml_method)	type of ml method
e(technique)	maximization technique

e(singularHmethod)	m-marquardt or hybrid; method used when Hessian is singular[1]
e(crittype)	optimization criterion[1]
e(properties)	b V
e(estat_cmd)	program used to implement estat
e(predict)	program used to implement predict
e(marginsok)	predictions allowed by margins
e(marginsnotok)	predictions disallowed by margins
e(marginsprop)	signals to the margins command
e(asbalanced)	factor variables fvset as asbalanced
e(asobserved)	factor variables fvset as asobserved

Matrices

e(b)	coefficient vector
e(Cns)	constraints matrix
e(ilog)	iteration log (up to 20 iterations)
e(init)	initial values
e(gradient)	gradient vector
e(V)	variance–covariance matrix of the estimators
e(V_modelbased)	model-based variance

Functions

| e(sample) | marks estimation sample |

1. Type ereturn list, all to view these results; see [P] **return**.

In addition to the above, the following is stored in r():

Matrices

| r(table) | matrix containing the coefficients with their standard errors, test statistics, p-values, and confidence intervals |

Note that results stored in r() are updated when the command is replayed and will be replaced when any r-class command is run after the estimation command.

Methods and formulas

Optimization is carried out using moptimize(); see [M-5] **moptimize( )**.

B Syntax of ml

Syntax

ml model in interactive mode

    `ml model`   *method progname eq* [*eq* ...] [*if*] [*in*] [*weight*]
                           [, *model_options* `svy` *diparm_options*]

    `ml model`   *method funcname() eq* [*eq* ...] [*if*] [*in*] [*weight*]
                           [, *model_options* `svy` *diparm_options*]

ml model in noninteractive mode

    `ml model`   *method progname eq* [*eq* ...] [*if*] [*in*] [*weight*] , `maximize`
                         [*model_options* `svy` *diparm_options* *noninteractive_options*]

    `ml model`   *method funcname() eq* [*eq* ...] [*if*] [*in*] [*weight*] ,
                         `maximize` [*model_options* `svy` *diparm_options*
                         noninteractive_options]

Noninteractive mode is invoked by specifying the `maximize` option. Use `maximize` when `ml` will be used as a subroutine of another ado-file or program and you want to carry forth the problem, from definition to posting of final results, in one command.

```
ml clear

ml query

ml check

ml search     [ [/]eqname[:] #lb #ub ] [ ... ] [ , search_options ]

ml plot       [eqname:]name [ # [ # [ #]]]
              [ , saving(filename[ , replace]) ]

ml init       { [eqname:]name=# | /eqname=# } [ ... ]
ml init       # [ # ... ] , copy
ml init       matname [ , skip copy ]

ml report

ml trace      {on | off}

ml count      [ clear | on | off ]

ml maximize   [ , ml_maximize_options display_options eform_option ]

ml graph      [#] [ , saving(filename[ , replace]) ]

ml display    [ , display_options eform_option ]

ml footnote

ml score      newvar [if] [in] [ , equation(eqname) missing ]

ml score      newvarlist [if] [in] [ , missing ]

ml score      [type] stub* [if] [in] [ , missing ]
```

where *method* is one of

lf	d0	lf0	gf0
	d1	lf1	
	d1debug	lf1debug	
	d2	lf2	
	d2debug	lf2debug	

method can be specified using one of the longer, more descriptive names if desired:

method	Longer name
lf	linearform
d0	derivative0
d1	derivative1
d1debug	derivative1debug
d2	derivative2
d2debug	derivative2debug
lf0	linearform0
lf1	linearform1
lf1debug	linearform1debug
lf2	linearform2
lf2debug	linearform2debug
gf0	generalform0

eq is the equation to be estimated, enclosed in parentheses, and optionally with a name to be given to the equation, preceded by a colon:

$$([\,eqname\!:\,]\ [\,varlist_y\ =\,]\ [\,varlist_x\,]\ [\,,\ eq\_options\,])$$

Or *eq* is the name of a parameter, such as `sigma`, with a slash in front:

$$/eqname \quad \text{which is equivalent to} \quad (eqname\!:\,,\ \texttt{freeparm})$$

diparm_options is one or more `diparm(`*diparm_args*`)` options where *diparm_args* is either `__sep__` or anything accepted by the "undocumented" `_diparm` command; see `help _diparm`.

eq_options	Description
noconstant	do not include intercept in the equation
offset(*varname_o*)	include *varname_o* in model with coefficient constrained to be 1
exposure(*varname_e*)	include ln(*varname_e*) in model with coefficient constrained to be 1
freeparm	*eqname* is a free parameter

model_options	Description
group(*varname*)	use *varname* to identify groups
vce(*vcetype*)	*vcetype* may be <u>robust</u>, <u>cl</u>uster *clustvar*, oim, or opg
<u>constraints</u>(*numlist*)	constraints by number to be applied
<u>constraints</u>(*matname*)	matrix that contains the constraints to be applied
<u>nocnsnotes</u>	do not display notes when constraints are dropped
<u>title</u>(*string*)	place title on the estimation output
<u>nopreserve</u>	do not preserve the estimation subsample in memory
<u>collinear</u>	keep collinear variables within equations
<u>missing</u>	keep observations containing variables with missing values
lf0($\#_k$ $\#_{ll}$)	number of parameters and log-likelihood value of "constant-only" model
<u>contin</u>ue	used to specify that model has been fit and set initial values $\mathbf{b}_0$ for model to be fit based on those results
<u>waldt</u>est($\#$)	perform Wald test; see *Options for use with ml model in interactive or noninteractive mode* below
obs($\#$)	number of observations
crittype(*string*)	describe the criterion optimized by ml
<u>subpop</u>(*varname*)	compute estimates for the single subpopulation
<u>nosvy</u>adjust	carry out Wald test as $W/k \sim F(k,d)$
<u>techni</u>que(*algorithm_spec*)	specify how the likelihood function is to be maximized

noninteractive_options	Description
<u>init</u>(*ml_init_args*)	set the initial values $\mathbf{b}_0$
<u>search</u>(on)	equivalent to ml search, repeat(0); default
<u>search</u>(norescale)	equivalent to ml search, repeat(0) norescale
<u>search</u>(quietly)	same as search(on) except that output is suppressed
<u>search</u>(off)	prevents calling ml search
<u>repeat</u>($\#$)	ml search's repeat() option; see below
<u>bounds</u>(*ml_search_bounds*)	specify bounds for ml search
<u>nowarn</u>ing	suppress "convergence not achieved" message of iterate(0)
novce	substitute the zero matrix for the variance matrix
negh	indicate that the evaluator return the negative Hessian matrix
<u>sc</u>ore(*newvars*)	new variables containing the contribution to the score
maximize_options	control the maximization process; seldom used

search_options	Description
<u>r</u>epeat(#)	number of random attempts to find better initial-value vector; default is repeat(10) in interactive mode and repeat(0) in noninteractive mode
<u>rest</u>art	use random actions to find starting values; not recommended
<u>noresca</u>le	do not rescale to improve parameter vector; not recommended
maximize_options	control the maximization process; seldom used

ml_maximize_options	Description
<u>nowarn</u>ing	suppress "convergence not achieved" message of iterate(0)
novce	substitute the zero matrix for the variance matrix
negh	indicate that the evaluator return the negative Hessian matrix
<u>score</u>(newvars | stub*)	new variables containing the contribution to the score
<u>nooutp</u>ut	suppress the display of the final results
<u>noskip</u>line	suppress the display of the blank line before the iteration log
noclear	do not clear ml problem definition after model has converged
maximize_options	control the maximization process; seldom used

display_options	Description
<u>noh</u>eader	suppress display of the header above the coefficient table
<u>nof</u>ootnote	suppress display of the footnote below the coefficient table
<u>l</u>evel(*#*)	set confidence level; default is level(95)
<u>f</u>irst	display coefficient table reporting results for first equation only
neq(*#*)	display coefficient table reporting first *#* equations
showeqns	display equation names in the coefficient table
<u>p</u>lus	display coefficient table ending in dashes–plus-sign–dashes
nocnsreport	suppress constraints display above the coefficient table
noomitted	suppress display of omitted variables
vsquish	suppress blank space separating factor-variable terms or time-series–operated variables from other variables
noemptycells	suppress empty cells for interactions of factor variables
<u>base</u>levels	report base levels of factor variables and interactions
<u>allbase</u>levels	display all base levels of factor variables and interactions
cformat(%*fmt*)	format the coefficients, standard errors, and confidence limits in the coefficient table
pformat(%*fmt*)	format the *p*-values in the coefficient table
sformat(%*fmt*)	format the test statistics in the coefficient table
nolstretch	do not automatically widen the coefficient table to accommodate longer variable names
<u>coef</u>legend	display legend instead of statistics

eform_option	Description
<u>ef</u>orm(*string*)	display exponentiated coefficients; column title is "*string*"
<u>ef</u>orm	display exponentiated coefficients; column title is "exp(b)"
hr	report hazard ratios
shr	report subhazard ratios
<u>ir</u>r	report incidence-rate ratios
or	report odds ratios
<u>rr</u>r	report relative-risk ratios

`fweights`, `pweights`, `aweights`, and `iweights` are allowed; see [U] **11.1.6 weight**. With all but method `lf`, you must write your likelihood-evaluation program carefully if `pweights` are to be specified, and `pweights` may not be specified with method `d0`, `d1`, `d1debug`, `d2`, or `d2debug`.

See [U] **20 Estimation and postestimation commands** for more capabilities of estimation commands.

To redisplay results, type `ml display`.

Syntax of subroutines for use by evaluator programs

`mleval`	$newvar$ = $vecname$ $\big[$, `eq(#)` $\big]$
`mleval`	$scalarname$ = $vecname$, `scalar` $\big[$ `eq(#)` $\big]$
`mlsum`	$scalarname_{\mathrm{lnf}}$ = exp $\big[$ if $\big]$ $\big[$, <u>`noweight`</u> $\big]$
`mlvecsum`	$scalarname_{\mathrm{lnf}}$ $rowvecname$ = exp $\big[$ if $\big]$ $\big[$, `eq(#)` $\big]$
`mlmatsum`	$scalarname_{\mathrm{lnf}}$ $matrixname$ = exp $\big[$ if $\big]$ $\big[$, `eq(#`$\big[$ `,#`$\big]$`)` $\big]$
`mlmatbysum`	$scalarname_{\mathrm{lnf}}$ $matrixname$ $varname_a$ $varname_b$ $\big[$ $varname_c$ $\big]$ $\big[$ if $\big]$, `by(`$varname$`)` $\big[$ `eq(#`$\big[$ `,#`$\big]$`)` $\big]$

Syntax of user-written evaluator

Summary of notation

The log-likelihood function is $\ln L(\theta_{1j}, \theta_{2j}, \ldots, \theta_{Ej})$, where $\theta_{ij} = \mathbf{x}_{ij}\mathbf{b}_i$, $j = 1, \ldots, N$ indexes observations, and $i = 1, \ldots, E$ indexes the linear equations defined by `ml model`. If the likelihood satisfies the linear-form restrictions, it can be decomposed as $\ln L = \sum_{j=1}^{N} \ln \ell(\theta_{1j}, \theta_{2j}, \ldots, \theta_{Ej})$.

Method-lf evaluators

```
program progname
        version 18
        args lnfj theta1 theta2 ...

        // if you need to create any intermediate results:
        tempvar tmp1 tmp2 ...
        quietly generate double 'tmp1' = ...
        ...
        quietly replace 'lnfj' = ...
end
```

where

'lnfj'	variable to be filled in with the values of $\ln \ell_j$
'theta1'	variable containing the evaluation of the first equation $\theta_{1j} = \mathbf{x}_{1j}\boldsymbol{\beta}_1$
'theta2'	variable containing the evaluation of the second equation $\theta_{2j} = \mathbf{x}_{2j}\boldsymbol{\beta}_2$
...	

Method-d0 evaluators

```
program progname
        version 18
        args todo b lnf

        tempvar theta1 theta2 ...
        mleval 'theta1' = 'b', eq(1)
        mleval 'theta2' = 'b', eq(2) // if there is a θ_2
        ...

        // if you need to create any intermediate results:
        tempvar tmp1 tmp2 ...
        generate double 'tmp1' = ...
        ...
        mlsum 'lnf' = ...
end
```

where

'todo'	always contains 0 (may be ignored)
'b'	full parameter row vector $\mathbf{b} = (\mathbf{b}_1, \mathbf{b}_2, ..., \mathbf{b}_E)$
'lnf'	scalar to be filled in with overall $\ln L$

Method-d1 evaluators

```
program progname
        version 18
        args todo b lnf g

        tempvar theta1 theta2 ...
        mleval 'theta1' = 'b', eq(1)
        mleval 'theta2' = 'b', eq(2) // if there is a θ_2
        ...

        // if you need to create any intermediate results:
        tempvar tmp1 tmp2 ...
        generate double 'tmp1' = ...
        ...

        mlsum 'lnf' = ...
        if ('todo'==0 | 'lnf'>=.) exit

        tempname d1 d2 ...
        mlvecsum 'lnf' 'd1' = formula for ∂ln ℓ_j/∂θ_1j, eq(1)
        mlvecsum 'lnf' 'd2' = formula for ∂ln ℓ_j/∂θ_2j, eq(2)
        ...
        matrix 'g' = ('d1','d2', ... )
end
```

where

'todo'	contains 0 or 1
	$\quad$ 0 ⇒ 'lnf' to be filled in;
	$\quad$ 1 ⇒ 'lnf' and 'g' to be filled in
'b'	full parameter row vector $\mathbf{b}=(\mathbf{b}_1,\mathbf{b}_2,...,\mathbf{b}_E)$
'lnf'	scalar to be filled in with overall $\ln L$
'g'	row vector to be filled in with overall $\mathbf{g}=\partial \ln L/\partial \mathbf{b}$

Method-d2 evaluators

```
program progname
        version 18
        args todo b lnf g H

        tempvar theta1 theta2 ...
        mleval 'theta1' = 'b', eq(1)
        mleval 'theta2' = 'b', eq(2) // if there is a θ_2
        ...

        // if you need to create any intermediate results:
        tempvar tmp1 tmp2 ...
        generate double 'tmp1' = ...
        ...

        mlsum 'lnf' = ...
        if ('todo'==0 | 'lnf'>=.) exit

        tempname d1 d2 ...
        mlvecsum 'lnf' 'd1' = formula for ∂ ln ℓ_j/∂θ_1j, eq(1)
        mlvecsum 'lnf' 'd2' = formula for ∂ ln ℓ_j/∂θ_2j, eq(2)
        ...
        matrix 'g' = ('d1','d2', ... )
        if ('todo'==1 | 'lnf'>=.) exit

        tempname d11 d12 d22 ...
        mlmatsum 'lnf' 'd11' = formula for ∂^2 ln ℓ_j/∂θ_1j^2, eq(1)
        mlmatsum 'lnf' 'd12' = formula for ∂^2 ln ℓ_j/∂θ_1j∂θ_2j, eq(1,2)
        mlmatsum 'lnf' 'd22' = formula for ∂^2 ln ℓ_j/∂θ_2j^2, eq(2)
        ...
        matrix 'H' = ('d11','d12', ... \ 'd12'','d22', ... )
end
```

where

'todo'	contains 0, 1, or 2
	$0 \Rightarrow$ 'lnf' to be filled in;
	$1 \Rightarrow$ 'lnf' and 'g' to be filled in;
	$2 \Rightarrow$ 'lnf', 'g', and 'H' to be filled in
'b'	full parameter row vector $\mathbf{b}=(\mathbf{b}_1,\mathbf{b}_2,...,\mathbf{b}_E)$
'lnf'	scalar to be filled in with overall $\ln L$
'g'	row vector to be filled in with overall $\mathbf{g}=\partial \ln L/\partial \mathbf{b}$
'H'	matrix to be filled in with overall Hessian $\mathbf{H}=\partial^2 \ln L/\partial \mathbf{b} \partial \mathbf{b}'$

Method-lf0 evaluators

```
program progname
        version 18
        args todo b lnfj

        tempvar theta1 theta2 ...
        mleval 'theta1' = 'b', eq(1)
        mleval 'theta2' = 'b', eq(2) // if there is a θ_2
        ...
        // if you need to create any intermediate results:
        tempvar tmp1 tmp2 ...
        generate double 'tmp1' = ...
        ...
        quietly replace 'lnfj' = ...
end
```

where

'todo'	always contains 0 (may be ignored)
'b'	full parameter row vector $\mathbf{b}=(\mathbf{b}_1,\mathbf{b}_2,...,\mathbf{b}_E)$
'lnfj'	variable to be filled in with observation-by-observation values of $\ln \ell_j$

Method-lf1 evaluators

```
program progname
        version 18
        args todo b lnfj g1 g2 ...

        tempvar theta1 theta2 ...
        mleval 'theta1' = 'b', eq(1)
        mleval 'theta2' = 'b', eq(2) // if there is a θ_2
        ...
        // if you need to create any intermediate results:
        tempvar tmp1 tmp2 ...
        generate double 'tmp1' = ...
        ...
        quietly replace 'lnfj' = ...
        if ('todo'==0) exit
        quietly replace 'g1' = formula for ∂ ln ℓ_j/∂θ_1j
        quietly replace 'g2' = formula for ∂ ln ℓ_j/∂θ_2j
        ...
end
```

where

'todo'	contains 0 or 1
	$0 \Rightarrow$ 'lnfj' to be filled in;
	$1 \Rightarrow$ 'lnfj', 'g1', 'g2', ... to be filled in
'b'	full parameter row vector $\mathbf{b}=(\mathbf{b}_1,\mathbf{b}_2,...,\mathbf{b}_E)$
'lnfj'	variable to be filled in with observation-by-observation values of $\ln \ell_j$
'g1'	variable to be filled in with $\partial \ln \ell_j/\partial\theta_{1j}$
'g2'	variable to be filled in with $\partial \ln \ell_j/\partial\theta_{2j}$

...

Method-lf2 evaluators

```
program progname
        version 18
        args todo b lnfj g1 g2 ... H

        tempvar theta1 theta2 ...
        mleval 'theta1' = 'b', eq(1)
        mleval 'theta2' = 'b', eq(2) // if there is a θ_2
        ...

        // if you need to create any intermediate results:
        tempvar tmp1 tmp2 ...
        generate double 'tmp1' = ...
        ...

        quietly replace 'lnfj' = ...
        if ('todo'==0) exit

        quietly replace 'g1' = formula for ∂ln ℓ_j/∂θ_1j
        quietly replace 'g2' = formula for ∂ln ℓ_j/∂θ_2j
        ...

        if ('todo'==1) exit

        tempname d11 d12 d22 lnf ...
        mlmatsum 'lnf' 'd11' = formula for ∂^2 ln ℓ_j/∂θ_1j^2, eq(1)
        mlmatsum 'lnf' 'd12' = formula for ∂^2 ln ℓ_j/∂θ_1j∂θ_2j, eq(1,2)
        mlmatsum 'lnf' 'd22' = formula for ∂^2 ln ℓ_j/∂θ_2j^2, eq(2)
        ...
        matrix 'H' = ('d11','d12', ... \ 'd12'','d22', ... )
end
```

where

'todo'	contains 0 or 1
	$0 \Rightarrow$ 'lnfj'to be filled in;
	$1 \Rightarrow$ 'lnfj', 'g1', 'g2', ... to be filled in
	$2 \Rightarrow$ 'lnfj', 'g1', 'g2', ... and 'H' to be filled in
'b'	full parameter row vector $\mathbf{b}=(\mathbf{b}_1, \mathbf{b}_2, ..., \mathbf{b}_E)$
'lnfj'	scalar to be filled in with observation-by-observation $\ln L$
'g1'	variable to be filled in with $\partial \ln \ell_j / \partial \theta_{1j}$
'g2'	variable to be filled in with $\partial \ln \ell_j / \partial \theta_{2j}$
...	
'H'	matrix to be filled in with overall Hessian $\mathbf{H}=\partial^2 \ln L/\partial \mathbf{b}\partial \mathbf{b}'$

Method-gf0 evaluators

```
program progname
        version 18
        args todo b lnfj

        tempvar theta1 theta2 ...
        mleval 'theta1' = 'b', eq(1)
        mleval 'theta2' = 'b', eq(2) // if there is a θ_2
        ...

        // if you need to create any intermediate results:
        tempvar tmp1 tmp2 ...
        generate double 'tmp1' = ...
        ...

        quietly replace 'lnfj' = ...
end
```

where

'todo'	always contains 0 (may be ignored)
'b'	full parameter row vector $\mathbf{b}=(\mathbf{b}_1,\mathbf{b}_2,...,\mathbf{b}_E)$
'lnfj'	variable to be filled in with the values of the log likelihood $\ln \ell_j$

Global macros for use by all evaluators

$ML_y1	name of first dependent variable
$ML_y2	name of second dependent variable, if any
...	
$ML_samp	variable containing 1 if observation to be used; 0 otherwise
$ML_w	variable containing weight associated with observation or 1 if no weights specified

Method lf can ignore $ML_samp, but restricting calculations to the $ML_samp==1 subsample will speed execution. Method-lf evaluators must ignore $ML_w; application of weights is handled by the method itself. Method-lf assumes that the weights enter multiplicatively. For weights that do not enter multiplicatively, another method that allows you to specify weights in the likelihood computation should be used. For instance, to properly incorporate aweights in a tobit model, you would need to use a d or gf evaluator instead of lf and incorporate the aweights directly in the likelihood computation.

Methods d0, d1, d2, lf0, lf1, lf2, and gf0 can ignore $ML_samp as long as ml model's nopreserve option is not specified. These methods will run more quickly if nopreserve is specified. These evaluators can ignore $ML_w only if they use mlsum, mlvecsum, mlmatsum, and mlmatbysum to produce all final results.

Description

ml model defines the current problem.

ml clear clears the current problem definition. This command is rarely used because when you type ml model, any previous problem is automatically cleared.

ml query displays a description of the current problem.

ml check verifies that the log-likelihood evaluator you have written works. We strongly recommend using this command.

ml search searches for (better) initial values. We recommend using this command.

ml plot provides a graphical way of searching for (better) initial values.

ml init provides a way to specify initial values.

ml report reports the values of $\ln L$, its gradient, and its Hessian at the initial values or current parameter estimates β_0.

ml trace traces the execution of the user-defined log-likelihood–evaluation program.

ml count counts the number of times the user-defined log-likelihood–evaluation program is called; the command is seldom used. ml count clear clears the counter. ml count on turns on the counter. ml count without arguments reports the current values of the counters. ml count off stops counting calls.

ml maximize maximizes the likelihood function and reports final results. Once ml maximize has successfully completed, the previously mentioned ml commands may no longer be used unless noclear is specified. ml graph and ml display may be used whether or not noclear is specified.

ml graph graphs the log-likelihood values against the iteration number.

ml display redisplays final results.

ml footnote displays a warning message when the model did not converge within the specified number of iterations.

ml score creates new variables containing the equation-level scores. The variables generated by ml score are equivalent to those generated by specifying the score() option of ml maximize (and ml model ..., ... maximize).

progname is the name of a program you write to evaluate the log-likelihood function.

funcname() is the name of a Mata function you write to evaluate the log-likelihood function.

In this documentation, *progname* and *funcname()* are referred to as the *user-written evaluator*, the *likelihood evaluator*, or sometimes simply as the *evaluator*. The program you write is written in the style required by the method you choose. The methods are lf, d0, d1, d2, lf0, lf1, lf2, and gf0. Thus, if you choose to use method lf, your program is called a method-lf evaluator.

Method-lf evaluators are required to evaluate the observation-by-observation log likelihood $\ln \ell_j$, $j = 1, \ldots, N$.

Method-d0 evaluators are required to evaluate the overall log likelihood $\ln L$. Method-d1 evaluators are required to evaluate the overall log likelihood and its gradient vector $\mathbf{g} = \partial \ln L / \partial \mathbf{b}$. Method-d2 evaluators are required to evaluate the overall log likelihood, its gradient, and its Hessian matrix $H = \partial^2 \ln L / \partial \mathbf{b} \partial \mathbf{b}'$.

Method-lf0 evaluators are required to evaluate the observation-by-observation log likelihood $\ln \ell_j$, $j = 1, \ldots, N$. Method-lf1 evaluators are required to evaluate the observation-by-observation log likelihood and its equation-level scores $g_{ji} = \partial \ln \ell / \partial \mathbf{x}_{ji} \mathbf{b}_i$. Method-lf2 evaluators are required to evaluate the observation-by-observation log likelihood, its equation-level scores, and its Hessian matrix $H = \partial^2 \ln \ell / \partial \mathbf{b} \partial \mathbf{b}'$.

Method-gf0 evaluators are required to evaluate the summable pieces of the log likelihood $\ln \ell_k$, $k = 1, \ldots, K$.

mleval is a subroutine used by evaluators of methods d0, d1, and d2 to evaluate the coefficient vector $\boldsymbol{\beta}$ that they are passed.

mlsum is a subroutine used by evaluators of methods d0, d1, and d2 to define the value $\ln L$ that is to be returned.

mlvecsum is a subroutine used by evaluators of methods d1 and d2 to define the gradient vector, $\mathbf{g}$, that is to be returned. It is suitable for use only when the likelihood function meets the linear-form restrictions.

mlmatsum is a subroutine used by evaluators of methods d2 and lf2 to define the Hessian matrix, $\mathbf{H}$, that is to be returned. It is suitable for use only when the likelihood function meets the linear-form restrictions.

mlmatbysum is a subroutine used by evaluators of method d2 to help define the Hessian matrix, $\mathbf{H}$, that is to be returned. It is suitable for use when the likelihood function contains terms made up of grouped sums, such as in panel-data models. For such models, use mlmatsum to compute the observation-level outer products and use mlmatbysum to compute the group-level outer products. mlmatbysum requires that the data be sorted by the variable identified in the by() option.

Options for use with ml model in interactive or noninteractive mode

group(*varname*) specifies the numeric variable that identifies groups. This option is typically used to identify panels for panel-data models.

vce(*vcetype*) specifies the type of standard error reported, which includes types that are robust to some kinds of misspecification, that allow for intragroup correlation, and that are derived from asymptotic theory; see [R] **vce_option**.

> vce(robust), vce(cluster *clustvar*), pweight, and svy will work with evaluators of methods lf, lf0, lf1, lf2, and gf0; all you need to do is specify them.

> These options will not work with evaluators of methods d0, d1, or d2, and specifying these options will produce an error message.

constraints(*numlist | matname*) specifies the linear constraints to be applied during estimation.

> constraints(*numlist*) specifies the constraints by number. Constraints are defined using the constraint command; see [R] **constraint**.

> constraint(*matname*) specifies a matrix that contains the constraints.

nocnsnotes prevents notes from being displayed when constraints are dropped. A constraint will be dropped if it is inconsistent, contradicts other constraints, or causes some other error when the constraint matrix is being built. Constraints are checked in the order they are specified.

title(*string*) specifies the title to be placed on the estimation output when results are complete.

nopreserve specifies that it is not necessary for ml to ensure that only the estimation subsample is in memory when the user-written likelihood evaluator is called. nopreserve is irrelevant when you use method lf.

> For the other methods, if nopreserve is not specified, ml saves the data in a file (preserves the original dataset) and drops the irrelevant observations before calling the user-written evaluator. This way, even if the evaluator does not restrict its attentions to the $ML_samp==1 subsample, results will still be correct. Later, ml automatically restores the original dataset.

> ml need not go through these machinations in the case of method lf because the user-written evaluator calculates observation-by-observation values, and ml itself sums the components.

> ml goes through these machinations if and only if the estimation sample is a subsample of the data in memory. If the estimation sample includes every observation in memory, ml does not preserve the original dataset. Thus, programmers must not alter the original dataset unless they preserve the data themselves.

We recommend that interactive users of `ml` not specify `nopreserve`; the speed gain is not worth the possibility of getting incorrect results.

We recommend that programmers specify `nopreserve`, but only after verifying that their evaluator really does restrict its attentions solely to the `$ML_samp==1` subsample.

`collinear` specifies that `ml` not remove the collinear variables within equations. There is no reason you would want to leave collinear variables in place, but this option is of interest to programmers who, in their code, have already removed collinear variables and do not want `ml` to waste computer time checking again.

`missing` specifies that observations containing variables with missing values not be eliminated from the estimation sample. There are two reasons you might want to specify `missing`:

Programmers may wish to specify `missing` because, in other parts of their code, they have already eliminated observations with missing values and do not want `ml` to waste computer time looking again.

You may wish to specify `missing` if your model explicitly deals with missing values. Stata's `heckman` command is a good example of this. In such cases, there will be observations where missing values are allowed and other observations where they are not—where their presence should cause the observation to be eliminated. If you specify `missing`, it is your responsibility to specify an `if` *exp* that eliminates the irrelevant observations.

`lf0(#_k #_ll)` is typically used by programmers. It specifies the number of parameters and log-likelihood value of the "constant-only" model so that `ml` can report a likelihood-ratio test rather than a Wald test. These values may have been analytically determined or they may have been determined by a previous fitting of the constant-only model on the estimation sample.

Also see the `continue` option directly below.

If you specify `lf0()`, it must be safe for you to specify the `missing` option, too, else how did you calculate the log likelihood for the constant-only model on the same sample? You must have identified the estimation sample, and done so correctly, so there is no reason for `ml` to waste time rechecking your results. All of which is to say, do not specify `lf0()` unless you are certain your code identifies the estimation sample correctly.

`lf0()`, even if specified, is ignored if `vce(robust)`, `vce(cluster varname)`, `pweight`, or `svy` is specified because, in that case, a likelihood-ratio test would be inappropriate.

Do not confuse the `lf0()` option with evaluator method `lf0`. They are unrelated.

`continue` is typically specified by programmers and does two things:

> First, it specifies that a model has just been fit by either `ml` or some other estimation command, such as `logit`, and that the likelihood value stored in `e(ll)` and the number of parameters stored in `e(b)` as of that instant are the relevant values of the constant-only model. The current value of the log likelihood is used to present a likelihood-ratio test unless `pweight`, `svy`, `vce(robust)`, `vce(cluster` *clustvar*`)`, or `constraints()` is specified. A likelihood-ratio test is inappropriate when `pweight`, `svy`, `vce(robust)`, or `vce(cluster` *clustvar*`)` is specified. We suggest using `lrtest` when `constraints()` is specified; see [R] **lrtest**.

> Second, `continue` sets the initial values, $\mathbf{b}_0$, for the model about to be fit according to the `e(b)` currently stored.

> The comments made about specifying `missing` with `lf0()` apply equally well here.

`waldtest(#)` is typically specified by programmers. By default, `ml` presents a Wald test, but that is overridden if the `lf0()` or `continue` option is specified. A Wald test is performed if `vce(robust)`, `vce(cluster` *clustvar*`)`, or `pweight` is specified.

> `waldtest(0)` prevents even the Wald test from being reported.

> `waldtest(-1)` is the default. It specifies that a Wald test be performed by constraining all coefficients except the intercept to 0 in the first equation. Remaining equations are to be unconstrained. A Wald test is performed if neither `lf0()` nor `continue` was specified, and a Wald test is forced if `vce(robust)`, `vce(cluster` *clustvar*`)`, or `pweight` was specified.

> `waldtest(k)` for $k \leq -1$ specifies that a Wald test be performed by constraining all coefficients except intercepts to 0 in the first $|k|$ equations; remaining equations are to be unconstrained. A Wald test is performed if neither `lf0()` nor `continue` was specified, and a Wald test is forced if `vce(robust)`, `vce(cluster` *clustvar*`)`, or `pweight` was specified.

> `waldtest(k)` for $k \geq 1$ works like the options above, except that it forces a Wald test to be reported even if the information to perform the likelihood-ratio test is available and even if none of `vce(robust)`, `vce(cluster` *clustvar*`)`, or `pweight` was specified. `waldtest(k)`, $k \geq 1$, may not be specified with `lf0()`.

`obs(#)` is used mostly by programmers. It specifies that the number of observations reported and ultimately stored in `e(N)` be #. Ordinarily, `ml` works that out for itself. Programmers may want to specify this option when, for the likelihood-evaluator to work for N observations, they first had to modify the dataset so that it contained a different number of observations.

`crittype(string)` is used mostly by programmers. It allows programmers to supply a string (up to 32 characters long) that describes the criterion that is being optimized by `ml`. The default is `"Log likelihood"` for nonrobust and `"Log pseudolikelihood"` for robust estimation.

svy indicates that `ml` is to pick up the `svy` settings set by `svyset` and use the robust variance estimator. This option requires the data to be `svyset`; see [SVY] **svyset**. `svy` may not be specified with *weights*, `vce(robust)`, `vce(opg)`, or `vce(cluster varname)`.

subpop(*varname*) specifies that estimates be computed for the single subpopulation defined by the observations for which *varname* $\neq 0$. Typically, *varname* $= 1$ defines the subpopulation, and *varname* $= 0$ indicates observations not belonging to the subpopulation. For observations whose subpopulation status is uncertain, *varname* should be set to missing ('.'). This option requires the `svy` option.

nosvyadjust specifies that the model Wald test be carried out as $W/k \sim F(k, d)$, where W is the Wald test statistic, k is the number of terms in the model excluding the constant term, d is the total number of sampled primary sampling units minus the total number of strata, and $F(k, d)$ is an F distribution with k numerator degrees of freedom and d denominator degrees of freedom. By default, an adjusted Wald test is conducted: $(d - k + 1)W/(kd) \sim F(k, d - k + 1)$. See Korn and Graubard (1990) for a discussion of the Wald test and the adjustments thereof. This option requires the `svy` option.

technique(*algorithm_spec*) specifies how the likelihood function is to be maximized. The following algorithms are currently implemented in `ml`.

technique(nr) specifies Stata's modified Newton–Raphson (NR) algorithm.

technique(bhhh) specifies the Berndt–Hall–Hall–Hausman (BHHH) algorithm.

technique(dfp) specifies the Davidon–Fletcher–Powell (DFP) algorithm.

technique(bfgs) specifies the Broyden–Fletcher–Goldfarb–Shanno (BFGS) algorithm.

The default is technique(nr).

You can switch between algorithms by specifying more than one in the `technique()` option. By default, `ml` will use an algorithm for five iterations before switching to the next algorithm. To specify a different number of iterations, include the number after the technique in the option. For example, specifying `technique(bhhh 10 nr 1000)` requests that `ml` perform 10 iterations using the BHHH algorithm followed by 1,000 iterations using the NR algorithm, and then switch back to BHHH for 10 iterations, and so on. The process continues until convergence or until the maximum number of iterations is reached.

Options for use with ml model in noninteractive mode

The following additional options are for use with `ml model` in noninteractive mode. Noninteractive mode is for programmers who use `ml` as a subroutine and want to issue a single command that will carry forth the estimation from start to finish.

`maximize` is required. It specifies noninteractive mode.

`init(`*ml_init_args*`)` sets the initial values $\mathbf{b}_0$. *ml_init_args* are whatever you would type after the `ml init` command.

`search(on|norescale|quietly|off)` specifies whether `ml search` is to be used to improve the initial values. `search(on)` is the default and is equivalent to running separately `ml search, repeat(0)`. `search(norescale)` is equivalent to running separately `ml search, repeat(0) norescale`. `search(quietly)` is equivalent to `search(on)`, except that it suppresses `ml search`'s output. `search(off)` prevents the calling of `ml search` altogether.

`repeat(#)` is `ml search`'s `repeat()` option. `repeat(0)` is the default.

`bounds(`*ml_search_bounds*`)` specifies the search bounds. *ml_search_bounds* is specified as

$$\big[\, eqn\_name \,\big]\ lower\_bound\ upper\_bound\ \ldots\ \big[\, eqn\_name \,\big]\ lower\_bound\ upper\_bound$$

for instance, `bounds(100 100 lnsigma 0 10)`. The command `ml model` issues is `ml search` *ml_search_bounds*`, repeat(#)`. Specifying search bounds is optional.

`nowarning`, `novce`, `negh`, and `score()` are `ml maximize`'s equivalent options.

maximize_options: `difficult`, `technique(`*algorithm_spec*`)`, `iterate(#)`, `[no]log`, `trace`, `gradient`, `showstep`, `hessian`, `showtolerance`, `tolerance(#)`, `ltolerance(#)`, `nrtolerance(#)`, `nonrtolerance`, and `from(`*init_specs*`)`; see [R] **Maximize**. These options are seldom used.

Options for use when specifying equations

`noconstant` specifies that the equation not include an intercept.

`offset(`*varname_o*`)` specifies that the equation be $\mathbf{x}\boldsymbol{\beta}$ + *varname_o*—that it include *varname_o* with the coefficient constrained to be 1.

`exposure(`*varname_e*`)` is an alternative to `offset(`*varname_o*`)`; it specifies that the equation be $\mathbf{x}\boldsymbol{\beta} + \ln(varname\_e)$. The equation is to include $\ln(varname\_e)$ with the coefficient constrained to be 1.

`freeparm` specifies that the associated *eqname* is a free parameter. The corresponding full column name on `e(b)` will be `/`*eqname* instead of *eqname*`:_cons`. This option is not allowed with *varlist*$_x$.

Options for use with ml search

`repeat(#)` specifies the number of random attempts that are to be made to find a better initial-value vector. The default is `repeat(10)`.

repeat(0) specifies that no random attempts be made. More correctly, repeat(0) specifies that no random attempts be made if the first initial-value vector is a feasible starting point. If it is not, ml search will make random attempts, even if you specify repeat(0), because it has no alternative. The repeat() option refers to the number of random attempts to be made to improve the initial values. When the initial starting value vector is not feasible, ml search will make up to 1,000 random attempts to find starting values. It stops when it finds one set of values that works and then moves into its improve-initial-values logic.

repeat(k), $k > 0$, specifies the number of random attempts to be made to improve the initial values.

restart specifies that random actions be taken to obtain starting values and that the resulting starting values not be a deterministic function of the current values. Generally, you should not specify this option because, with restart, ml search intentionally does not produce as good a set of starting values as it could. restart is included for use by the optimizer when it gets into serious trouble. The random actions ensure that the actions of the optimizer and ml search, working together, do not result in a long, endless loop.

restart implies norescale, which is why we recommend that you do not specify restart. In testing, sometimes rescale worked so well that, even after randomization, the rescaler would bring the starting values right back to where they had been the first time and thus defeat the intended randomization.

norescale specifies that ml search not engage in its rescaling actions to improve the parameter vector. We do not recommend specifying this option because rescaling tends to work so well.

maximize_options: [no]log, <u>trace</u>; see [R] **Maximize**. These options are seldom used.

Options for use with ml plot

saving(*filename*[, replace]) specifies that the graph be saved in *filename*.gph. See [G-3] ***saving_option***.

Options for use with ml init

copy specifies that the list of numbers or the initialization vector be copied into the initial-value vector by position rather than by name.

skip specifies that any parameters found in the specified initialization vector that are not also found in the model be ignored. The default action is to issue an error message.

Options for use with ml maximize

nowarning is allowed only with iterate(0). nowarning suppresses the "convergence not achieved" message. Programmers might specify iterate(0) nowarning when they have a vector **b** already containing the final estimates and want ml to calculate the variance matrix and post final estimation results. Then, specify init(b) search(off) iterate(0) nowarning nolog.

novce is allowed only with iterate(0). novce substitutes the zero matrix for the variance matrix, which in effect posts estimation results as fixed constants.

negh specifies that the evaluator return the negative Hessian matrix. By default, ml assumes d2 and lf2 evaluators return the Hessian matrix.

score(*newvars* | *stub**) creates new variables containing the contributions to the score for each equation and ancillary parameter in the model; see [U] **20.23 Obtaining scores**.

If score(*newvars*) is specified, the *newvars* must contain k new variables. For evaluators of methods lf, lf0, lf1, and lf2, k is the number of equations. For evaluators of method gf0, k is the number of parameters. If score(*stub**) is specified, variables named *stub*1, *stub*2, ..., *stubk* are created.

For evaluators of methods lf, lf0, lf1, and lf2, the first variable contains $\partial \ln \ell_j / \partial(\mathbf{x}_{1j}\mathbf{b}_1)$, the second variable contains $\partial \ln \ell_j / \partial(\mathbf{x}_{2j}\mathbf{b}_2)$, and so on.

For evaluators of method gf0, the first variable contains $\partial \ln \ell_j / \partial \mathbf{b}_1$, the second variable contains $\partial \ln \ell_j / \partial \mathbf{b}_2$, and so on.

nooutput suppresses display of the final results. This is different from prefixing ml maximize with quietly in that the iteration log is still displayed (assuming that nolog is not specified).

noskipline suppresses display of a blank line before the iteration log. This is useful in programs when there is a header preceding the iteration log and a blank line is not wanted after the header.

noclear specifies that after the model has converged, the ml problem definition not be cleared. Perhaps you are having convergence problems and intend to run the model to convergence. If so, use ml search to see if those values can be improved, and then start the estimation again.

maximize_options: difficult, iterate(#), [no]log, trace, gradient, showstep, hessian, showtolerance, tolerance(#), ltolerance(#), nrtolerance(#), and nonrtolerance; see [R] **Maximize**. These options are seldom used.

display_options; see *Options for use with ml display* below.

eform_option; see *Options for use with ml display* below.

Options for use with ml graph

saving(*filename*[, replace]) specifies that the graph be saved in *filename*.gph.
See [G-3] *saving_option*.

Options for use with ml display

noheader suppresses, above the coefficient table, the header display that displays the
final log-likelihood value, the number of observations, and the model significance
test.

nofootnote suppresses the footnote display below the coefficient table. The footnote
displays a warning if the model fit did not converge within the specified number of
iterations. Use ml footnote to display the warning if 1) you add to the coefficient
table using the plus option or 2) you have your own footnotes and want the warning
to be last.

level(#) is the standard confidence-level option. It specifies the confidence level, as a
percentage, for confidence intervals of the coefficients. The default is level(95) or
as set by set level; see [U] **20.8 Specifying the width of confidence intervals**.

first displays a coefficient table reporting results for the first equation only, and the report makes it appear that the first equation is the only equation. This option is used
by programmers who estimate ancillary parameters in the second and subsequent
equations and who wish to report the values of such parameters themselves.

neq(#) is an alternative to first. neq(#) displays a coefficient table reporting results
for the first # equations. This option is used by programmers who estimate ancillary
parameters in the #+1 and subsequent equations and who wish to report the values
of such parameters themselves.

showeqns is a seldom-used option that displays the equation names in the coefficient
table. ml display uses the numbers stored in e(k_eq) and e(k_aux) to determine
how to display the coefficient table. e(k_eq) identifies the number of equations,
and e(k_aux) identifies how many of these are for ancillary parameters. The first
option is implied when showeqns is not specified and all but the first equation are
for ancillary parameters.

plus displays the coefficient table, but rather than ending the table in a line of dashes,
ends it in dashes–plus-sign–dashes. This is so that programmers can write additional
display code to add more results to the table and make it appear as if the combined
result is one table. Programmers typically specify plus with the first or neq()
option. plus implies nofootnote.

nocnsreport suppresses the display of constraints above the coefficient table. This
option is ignored if constraints were not used to fit the model.

`noomitted` specifies that variables that were omitted because of collinearity not be displayed. The default is to include in the table any variables omitted because of collinearity and to label them as "(omitted)".

`vsquish` specifies that the blank space separating factor-variable terms or time-series–operated variables from other variables in the model be suppressed.

`noemptycells` specifies that empty cells for interactions of factor variables not be displayed. The default is to include in the table interaction cells that do not occur in the estimation sample and to label them as "(empty)".

`baselevels` and `allbaselevels` control whether the base levels of factor variables and interactions are displayed. The default is to exclude from the table all base categories.

> `baselevels` specifies that base levels be reported for factor variables and for interactions whose bases cannot be inferred from their component factor variables.

> `allbaselevels` specifies that all base levels of factor variables and interactions be reported.

`cformat(%`*fmt*`)` specifies how to format coefficients, standard errors, and confidence limits in the coefficient table.

`pformat(%`*fmt*`)` specifies how to format p-values in the coefficient table.

`sformat(%`*fmt*`)` specifies how to format test statistics in the coefficient table.

`nolstretch` specifies that the width of the coefficient table not be automatically widened to accommodate longer variable names. The default, `lstretch`, is to automatically widen the coefficient table up to the width of the Results window. Specifying `lstretch` or `nolstretch` overrides the setting given by `set lstretch`. If `set lstretch` has not been set, the default is `lstretch`.

`coeflegend` specifies that the legend of the coefficients and how to specify them in an expression be displayed rather than the coefficient table.

eform_option: `eform(`*string*`)`, `eform`, `hr`, `shr`, `irr`, `or`, and `rrr` display the coefficient table in exponentiated form: for each coefficient, $\exp(b)$ rather than b is displayed, and standard errors and confidence intervals are transformed. *string* is the table header that will be displayed above the transformed coefficients and must be 11 characters or shorter in length—for example, `eform("Odds ratio")`. The options `eform`, `hr`, `shr`, `irr`, `or`, and `rrr` provide a default *string* equivalent to "exp(b)", "Haz. ratio", "SHR", "IRR", "Odds ratio", and "RRR", respectively. These options may not be combined.

> `ml display` looks at `e(k_eform)` to determine how many equations are affected by an *eform_option*; by default, only the first equation is affected. Type `ereturn list, all` to view `e(k_eform)`; see [P] **ereturn**.

Options for use with mleval

eq($\#$) specifies the equation number i for which $\theta_{ij} = \mathbf{x}_{ij}\boldsymbol{\beta}_i$ is to be evaluated. eq(1) is assumed if eq() is not specified.

scalar asserts that the ith equation is known to evaluate to a constant; the equation was specified as (), (*name*:), or /*name* on the ml model statement. If you specify this option, the new variable is created as a scalar. If the ith equation does not evaluate to a scalar, an error message is issued.

Options for use with mlsum

noweight specifies that weights ($ML_w) be ignored when summing the likelihood function.

Options for use with mlvecsum

eq($\#$) specifies the equation for which a gradient vector $\partial \ln L/\partial \boldsymbol{\beta}_i$ is to be constructed. The default is eq(1).

Options for use with mlmatsum

eq($\#[\,,\#\,]$) specifies the equations for which the Hessian matrix is to be constructed. The default is eq(1), which is the same as eq(1,1), which means $\partial^2 \ln L/\partial \boldsymbol{\beta}_1 \partial \boldsymbol{\beta}_1'$. Specifying eq($i,j$) results in $\partial^2 \ln L/\partial \boldsymbol{\beta}_i \partial \boldsymbol{\beta}_j'$.

Options for use with mlmatbysum

by(*varname*) is required and specifies the group variable.

eq($\#[\,,\#\,]$) specifies the equations for which the Hessian matrix is to be constructed. The default is eq(1), which is same as eq(1,1), which means $\partial^2 \ln L/\partial \boldsymbol{\beta}_1 \partial \boldsymbol{\beta}_1'$. Specifying eq($i,j$) results in $\partial^2 \ln L/\partial \boldsymbol{\beta}_i \partial \boldsymbol{\beta}_j'$.

Options for use with ml score

equation(*eqname*) identifies from which equation the observation-level scores should come. This option may be used only when generating one variable.

missing specifies that observations containing variables with missing values not be eliminated from the estimation sample.

Stored results

For results stored by ml without the svy option, see [R] **Maximize**.

For results stored by ml with the svy option, see [SVY] **svy**.

C Syntax of moptimize()

Syntax

If you are reading this entry for the first time, skip down to *Description* and to *Remarks and examples*, and more especially, we suggest you read *Mathematical statement of the moptimize( ) problem* under *Remarks and examples* in [M-5] **moptimize( )**.

The syntax is presented under the following headings:

Step 1: Initialization

$$M = \texttt{moptimize\_init()}$$

Step 2: Definition of maximization or minimization problem

In each of the functions, the last argument is optional. If specified, the function sets the value and returns *void*. If not specified, no change is made, and instead what is currently set is returned.

(varies)	`moptimize_init_which(`M`, {` `"max"` `|` `"min"` `})`
(varies)	`moptimize_init_evaluator(`M`, &`*functionname*`())`
(varies)	`moptimize_init_evaluator(`M`, "`*programname*`")`
(varies)	`moptimize_init_evaluatortype(`M`, `*evaluatortype*`)`
(varies)	`moptimize_init_negH(`M`, {` `"off"` `|` `"on"` `})`
(varies)	`moptimize_init_touse(`M`, "`*tousevarname*`")`
(varies)	`moptimize_init_ndepvars(`M`, `D`)`
(varies)	`moptimize_init_depvar(`M`, `j`, `y`)`
(varies)	`moptimize_init_eq_n(`M`, `m`)`
(varies)	`moptimize_init_eq_indepvars(`M`, `i`, `X`)`
(varies)	`moptimize_init_eq_cons(`M`, `i`, {` `"on"` `|` `"off"` `})`
(varies)	`moptimize_init_eq_offset(`M`, `i`, `o`)`
(varies)	`moptimize_init_eq_exposure(`M`, `i`, `t`)`
(varies)	`moptimize_init_eq_name(`M`, `i`, `*name*`)`
(varies)	`moptimize_init_eq_colnames(`M`, `i`, `*names*`)`
(varies)	`moptimize_init_eq_freeparm(`M`, `i`, {` `"on"` `|` `"off"` `})`
(varies)	`moptimize_init_eq_coefs(`M`, `i`, `*b0*`)`
(varies)	`moptimize_init_constraints(`M`, `Cc`)`

(varies)   `moptimize_init_search(`M`, {"on"|"off"})`

(varies)   `moptimize_init_search_random(`M`, {"off"|"on"})`

(varies)   `moptimize_init_search_repeat(`M`, `nr`)`

(varies)   `moptimize_init_search_bounds(`M`, `i`, `$minmax$`)`

(varies)   `moptimize_init_search_rescale(`M`, {"on"|"off"})`

(varies)   `moptimize_init_weight(`M`, `w`)`

(varies)   `moptimize_init_weighttype(`M`, `$weighttype$`)`

(varies)   `moptimize_init_cluster(`M`, `c`)`

(varies)   `moptimize_init_svy(`M`, {"off"|"on"})`

(varies)   `moptimize_init_by(`M`, `by`)`

(varies)   `moptimize_init_nuserinfo(`M`, `n_user`)`

(varies)   `moptimize_init_userinfo(`M`, `l`, `Z`)`

(varies)   `moptimize_init_technique(`M`, `$technique$`)`

(varies)   `moptimize_init_vcetype(`M`, `$vcetype$`)`

(varies)   `moptimize_init_nmsimplexdeltas(`M`, `$delta$`)`

(varies)   `moptimize_init_gnweightmatrix(`M`, `W`)`

(varies)   `moptimize_init_singularHmethod(`M`, `$singularHmethod$`)`

(varies)   `moptimize_init_conv_maxiter(`M`, `$maxiter$`)`

(varies)   `moptimize_init_conv_warning(`M`, {"on"|"off"})`

(varies)   `moptimize_init_conv_ptol(`M`, `$ptol$`)`

(varies)   `moptimize_init_conv_vtol(`M`, `$vtol$`)`

(varies)   `moptimize_init_conv_nrtol(`M`, `$nrtol$`)`

(varies)   `moptimize_init_conv_ignorenrtol(`M`, {"off"|"on"})`

(varies)   `moptimize_init_iterid(`M`, `id`)`

(varies)   `moptimize_init_valueid(`M`, `id`)`

(varies)   `moptimize_init_tracelevel(`M`, `$tracelevel$`)`

(varies)   `moptimize_init_trace_ado(`M`, {"off"|"on"})`

(varies)	`moptimize_init_trace_dots(`M`, {"off"|"on"})`
(varies)	`moptimize_init_trace_value(`M`, {"on"|"off"})`
(varies)	`moptimize_init_trace_tol(`M`, {"off"|"on"})`
(varies)	`moptimize_init_trace_step(`M`, {"off"|"on"})`
(varies)	`moptimize_init_trace_coefdiffs(`M`, {"off"|"on"})`
(varies)	`moptimize_init_trace_coefs(`M`, {"off"|"on"})`
(varies)	`moptimize_init_trace_gradient(`M`, {"off"|"on"})`
(varies)	`moptimize_init_trace_Hessian(`M`, {"off"|"on"})`
(varies)	`moptimize_init_evaluations(`M`, {"off"|"on"})`
(varies)	`moptimize_init_verbose(`M`, {"on"|"off"})`

Step 3: Perform optimization or perform a single function evaluation

void	`moptimize(`M`)`
real scalar	`_moptimize(`M`)`
void	`moptimize_evaluate(`M`)`
real scalar	`_moptimize_evaluate(`M`)`

Step 4: Post, display, or obtain results

void	`moptimize_result_post(`M`[, `*vcetype*`])`
void	`moptimize_result_display([`M`[, `*vcetype*`]])`
real scalar	`moptimize_result_value(`M`)`
real scalar	`moptimize_result_value0(`M`)`
real rowvector	`moptimize_result_eq_coefs(`M`[, `i`])`
real rowvector	`moptimize_result_coefs(`M`)`
string matrix	`moptimize_result_colstripe(`M`[, `i`])`
real matrix	`moptimize_result_scores(`M`)`
real rowvector	`moptimize_result_gradient(`M`[, `i`])`

real matrix	moptimize_result_Hessian(M[, i])
real matrix	moptimize_result_V(M[, i])
string scalar	moptimize_result_Vtype(M)
real matrix	moptimize_result_V_oim(M[, i])
real matrix	moptimize_result_V_opg(M[, i])
real matrix	moptimize_result_V_robust(M[, i])
real scalar	moptimize_result_iterations(M)
real scalar	moptimize_result_converged(M)
real colvector	moptimize_result_iterationlog(M)
real rowvector	moptimize_result_evaluations(M)
real scalar	moptimize_result_errorcode(M)
string scalar	moptimize_result_errortext(M)
real scalar	moptimize_result_returncode(M)
void	moptimize_ado_cleanup(M)

Utility functions for use in all steps

void   moptimize_query(M)

real matrix   moptimize_util_eq_indices(M, i [, $i2$])

(varies)   moptimize_util_depvar(M, j)
      returns y set by moptimize_init_depvar(M, j, y), which is usually a real
            colvector

real colvector moptimize_util_xb(M, b, i)

real scalar   moptimize_util_sum(M, *real colvector v*)

real rowvector moptimize_util_vecsum(M, i, *real colvector s*,
            real scalar value)

real matrix   moptimize_util_matsum(M, i, $i2$, *real colvector s*,
            real scalar value)

real matrix	moptimize_util_matbysum(*M*, *i*, *real colvector a*, *real colvector b*, *real scalar value*)
real matrix	moptimize_util_matbysum(*M*, *i*, *i2*, *real colvector a*, *real colvector b*, *real colvector c*, *real scalar value*)
pointer scalar	moptimize_util_by(*M*)

Definition of M

M, if it is declared, should be declared transmorphic. *M* is obtained from the function moptimize_init() and then passed as an argument to the other moptimize() functions.

> moptimize_init() returns *M*, called an moptimize() problem handle. The function takes no arguments. *M* holds the information about the optimization problem.

Setting the sample

Various moptimize_init_*() functions set values for dependent variables, independent variables, etc. When you set those values, you do that either by specifying Stata variable names or by specifying Mata matrices containing the data themselves. Function moptimize_init_touse() specifies the sample to be used when you specify Stata variable names.

> moptimize_init_touse(*M*, "*tousevarname*") specifies the name of the variable in the Stata dataset that marks the observations to be included. Observations for which the Stata variable is nonzero are included. The default is "", meaning all observations are to be used.

> You need to specify *tousevarname* only if you specify Stata variable names in the other moptimize_init_*() functions, and even then it is not required. Setting tousevar when you specify the data themselves via Mata matrices, whether views or not, has no effect.

Specifying dependent variables

D and *j* index dependent variables:

index	Description
D	number of dependent variables, $D \geq 0$
j	dependent variable index, $1 \leq j \leq D$

D and *j* are real scalars.

You set the dependent variables one at a time. In a particular optimization problem, you may have no dependent variables or have more dependent variables than equations.

moptimize_init_depvar(M, j, y) sets the jth dependent variable to be y. y may be a string scalar containing a Stata variable name that in turn contains the values of the jth dependent variable, or y may be a real colvector directly containing the values.

moptimize_init_ndepvars(M, D) sets the total number of dependent variables. You can set D before defining dependent variables, and that speeds execution slightly, but it is not necessary because D is automatically set to the maximum j.

Specifying independent variables

Independent variables are defined within parameters or, equivalently, equations. The words parameter and equation mean the same thing. m, i, and $i2$ index parameters:

index	Description
m	number of parameters (equations), $m \geq 1$
i	equation index, $1 \leq i \leq m$
$i2$	equation index, $1 \leq i2 \leq m$

m, i, and $i2$ are real scalars.

The function to be optimized is $f(p_1, p_2, \ldots, p_m)$. The ith parameter (equation) is defined as

$$pi = Xi \times bi' + oi + \ln(ti) \;{:}{+}\; b0i$$

where

pi:	$Ni \times 1$	(ith parameter)
Xi:	$Ni \times ki$	(Ni observations on ki independent variables)
bi:	$1 \times ki$	(coefficients to be fit)
oi:	$Ni \times 1$	(exposure/offset in offset form, optional)
ti:	$Ni \times 1$	(exposure/offset in exposure form, optional)
$b0i$:	1×1	(constant or intercept, optional)

Any of the terms may be omitted. The most common forms for a parameter are $pi = Xi \times bi' + b0i$ (standard model), $pi = Xi \times bi'$ (no-constant model), and $pi = b0i$ (constant-only model).

In addition, define b: $1 \times K$ as the entire coefficient vector, which is to say,

$$b = (b1, [b01,] \quad b2, [b02,] \quad \ldots)$$

That is, because bi is $1 \times ki$ for $i = 1, 2, \ldots, m$, then b is $1 \times K$, where $K = \sum_i ki + ci$, where ci is 1 if equation i contains an intercept and is 0 otherwise. Note that bi does not contain the constant or intercept, if there is one, but b contains all the coefficients, including the intercepts. b is called *the full set of coefficients*.

Parameters are defined one at a time by using the following functions:

moptimize_init_eq_n(M, m) sets the number of parameters. The use of this function is optional. m will be automatically determined from the other moptimize_init_eq_*() functions you issue.

moptimize_init_eq_indepvars(M, i, X) sets X to be the data (independent variables) for the ith parameter. X may be a $1 \times ki$ string rowvector containing Stata variable names, or X may be a string scalar containing the same names in space-separated format, or X may be an $Ni \times ki$ real matrix containing the data for the independent variables. Specify X as "" to omit term $Xi \times bi'$, for instance, as when fitting a constant-only model. The default is "".

moptimize_init_eq_cons(M, i, {"on" | "off"}) specifies whether the equation for the ith parameter includes $b0i$, a constant or intercept. Specify "on" to include $b0i$, "off" to exclude it. The default is "on".

moptimize_init_eq_offset(M, i, o) specifies oi in the equation for the ith parameter. o may be a string scalar containing a Stata variable name, or o may be an $Ni \times 1$ real colvector containing the offsets. The default is "", meaning term oi is omitted. Parameters may not have both oi and $\ln(ti)$ terms.

moptimize_init_eq_exposure(M, i, t) specifies ti in term $\ln(ti)$ of the equation for the ith parameter. t may be a string scalar containing a Stata variable name, or t may be an $Ni \times 1$ real colvector containing the exposure values. The default is "", meaning term $\ln(ti)$ is omitted.

moptimize_init_eq_name(M, i, *name*) specifies a string scalar, *name*, to be used in the output to label the ith parameter. The default is to use an automatically generated name.

moptimize_init_eq_colnames(M, i, *names*) specifies a $1 \times ki$ string rowvector, *names*, to be used in the output to label the coefficients for the ith parameter. The default is to use automatically generated names.

moptimize_init_eq_freeparm(M, i, {"on" | "off"}) specifies whether the equation for the ith parameter is to be treated as a free parameter. This setting is ignored if there are independent variables or an offset attached to the ith parameter. Free parameters have a shortcut notation that distinguishes them from constant linear equations. The free parameter notation for an equation labeled *name* is /*name*. The corresponding notation for a constant linear equation is *name*:_cons.

Specifying constraints

Linear constraints may be placed on the coefficients, b, which may be either within equation or between equations.

> $\texttt{moptimize\_init\_constraints}(M,\ Cc)$ specifies an $R \times K + 1$ real matrix, Cc, that places R linear restrictions on the $1 \times K$ full set of coefficients, b. Think of Cc as being (C,c), C: $R \times K$ and c: $R \times 1$. Optimization will be performed subject to the constraint $Cb' = c$. The default is no constraints.

Specifying weights or survey data

You may specify weights, and once you do, everything is automatic, assuming you implement your evaluator by using the provided utility functions.

> $\texttt{moptimize\_init\_weight}(M,\ w)$ specifies the weighting variable or data. w may be a string scalar containing a Stata variable name, or w may be a real colvector directly containing the weight values. The default is "", meaning no weights.

> $\texttt{moptimize\_init\_weighttype}(M,\ weighttype)$ specifies how w is to be treated. *weighttype* may be "fweight", "aweight", "pweight", or "iweight". You may set w first and then *weighttype*, or the reverse. If you set w without setting *weighttype*, then "fweight" is assumed. If you set *weighttype* without setting w, then *weighttype* is ignored. The default *weighttype* is "fweight".

Alternatively, you may inherit the full set of survey settings from Stata by using $\texttt{moptimize\_init\_svy}()$. If you do this, do not use $\texttt{moptimize\_init\_weight}()$, $\texttt{moptimize\_init\_weighttype}()$, or $\texttt{moptimize\_init\_cluster}()$. When you use the survey settings, everything is nearly automatic, assuming you use the provided utility functions to implement your evaluator. The proviso is that your evaluator must be of *evaluatortype* lf, lf*, gf, or q.

> $\texttt{moptimize\_init\_svy}(M,\ \{\,\texttt{"off"}\,|\,\texttt{"on"}\,\})$ specifies whether Stata's survey settings should be used. The default is "off". Using the survey settings changes the default *vcetype* to "svy", which is equivalent to "robust".

Specifying clusters and panels

Clustering refers to possible nonindependence of the observations within groups called clusters. A cluster variable takes on the same value within a cluster and different values across clusters. After setting the cluster variable, there is nothing special you have to do, but be aware that clustering is allowed only if you use a type lf, lf*, gf, or q evaluator. $\texttt{moptimize\_init\_cluster}()$ allows you to set a cluster variable.

Panels refer to likelihood functions or other objective functions that can only be calculated at the panel level, for which there is no observation-by-observation decomposition. Unlike clusters, these panel likelihood functions are difficult to calculate and require the use of type d or gf evaluator. A panel variable takes on the same value within a panel and different values across panels. moptimize_init_by() allows you to set a panel variable.

You may set both a cluster variable and a panel variable, but be careful because, for most likelihood functions, panels are mathematically required to be nested within cluster.

moptimize_init_cluster(M, c) specifies a cluster variable. c may be a string scalar containing a Stata variable name, or c may be a real colvector directly containing the cluster values. The default is "", meaning no clustering. If clustering is specified, the default *vcetype* becomes "robust".

moptimize_init_by(M, by) specifies a panel variable and specifies that only panel-level calculations are meaningful. by may be a string scalar containing a Stata variable name, or by may be a real colvector directly containing the panel ID values. The default is "", meaning no panels. If panels are specified, the default *vcetype* remains unchanged, but if the opg variance estimator is used, the opg calculation is modified so that it is clustered at the panel level.

Specifying optimization technique

Technique refers to the numerical methods used to solve the optimization problem. The default is Newton–Raphson maximization.

moptimize_init_which(M, {"max" | "min"}) sets whether the maximum or minimum of the objective function is to be found. The default is "max".

moptimize_init_technique(M, *technique*) specifies the technique to be used to find the coefficient vector b that maximizes or minimizes the objective function. Allowed values are

technique	Description
"nr"	modified Newton–Raphson
"dfp"	Davidon–Fletcher–Powell
"bfgs"	Broyden–Fletcher–Goldfarb–Shanno
"bhhh"	Berndt–Hall–Hall–Hausman
"nm"	Nelder–Mead
"gn"	Gauss–Newton (quadratic optimization)

The default is "nr".

You can switch between "nr", "dfp", "bfgs", and "bhhh" by specifying two or more of them in a space-separated list. By default, moptimize() will use an algorithm for five iterations before switching to the next algorithm. To specify a different number of iterations, include the number after the technique. For example, specifying moptimize_init_technique(M, "bhhh 10 nr 1000") requests that moptimize() perform 10 iterations using the Berndt–Hall–Hall–Hausman algorithm, followed by 1,000 iterations using the modified Newton–Raphson algorithm, and then switch back to Berndt–Hall–Hall–Hausman for 10 iterations, and so on. The process continues until convergence or until *maxiter* is exceeded.

`moptimize_init_singularHmethod(`*M*, *singularHmethod*`)` specifies the action to be taken during optimization if the Hessian is found to be singular and the *technique* requires the Hessian be of full rank. Allowed values are

singularHmethod	Description
`"m-marquardt"`	modified Marquardt algorithm
`"hybrid"`	mixture of steepest descent and Newton

The default is `"m-marquardt"`.
`"hybrid"` is equivalent to `ml`'s `difficult` option; see [R] **ml**.

`moptimize_init_nmsimplexdeltas(`*M*, *delta*`)` is for use with Nelder–Mead, also known as technique `nm`. This function sets the values of *delta* to be used, along with the initial parameters, to build the simplex required by Nelder–Mead. Use of this function is required only in the Nelder–Mead case. The values in *delta* must be at least 10 times larger than *ptol*. The initial simplex will be $\{p, p + (d_1, 0, \ldots, 0), p + (0, d_2, 0, \ldots, 0), \ldots, p + (0, 0, \ldots, 0, d_K)\}$.

Specifying initial values

Initial values are values you optionally specify that via a search procedure result in starting values that are then used for the first iteration of the optimization technique. That is,

$$\text{initial values} \xrightarrow{\text{(searching)}} \text{starting values} \xrightarrow{\substack{\text{(optimization} \\ \text{technique)}}} \text{final results}$$

Initial values are specified parameter by parameter.

`moptimize_init_eq_coefs(`*M*, *i*, *b0*`)` sets the initial values of the coefficients for the *i*th parameter to be *b0*: $1 \times (ki + ci)$. The default is $(0, 0, \ldots, 0)$.

The following functions control whether searching is used to improve on the initial values to produce better starting values. In addition to searching a predetermined set of hardcoded starting values, there are two other methods that can be used to improve on the initial values: random and rescaling. By default, random is off and rescaling is on. You can use one, the other, or both.

`moptimize_init_search(`*M*, { `"on"` | `"off"` }`)` determines whether any attempts are to be made to improve on the initial values via a search technique. The default is `"on"`. If you specify `"off"`, the initial values become the starting values.

`moptimize_init_search_random(`M`, {"off"|"on"})` determines whether the random method of improving initial values is to be attempted. The default is `"off"`. Use of the random method is recommended when the initial values are or might be infeasible. Infeasible means that the function cannot be evaluated, which mechanically corresponds to the user-written evaluator returning a missing value. The random method is seldom able to improve on feasible initial values. It works well when the initial values are or might be infeasible.

`moptimize_init_search_repeat(`M`, `nr`)` controls how many times random values are tried if the random method is turned on. The default is 10.

`moptimize_init_search_bounds(`M`, `i`, `*minmax*`)` specifies the bounds for the random search. *minmax* is a 1×2 real rowvector containing the minimum and maximum values for the ith parameter (equation). The default is $(.,.)$, meaning no lower and no upper bounds.

`moptimize_init_search_rescale(`M`, {"on"|"off"})` determines whether rescaling is attempted. The default is `"on"`. Rescaling is a deterministic (not random) method. It also usually improves initial values, and usually reduces the number of subsequent iterations required by the optimization technique.

Performing one evaluation of the objective function

`moptimize_evaluate(`M`)` and `_moptimize_evaluate(`M`)` perform one evaluation of the function evaluated at the initial values. Results can be accessed by using `moptimize_result_*()`, including first- and second-derivative-based results.

`moptimize_evaluate()` and `_moptimize_evaluate()` do the same thing, differing only in that `moptimize_evaluate()` aborts with a nonzero return code if things go badly, whereas `_moptimize_evaluate()` returns the real scalar error code. An infeasible initial value is an error.

The evaluation is performed at the initial values, not the starting values, and this is true even if search is turned on. If you want to perform an evaluation at the starting values, then perform optimization with *maxiter* set to 0.

Performing optimization of the objective function

`moptimize(`M`)` and `_moptimize(`M`)` perform optimization. Both routines do the same thing; they differ only in their behavior when things go badly. `moptimize()` returns nothing and aborts with error. `_moptimize()` returns a real scalar error code. `moptimize()` is best for interactive use and often adequate for use in programs that do not want to consider the possibility that optimization might fail.

The optimization process is as follows:

1. The initial values are used to create starting values. The value of the function at the starting values is calculated. If that results in a missing value, the starting values are declared infeasible. `moptimize()` aborts with return code 430; `_moptimize()` returns a nonzero error code, which maps to 430 via `moptimize_result_returncode()`. This step is called iteration 0.

2. The starting values are passed to the technique to produce better values. Usually this involves the technique calculating first and second derivatives, numerically or analytically, and then stepping multiple times in the appropriate direction, but techniques can vary on this. In general, the technique performs what it calls one iteration, the result of which is to produce better values. Those new values then become the starting values and the process repeats.

   An iteration is said to fail if the new coefficient vector is infeasible (results in a missing value). Then attempts are made to recover and, if those attempts are successful, optimization continues. If they fail, `moptimize()` aborts with error and `_moptimize()` returns a nonzero error code.

   Other problems may arise, such as singular Hessians or the inability to find better values. Various fix-ups are made and optimization continues. These are not failures.

   This step is called iterations 1, 2, and so on.

3. Step 2 continues until either the process converges or the maximum number of iterations (*maxiter*) is exceeded. Stopping due to *maxiter* is not considered an error. Upon completion, programmers should check `moptimize_result_converged()`.

If optimization succeeds, which is to say, if `moptimize()` does not abort or if `_moptimize()` returns 0, you can use the `moptimize_result_*()` functions to access results.

Tracing optimization

`moptimize()` and `_moptimize()` will produce output like

```
Iteration 0:  f(p) = ........
Iteration 1:  f(p) = ........
```

You can change the $f(p)$ to be "Log likelihood" or whatever else you want. You can also change "Iteration".

`moptimize_init_iterid(M, id)` sets the string to be used to label the iterations in the iteration log. *id* is a string scalar. The default is `"Iteration"`.

`moptimize_init_valueid(M, id)` sets the string to be used to label the objective function value in the iteration log. *id* is a string scalar. The default is `"f(p)"`.

Additional results can be displayed during optimization, which can be useful when you are debugging your evaluator. This is called tracing the execution.

moptimize_init_tracelevel(M, *tracelevel*) specifies the output to be displayed during the optimization process. Allowed values are

tracelevel	To be displayed each iteration
"none"	nothing
"value"	function value
"tolerance"	previous + convergence values
"step"	previous + stepping information
"coefdiffs"	previous + parameter relative differences
"paramdiffs"	same as "coefdiffs"
"coefs"	previous + parameter values
"params"	same as "coefs"
"gradient"	previous + gradient vector
"hessian"	previous + Hessian matrix

The default is "value" when set iterlog is on and "none" when set iterlog is off; see [R] *set iter*.

Setting *tracelevel* is a shortcut. The other trace functions allow you to turn on and off individual features. In what follows, the documented defaults are the defaults when *tracelevel* is "value".

moptimize_init_trace_ado(M, {"off"|"on"}) traces the execution of evaluators written as ado-files. This topic is not discussed in this manual entry. The default is "off".

moptimize_init_trace_dots(M, {"off"|"on"}) displays a dot each time your evaluator is called. The default is "off".

moptimize_init_trace_value(M, {"on"|"off"}) displays the function value at the start of each iteration. The default is "on".

moptimize_init_trace_tol(M, {"off"|"on"}) displays the value of the calculated result that is compared with the effective convergence criterion at the end of each iteration. The default is "off".

moptimize_init_trace_step(M, {"off"|"on"}) displays the steps within iteration. Listed are the value of objective function along with the word forward or backward. The default is "off".

moptimize_init_trace_coefdiffs(M, {"off"|"on"}) displays the coefficient relative differences from the previous iteration that are greater than the coefficient tolerance *ptol*. The default is "off".

moptimize_init_trace_coefs(M, {"off"|"on"}) displays the coefficients at the start of each iteration. The default is "off".

`moptimize_init_trace_gradient(`M`, {"off"|"on"})` displays the gradient vector at the start of each iteration. The default is `"off"`.

`moptimize_init_trace_Hessian(`M`, {"off"|"on"})` displays the Hessian matrix at the start of each iteration. The default is `"off"`.

Specifying convergence criteria

Convergence is based on several rules controlled by four parameters: *maxiter*, *ptol*, *vtol*, and *nrtol*. The first rule is not a convergence rule, but a stopping rule, and it is controlled by *maxiter*.

`moptimize_init_conv_maxiter(`M`, `*maxiter*`)` specifies the maximum number of iterations. If this number is exceeded, optimization stops and results are posted where they are accessible by using the `moptimize_result_*()` functions, just as if convergence had been achieved. `moptimize_result_converged()`, however, is set to 0 rather than 1. The default *maxiter* is Stata's `c(maxiter)`, which is 300 by default.

`moptimize_init_conv_warning(`M`, {"on"|"off"})` specifies whether the warning message "convergence not achieved" is to be displayed when this stopping rule is invoked. The default is `"on"`.

Usually, convergence occurs before the stopping rule comes into effect. The convergence criterion is a function of three real scalar values: *ptol*, *vtol*, and *nrtol*. Let

$$
\begin{aligned}
b &= \text{full set of coefficients} \\
b\_prior &= \text{value of } b \text{ from prior iteration} \\
v &= \text{value of objective function} \\
v\_prior &= \text{value of } v \text{ from prior iteration} \\
g &= \text{gradient vector from this iteration} \\
H &= \text{Hessian matrix from this iteration}
\end{aligned}
$$

Define, for maximization,

$$
\begin{array}{rll}
C\_ptol\!: & \texttt{mreldif}(b,\ b\_prior) & \leq & ptol \\
C\_vtol\!: & \texttt{reldif}(v,\ v\_prior) & \leq & vtol \\
C\_nrtol\!: & g \times \texttt{invsym}(-H) \times g' & < & nrtol \\
C\_concave\!: & -H \text{ is positive semidefinite}
\end{array}
$$

For minimization, think in terms of maximization of $-f(p)$. Convergence is declared when

$$
(C\_ptol \,|\, C\_vtol) \ \& \ C\_nrtol \ \& \ C\_concave
$$

The above applies in cases of derivative-based optimization, which currently is all techniques except `"nm"` (Nelder–Mead). In the Nelder–Mead case, the criterion is

$$C\_ptol\text{:}\quad \mathtt{mreldif}(\text{vertices of } R) \;\leq\; ptol$$
$$C\_vtol\text{:}\quad\qquad\qquad \mathtt{reldif}(R) \;\leq\; vtol$$

where R is the minimum and maximum values on the simplex. Convergence is declared when $C\_ptol \,|\, C\_vtol$.

The values of *ptol*, *vtol*, and *nrtol* are set by the following functions:

> `moptimize_init_conv_ptol(`M`, ` *ptol*`)` sets *ptol*. The default is 1e–6.

> `moptimize_init_conv_vtol(`M`, ` *vtol*`)` sets *vtol*. The default is 1e–7.

> `moptimize_init_conv_nrtol(`M`, ` *nrtol*`)` sets *nrtol*. The default is 1e–5.

> `moptimize_init_conv_ignorenrtol(`M`, ` {`"off"`|`"on"`}`)` sets whether C_nrtol should always be treated as true, which in effect removes the *nrtol* criterion from the convergence rule. The default is `"off"`.

Accessing results

Once you have successfully performed optimization, or you have successfully performed a single function evaluation, you may display results, post results to Stata, or access individual results.

To display results, use `moptimize_result_display()`.

> `moptimize_result_display(`M`)` displays estimation results. Standard errors are shown using the default *vcetype*.

> `moptimize_result_display(`M`, ` *vcetype*`)` displays estimation results. Standard errors are shown using the specified *vcetype*.

Also there is a third syntax for use after results have been posted to Stata, which we will discuss below.

> `moptimize_result_display()` without arguments (not even M) displays the estimation results currently posted in Stata.

vcetype specifies how the variance–covariance matrix of the estimators (VCE) is to be calculated. Allowed values are

vcetype	Description
""	use default for technique
"oim"	observed information matrix
"opg"	outer product of gradients
"robust"	Huber/White/sandwich estimator
"svy"	survey estimator; equivalent to robust

The default *vcetype* is oim except for technique bhhh, where it is opg. If survey, pweights, or clusters are used, the default becomes robust or svy.

As an aside, if you set moptimize_init_vcetype() during initialization, that changes the default.

moptimize_init_vcetype(M, *vcetype*), *vcetype* being a string scalar, resets the default *vcetype*.

To post results to Stata, use moptimize_result_post().

moptimize_result_post(M) posts estimation results to Stata where they can be displayed with Mata function moptimize_result_post() (without arguments) or with Stata command ereturn display (see [P] **ereturn**). The posted VCE will be of the default *vcetype*.

moptimize_result_post(M, *vcetype*) does the same thing, except the VCE will be of the specified *vcetype*.

The remaining moptimize_result_*() functions simply return the requested result. It does not matter whether results have been posted or previously displayed.

moptimize_result_value(M) returns the real scalar value of the objective function.

moptimize_result_value0(M) returns the real scalar value of the objective function at the starting values.

moptimize_result_eq_coefs($M[$, $i]$) returns the $1 \times (ki + ci)$ coefficient rowvector for the ith equation. If $i \geq .$ or argument i is omitted, the $1 \times K$ full set of coefficients is returned.

moptimize_result_coefs(M) returns the $1 \times K$ full set of coefficients.

moptimize_result_colstripe($M[$, $i]$) returns a $(ki + ci) \times 2$ string matrix containing, for the ith equation, the equation names in the first column and the coefficient names in the second. If $i \geq .$ or argument i is omitted, the result is $K \times 2$.

`moptimize_result_scores(M)` returns an $N \times m$ (evaluator types `lf` and `lf*`), or an $N \times K$ (evaluator type `gf`), or an $L \times K$ (evaluator type `q`) real matrix containing the observation-by-observation scores. For all other evaluator types, `J(0,0,.)` is returned. For evaluator types `lf` and `lf*`, scores are defined as the derivative of the objective function with respect to the parameters. For evaluator type `gf`, scores are defined as the derivative of the objective function with respect to the coefficients. For evaluator type `q`, scores are defined as the derivatives of the L independent elements with respect to the coefficients.

`moptimize_result_gradient(M[, i])` returns the $1 \times (ki+ci)$ gradient rowvector for the ith equation. If $i \geq$. or argument i is omitted, the $1 \times K$ gradient corresponding to the full set of coefficients is returned. Gradient is defined as the derivative of the objective function with respect to the coefficients.

`moptimize_result_Hessian(M[, i])` returns the $(ki + ci) \times (ki + ci)$ Hessian matrix for the ith equation. If $i \geq$. or argument i is omitted, the $K \times K$ Hessian corresponding to the full set of coefficients is returned. The Hessian is defined as the second derivative of the objective function with respect to the coefficients.

`moptimize_result_V(M[, i])` returns the appropriate $(ki + ci) \times (ki + ci)$ submatrix of the full variance matrix calculated according to the default *vcetype*. If $i \geq$. or argument i is omitted, the full $K \times K$ variance matrix corresponding to the full set of coefficients is returned.

`moptimize_result_Vtype(M)` returns a string scalar containing the default *vcetype*.

`moptimize_result_V_oim(M[, i])` returns the appropriate $(ki + ci) \times (ki + ci)$ submatrix of the full variance matrix calculated as the inverse of the negative Hessian matrix (the observed information matrix). If $i \geq$. or argument i is omitted, the full $K \times K$ variance matrix corresponding to the full set of coefficients is returned.

`moptimize_result_V_opg(M[, i])` returns the appropriate $(ki + ci) \times (ki + ci)$ submatrix of the full variance matrix calculated as the inverse of the outer product of the gradients. If $i \geq$. or argument i is omitted, the full $K \times K$ variance matrix corresponding to the full set of coefficients is returned. If `moptimize_result_V_opg()` is used with evaluator types other than `lf`, `lf*`, `gf`, or `q`, an appropriately dimensioned matrix of zeros is returned.

`moptimize_result_V_robust(M[, i])` returns the appropriate $(ki+ci) \times (ki+ci)$ submatrix of the full variance matrix calculated via the sandwich estimator. If $i \geq$. or argument i is omitted, the full $K \times K$ variance matrix corresponding to the full set of coefficients is returned. If `moptimize_result_V_robust()` is used with evaluator types other than `lf`, `lf*`, `gf`, or `q`, an appropriately dimensioned matrix of zeros is returned.

`moptimize_result_iterations(M)` returns a real scalar containing the number of iterations performed.

moptimize_result_converged(*M*) returns a real scalar containing 1 if convergence was achieved and 0 otherwise.

moptimize_result_iterationlog(*M*) returns a real colvector containing the values of the objective function at the end of each iteration. Up to the last 20 iterations are returned, one to a row.

moptimize_result_errorcode(*M*) returns the real scalar containing the error code from the most recently run optimization or function evaluation. The error code is 0 if there are no errors. This function is useful only after _moptimize() or _moptimize_evaluate() because the nonunderscore versions aborted with error if there were problems.

moptimize_result_errortext(*M*) returns a string scalar containing the error text corresponding to moptimize_result_errorcode().

moptimize_result_returncode(*M*) returns a real scalar containing the Stata return code corresponding to moptimize_result_errorcode().

The following error codes and their corresponding Stata return codes are only for moptimize(). To see other error codes and their corresponding Stata return codes, see [M-5] **optimize( )**.

Error code	Return code	Error text
400	1400	could not find feasible values
401	491	Stata program evaluator returned an error
402	198	views are required when the evaluator is a Stata program
403	198	Stata program evaluators require a touse variable

Stata evaluators

The following function is useful only when your evaluator is a Stata program instead of a Mata function.

moptimize_ado_cleanup(*M*) removes all the global macros with the ML_ prefix. A temporary weight variable is also dropped if weights were specified.

Advanced functions

These functions are not really advanced, they are just seldomly used.

moptimize_init_verbose(*M*, {"on"|"off"}) specifies whether error messages are to be displayed. The default is "on".

moptimize_init_evaluations(*M*, {"off"|"on"}) specifies whether the system is to count the number of times the evaluator is called. The default is "off".

moptimize_result_evaluations(M) returns a 1×3 real rowvector containing the number of times the evaluator was called. moptimize_init_evaluations() must be set on. Contents are the number of times called for the purposes of 1) calculating the objective function, 2) calculating the objective function and its first derivative, and 3) calculating the objective function and its first and second derivatives. If moptimize_init_evaluations() was set off, returned is (0,0,0).

Syntax of evaluators

An evaluator is a program you write that calculates the value of the function being optimized and optionally calculates the function's first and second derivatives. The evaluator you write is called by the moptimize() functions.

There are five styles in which the evaluator can be written, known as types lf, d, lf*, gf, and q. *evaluatortype*, optionally specified in moptimize_init_evaluatortype(), specifies the style in which the evaluator is written. Allowed values are

evaluatortype	Description
"lf"	*function*() returns $N \times 1$ colvector value
"d0"	*function*() returns scalar value
"d1"	same as "d0" and returns gradient rowvector
"d2"	same as "d1" and returns Hessian matrix
"d1debug"	same as "d1" but checks gradient
"d2debug"	same as "d2" but checks gradient and Hessian
"lf0"	*function*() returns $N \times 1$ colvector value
"lf1"	same as "lf0" and return equation-level score matrix
"lf2"	same as "lf1" and returns Hessian matrix
"lf1debug"	same as "lf1" but checks gradient
"lf2debug"	same as "lf2" but checks gradient and Hessian
"gf0"	*function*() returns $N \times 1$ colvector value
"gf1"	same as "gf0" and returns score matrix
"gf2"	same as "gf1" and returns Hessian matrix
"gf1debug"	same as "gf1" but checks gradient
"gf2debug"	same as "gf2" but checks gradient and Hessian
"q0"	*function*() returns colvector value
"q1"	same as "q0" and returns score matrix
"q1debug"	same as "q1" but checks gradient

The default is "lf" if not set.
"q" evaluators are used with technique "gn".
Returned gradients are $1 \times K$ rowvectors.
Returned Hessians are $K \times K$ matrices.

Examples of each of the evaluator types are outlined below.

You must tell `moptimize()` the identity and type of your evaluator, which you do by using the `moptimize_init_evaluator()` and `moptimize_init_evaluatortype()` functions.

`moptimize_init_evaluator(M, &functionname())` sets the identity of the evaluator function that you write in Mata.

`moptimize_init_evaluator(M, "programname")` sets the identity of the evaluator program that you write in Stata.

`moptimize_init_evaluatortype(M, evaluatortype)` informs `moptimize()` of the style of evaluator you have written. *evaluatortype* is a string scalar from the table above. The default is `"lf"`.

`moptimize_init_negH(M, {"off"|"on"})` sets whether the evaluator you have written returns H or $-H$, the Hessian or the negative of the Hessian, if it returns a Hessian at all. This is for backward compatibility with prior versions of Stata's `ml` command (see [R] **ml**). Modern evaluators return H. The default is `"off"`.

Syntax of type lf evaluators

lfeval(M, b, fv):
    inputs:
        M: problem definition
        b: coefficient vector
    outputs:
        fv: $N \times 1$, $N = \#$ of observations

Notes:

1. The objective function is $f(\cdot) = \texttt{colsum}(\textit{fv})$.

2. In the case where $f(\cdot)$ is a log-likelihood function, the values of the log likelihood must be summable over the observations.

3. For use with any technique except `gn`.

4. May be used with robust, clustering, and survey.

5. Returns *fv* containing missing ($\textit{fv} = .$) if evaluation is not possible.

Syntax of type d evaluators

deval(M, todo, b, fv, g, H):
  inputs:

M:	problem definition
todo:	real scalar containing 0, 1, or 2
b:	coefficient vector

  outputs:

fv:	real scalar
g:	$1 \times K$, gradients, $K = \#$ of coefficients
H:	$K \times K$, Hessian

Notes:

1. The objective function is $f(\cdot) = fv$.

2. For use with any log-likelihood function, or any function.

3. For use with any technique except `gn` and `bhhh`.

4. Cannot be used with robust, clustering, or survey.

5. *deval*() must always fill in *fv*, and fill in *g* if *todo* ≥ 1, and fill in H if *todo* $= 2$. For type `d0`, *todo* will always be 0. For type `d1` and `d1debug`, *todo* will be 0 or 1. For type `d2` and `d2debug`, *todo* will be 0, 1, or 2.

6. Returns *fv* $= .$ if evaluation is not possible.

Syntax of type lf* evaluators

lfeval(M, *todo*, b, *fv*, S, H):
  inputs:

M:	problem definition
todo:	real scalar containing 0, 1, or 2
b:	coefficient vector

  outputs:

fv:	$N \times 1$, $N = \#$ of observations
S:	$N \times m$, scores, $m = \#$ of equations (parameters)
H:	$K \times K$, Hessian, $K = \#$ of coefficients

Notes:

1. The objective function is $f(\cdot) = $ `colsum`(fv).

2. Type `lf*` is a variation of type `lf` that allows the user to supply analytic derivatives. Although `lf*` could be used with an arbitrary function, it is intended for use when $f(\cdot)$ is a log-likelihood function and the log-likelihood values are summable over the observations.

3. For use with any technique except `gn`.

4. May be used with robust, clustering, and survey.

5. Always returns *fv*, returns S if $todo \geq 1$, and returns H if $todo = 2$. For type `lf0`, *todo* will always be 0. For type `lf1` and `lf1debug`, *todo* will be 0 or 1. For type `lf2` and `lf2debug`, *todo* will be 0, 1, or 2.

6. Returns *fv* containing missing $(fv = .)$ if evaluation is not possible.

Syntax of type gf evaluators

gfeval(M, *todo*, b, *fv*, S, H):
   inputs:

M:	problem definition
todo:	real scalar containing 0, 1, or 2
b:	coefficient vector

   outputs:

fv:	$L \times 1$, values, $L = \#$ of independent elements
S:	$L \times K$, scores, $K = \#$ of coefficients
H:	$K \times K$, Hessian

Notes:

1. The objective function is $f(\cdot) = \texttt{colsum}(fv)$.

2. Type `gf` is a variation on type `lf*` that relaxes the requirement that the log-likelihood function be summable over the observations. `gf` is especially useful for fitting panel-data models with technique `bhhh`. Then L is the number of panels.

3. For use with any `gf` is especially useful for fitting panel-data models with except `gn`.

4. May be used with robust, clustering, and survey.

5. Always returns *fv*, returns S if $todo \geq 1$, and returns H if $todo = 2$. For type `gf0`, *todo* will always be 0. For type `gf1` and `gf1debug`, *todo* will be 0 or 1. For type `gf2` and `gf2debug`, *todo* will be 0, 1, or 2.

6. Returns $fv = .$ if evaluation is not possible.

Syntax of type q evaluators

qeval(*M*, *todo*, *b*, *r*, *S*)

     inputs:
M:	problem definition
todo:	real scalar containing 0 or 1
b:	coefficient vector

     outputs:
r:	$L \times 1$ of independent elements
S:	$L \times K$, scores, $K = \#$ of coefficients

Notes:

1. Type q is for quadratic optimization. The objective function is $f(\cdot) = r'Wr$, where r is returned by *qeval*() and W has been previously set by using `moptimize_init_gnweightmatrix()`, described below.

2. For use only with techniques gn and nm.

3. Always returns r and returns S if $todo = 1$. For type q0, *todo* will always be 0. For type q1 and q1debug, *todo* will be 0 or 1. There is no type q2.

4. Returns r containing missing, or $r = .$ if evaluation is not possible.

Use `moptimize_init_gnweightmatrix()` during initialization to set matrix W.

`moptimize_init_gnweightmatrix(`*M*, *W*`)` sets real matrix W: $L \times L$, which is used only by type q evaluators. The objective function is $r'Wr$. If W is not set and if observation weights w are set by using `moptimize_init_weight()`, then $W = \text{diag}(w)$. If w is not set, then W is the identity matrix.

`moptimize()` does not produce a robust VCE when you set W with `moptimize_init_gnweight()`.

Passing extra information to evaluators

In addition to the arguments the evaluator receives, you may arrange that extra information be sent to the evaluator. Specify the extra information to be sent by using `moptimize_init_userinfo()`.

`moptimize_init_userinfo(`*M*, *l*, *Z*`)` specifies that the *l*th piece of extra information is *Z*. *l* is a real scalar. The first piece of extra information should be 1; the second piece, 2; and so on. *Z* can be anything. No copy of *Z* is made.

`moptimize_init_nuserinfo(`*M*, *n_user*`)` specifies the total number of pieces of information to be sent. Setting *n_user* is optional; it will be automatically determined from the `moptimize_init_userinfo()` calls you issue.

Inside your evaluator, you access the information by using
`moptimize_util_userinfo()`.

> `moptimize_util_userinfo(`M`, `l`)` returns the Z set by
> `moptimize_init_userinfo()`.

Utility functions

There are various utility functions that are helpful in writing evaluators and in processing results returned by the `moptimize_result_*()` functions.

The first set of utility functions are useful in writing evaluators, and the first set return results that all evaluators need.

> `moptimize_util_depvar(`M`, `j`)` returns an $Nj \times 1$ colvector containing the values of the jth dependent variable, the values set by `moptimize_init_depvar(`M`, `j`, ...)`.

> `moptimize_util_xb(`M`, `b`, `i`)` returns the $Ni \times 1$ colvector containing the value of the ith parameter, which is usually $Xi \times bi'$ `:+` $b0i$, but might be as complicated as $Xi \times bi' + oi + \ln(ti)$ `:+` $b0i$.

Once the inputs of an evaluator have been processed, the following functions assist in making the calculations required of evaluators.

> `moptimize_util_sum(`M`, `v`)` returns the "sum" of colvector v. This function is for use in evaluators that require you to return an overall objective function value rather than observation-by-observation results. Usually, `moptimize_util_sum()` returns `sum(`v`)`, but in the cases where you have specified a weight by using `moptimize_init_weight()` or where there is an implied weight due to the use of `moptimize_init_svy()`, the appropriately weighted sum is returned. Use `moptimize_util_sum()` to sum log-likelihood values.

> `moptimize_util_vecsum(`M`, `i`, `s`, `*value*`)` is like `moptimize_util_sum()`, but for use with gradients. The gradient is defined as the vector of partial derivatives of $f(\cdot)$ with respect to the coefficients bi. Some evaluator types require that your evaluator be able to return this vector. Nonetheless, it is usually easier to write your evaluator in terms of parameters rather than coefficients, and this function handles the mapping of parameter gradients to the required coefficient gradients.

> Input s is an $Ni \times 1$ colvector containing df/dpi for each observation. df/dpi is the partial derivative of the objective function, but with respect to the ith parameter rather than the ith set of coefficients. `moptimize_util_vecsum()` takes s and returns the $1 \times (ki + ci)$ summed gradient. Also weights, if any, are factored into the calculation.

> If you have $m < 1$ equation, you will need to call `moptimize_util_vecsum()` m times, once for each equation, and then concatenate the individual results into one vector.

value plays no role in `moptimize_util_vecsum()`'s calculations. *value*, however, should be specified as the result obtained from `moptimize_util_sum()`. If that is inconvenient, make *value* any nonmissing value. If the calculation from parameter space to vector space cannot be performed, or if your original parameter space derivatives have any missing values, *value* will be changed to missing. Remember, when a calculation cannot be made, the evaluator is to return a missing value for the objective function. Thus storing the value of the objective function in *value* ensures that your evaluator will return missing if it is supposed to.

`moptimize_util_matsum(`*M*, *i*, *i2*, *s*, *value*`)` is like `moptimize_util_vecsum()`, but for Hessians (matrix of second derivatives).

Input *s* is an $Ni \times 1$ colvector containing $d^2f/dpidpi2$ for each observation. `moptimize_util_matsum()` returns the $(ki + ci) \times (ki2 + ci2)$ summed Hessian. Also weights, if any, are factored into the calculation.

If you have $m > 1$ equations, you will need to call `moptimize_util_matsum()` $m \times (m+1)/2$ times and then join the results into one symmetric matrix.

value plays no role in the calculation and works the same way it does in `moptimize_util_vecsum()`.

`moptimize_util_matbysum()` is a helper for making `moptimize_util_matsum()` calculations in cases where you have panel data and the log-likelihood function's values exists only at the panel level. `moptimize_util_matbysum(`*M*, *i*, *a*, *b*, *value*`)` is for making diagonal calculations and `moptimize_util_matbysum(`*M*, *i*, *i2*, *a*, *b*, *c*, *value*`)` is for making off-diagonal calculations.

This is an advanced topic; see pages 268–269 for a full description of it. In applying the chain rule to translate results from parameter space to coefficient space, `moptimize_util_matsum()` can be used to make some of the calculations, and `moptimize_util_matbysum()` can be used to make the rest. *value* plays no role and works as it did in the other helper functions. `moptimize_util_matbysum()` is for use at times when *by* has been set, which is done by `moptimize_init_by(`*M*, *by*`)`. `moptimize_util_matbysum()` is never required unless *by* has been set.

The formula implemented in `moptimize_util_matbysum(`*M*, *i*, *a*, *b*, *value*`)` is

$$\sum_{j=1}^{N} \left(\sum_{t=1}^{T_j} a_{jt} \right) \left(\sum_{t=1}^{T_j} b_{jt} \mathbf{x}'_{1jt} \right) \left(\sum_{t=1}^{T_j} b_{jt} \mathbf{x}_{1jt} \right)$$

The formula implemented in `moptimize_util_matbysum(`*M*, *i*, *i2*, *a*, *b*, *c*, *value*`)` is

$$\sum_{j=1}^{N} \left(\sum_{t=1}^{T_j} a_{jt} \right) \left(\sum_{t=1}^{T_j} b_{jt} \mathbf{x}'_{1jt} \right) \left(\sum_{t=1}^{T_j} c_{jt} \mathbf{x}_{2jt} \right)$$

moptimize_util_by() returns a pointer to the vector of group identifiers that were set using moptimize_init_by(). This vector can be used with panelsetup() to perform panel level calculations.

The other utility functions are useful inside or outside of evaluators. One of the more useful is moptimize_util_eq_indices(), which allows two or three arguments.

moptimize_util_eq_indices(M, i) returns a 1×2 vector that can be used with range subscripts to extract the portion relevant for the ith equation from any $1 \times K$ vector, that is, from any vector conformable with the full coefficient vector.

moptimize_util_eq_indices(M, i, $i2$) returns a 2×2 matrix that can be used with range subscripts to exact the portion relevant for the ith and $i2$th equations from any $K \times K$ matrix, that is, from any matrix with rows and columns conformable with the full variance matrix.

For instance, let b be the $1 \times K$ full coefficient vector, perhaps obtained by being passed into an evaluator, or perhaps obtained from b = moptimize_result_coefs(M). Then b[|moptimize_util_eq_indices(M, i)|] is the $1 \times (ki + ci)$ vector of coefficients for the ith equation.

Let V be the $K \times K$ full variance matrix obtained by V = moptimize_result_V(M). Then V[|moptimize_util_eq_indices(M, i, i)|] is the $(ki + ci) \times (ki + ci)$ variance matrix for the ith equation. V[|moptimize_util_eq_indices(M, i, j)|] is the $(ki + ci) \times (kj + cj)$ covariance matrix between the ith and jth equations.

Finally, there is one more utility function that may help when you become confused: moptimize_query().

moptimize_query(M) displays in readable form everything you have set via the moptimize_init_*() functions, along with the status of the system.

Diagnostics

All functions abort with error when used incorrectly.

moptimize() aborts with error if it runs into numerical difficulties. _moptimize() does not; it instead returns a nonzero error code.

The moptimize_result*() functions abort with error if they run into numerical difficulties when called after moptimize() or moptimize_evaluate(). They do not abort when run after _moptimize() or _moptimize_evaluate(). They instead return a properly dimensioned missing result and set moptimize_result_errorcode() and moptimize_result_errortext().

D Likelihood-evaluator checklists

D.1 Method lf

1. To use method `lf`, the likelihood function must meet the linear-form restrictions; otherwise, you must use method `d0`.

2. Write a program to evaluate $\ln \ell_j$, the contribution to the overall likelihood for an observation. The outline for the program is

```
program myprog
        version 18
        args lnfj theta1 theta2 ...
        // if you need to create any intermediate results:
        tempvar tmp1 tmp2 ...
        quietly generate double 'tmp1' = ...
        ...
        quietly replace 'lnfj' = ...
end
```

   Access the dependent variables (if any) by typing $ML_y1, $ML_y2, ... Use $ML_y1, $ML_y2, ... just as you would any existing variable in the data.

   Final results are to be stored in 'lnfj'. ('lnfj' is just a double-precision variable `ml` created for you. When `ml` calls your program, 'lnfj' contains missing values.)

   You must create any temporary variables for intermediate results as `double`s: type `gen double` *name* = ...; do not omit the word `double`.

3. Issue the `ml model` statement as

   `. ml model lf` *myprog* (*equation for* θ_1) (*equation for* θ_2) ...

   Specify `if`, `in`, weights, and the `vce(opg)`, `vce(robust)`, `vce(cluster` *var-name*), ... options at this point.

4. Verify that your program works; type

   `. ml check`

5. Find starting values; type

   `. ml search`

   or use `ml init` or `ml plot`. We suggest you use `ml search`. In most cases, `ml search` can be omitted because `ml maximize` calls `ml search`, but in that case, `ml search` will not do as thorough a job.

6. Fit the model; type

   ```
   . ml maximize
   ```

7. If later you wish to redisplay results, type

   ```
   . ml display
   ```

D.2   Method d0

1. Method d0 may be used with any likelihood function.

2. Write a program to evaluate $\ln L$, the overall log-likelihood function. The outline for the evaluator is

```
program myprog
        version 18
        args todo b lnf

        tempvar theta1 theta2 ...
        mleval 'theta1' = 'b'
        mleval 'theta2' = 'b', eq(2)  // if there is a θ₂
        ...

        // if you need to create any intermediate results:
        tempvar tmp1 tmp2 ...
        generate double 'tmp1' = ...
        ...

        mlsum 'lnf' = ...
end
```

Argument 'todo' will always contain 0; it can be ignored.

You are passed the parameter row vector in 'b'. Use mleval to obtain the linear equation values $\theta_{ij} = \mathbf{x}_{ij}\boldsymbol{\beta}_i$ from it.

Access the dependent variables (if any) by typing $ML_y1, $ML_y2, ... Use $ML_y1, $ML_y2, ... just as you would any existing variable in the data.

Final results are to be stored in scalar 'lnf'. Use mlsum to produce it.

You must create any temporary variables for intermediate results as doubles: type gen double *name* = ...; do not omit the word double.

The estimation subsample is $ML_samp==1. You may safely ignore this if you do not specify ml model's nopreserve option. When your program is called, only relevant data will be in memory. If you do specify nopreserve, understand that mleval and mlsum automatically restrict themselves to the estimation subsample; it is not necessary to code if $ML_samp==1 on these commands. You merely need to restrict other commands to the $ML_samp==1 subsample.

The weights are stored in variable $ML_w, which contains 1 in every observation if no weights are specified. If you use mlsum to produce the log likelihood and mlvecsum to produce the gradient vector, you may ignore this because these commands handle that themselves.

3. Issue the `ml model` statement as

>   . ml model d0 *myprog* (*equation for θ_1*) (*equation for θ_2*) ...

Specify any `if` *exp* or weights at this point. Although method-d0 evaluators can be used with `fweights`, `aweights`, and `iweights`, `pweights` may not be specified. Use method `1f`, `1f0`, `1f1`, or `gf0` if you want to support `pweights`. Similarly, you may not specify `ml model`'s `vce(opg)`, `vce(robust)`, or `vce(cluster `*varname*`)` option. Use method `1f`, `1f0`, `1f1`, or `gf0` if you want to support these `vce()` options. The BHHH algorithm may not be used with method d0.

4. Verify that your program works; type

>   . ml check

5. Find starting values; type

>   . ml search

or use `ml init` or `ml plot`. We suggest that you use `ml search`. In most cases, `ml search` can be omitted because `ml maximize` calls `ml search`, but in that case `ml search` will not do as thorough a job.

6. Fit the model; type

>   . ml maximize

7. If later you wish to redisplay results, type

>   . ml display

D.3   Method d1

1. Method `d1` may be used with any likelihood function. This will be a considerable effort, however, if the function does not meet the linear-form restrictions.

2. Write a program to evaluate $\ln L$, the overall log-likelihood function and its derivatives. The outline for the evaluator is

```
program myprog
        version 18
        args todo b lnf g

        tempvar theta1 theta2 ...
        mleval 'theta1' = 'b'
        mleval 'theta2' = 'b', eq(2) // if there is a θ₂
        ...

        // if you need to create any intermediate results:
        tempvar tmp1 tmp2 ...
        generate double 'tmp1' = ...
        ...

        mlsum 'lnf' = ...
        if ('todo'==0 | 'lnf'==.) exit

        tempname d1 d2 ...
        mlvecsum 'lnf' 'd1' = formula for ∂ln ℓⱼ/∂θ₁ⱼ, eq(1)
        mlvecsum 'lnf' 'd2' = formula for ∂ln ℓⱼ/∂θ₂ⱼ, eq(2)
        ...

        matrix 'g' = ('d1', 'd2', ... )
end
```

Argument 'todo' will contain 0 or 1. You are to fill in the log likelihood if 'todo'==0 or fill in the log likelihood and gradient otherwise.

You are passed the parameter row vector in 'b'. Use mleval to obtain the linear equation values $\theta_{ij} = \mathbf{x}_{ij}\boldsymbol{\beta}_i$ from it.

Access the dependent variables (if any) by referring to \$ML_y1, \$ML_y2, ... Use \$ML_y1, \$ML_y2, ... just as you would any existing variable in the data.

The overall log-likelihood value is to be stored in scalar 'lnf'. Use mlsum to produce it.

The gradient vector is to be stored in vector 'g'. Use mlvecsum to produce it. You issue one mlvecsum per theta and then use matrix *name* = ... to put the results together.

You must create any temporary variables for intermediate results as doubles: type gen double *name* = ...; do not omit the word double.

The estimation subsample is \$ML_samp==1. You may safely ignore this if you do not specify ml model's nopreserve option. When your program is called, only relevant data will be in memory. If you do specify nopreserve, understand that mleval, mlsum, and mlvecsum automatically restrict themselves to the estimation subsample; it is not necessary to code if \$ML_samp==1 on these commands. You merely need to restrict other commands to the \$ML_samp==1 subsample.

The weights are stored in variable \$ML_w, which contains 1 if no weights are specified. If you use mlsum to produce the log likelihood and mlvecsum to produce the gradient vector, you may ignore this because these commands handle the weights themselves.

3. Issue the `ml model` statement as

   `. ml model d1debug` *myprog* (*equation for θ₁*) (*equation for θ₂*) ...

   Note: Begin with method **d1debug** and not method **d1**. This allows testing whether the program produces correct first derivatives.

4. Verify that your program works; type

   `. ml check`

5. Find starting values; type

   `. ml search`

   or use `ml init` or `ml plot`. We suggest that you use `ml search`. In most cases, `ml search` can be omitted because `ml maximize` calls `ml search`, but in that case `ml search` will not do as thorough a job.

6. Fit the model; type

   `. ml maximize`

   Review the reported `mreldif()`s in the iteration log. Even if you have programmed the derivatives correctly, it is typical to observe the `mreldif()`s falling and then rising and even observe the `mreldif()` being largest in the final iteration, when the gradient vector is near zero. This result should not concern you. The reported `mreldif()`s should be 1e–3 or smaller in the middle iterations. If you suspect problems, reissue the `ml model` statement and then run `ml maximize` with the **gradient** option.

7. Now that you have verified that your program is calculating derivatives correctly, switch to using method **d1** by typing

   `. ml model d1` *myprog* (*equation for θ₁*) (*equation for θ₂*) ...

   Repeat steps 4, 5, and 6. You should obtain approximately the same results. The iteration log will probably differ; method **d1** may require fewer or more iterations than method **d1debug**. This does not indicate problems. Method **d1debug** used the numerically calculated derivatives, and maximization is a chaotic process.

8. If later you wish to redisplay results, type

   `. ml display`

D.4   Method d2

1. Method **d2** may be used with any likelihood function. This will be a considerable effort, however, if the function does not meet the linear-form restrictions.

2. Write a program to evaluate $\ln L$, the overall log-likelihood function, and its first and second derivatives. The outline for the evaluator is

```
program myprog
        version 18
        args todo b lnf g H

        tempvar theta1 theta2 ...
        mleval 'theta1' = 'b'
        mleval 'theta2' = 'b', eq(2) // if there is a θ2
        ...

        // if you need to create any intermediate results:
        tempvar tmp1 tmp2 ...
        generate double 'tmp1' = ...
        ...

        mlsum 'lnf' = ...
        if ('todo'==0 | 'lnf'==.) exit

        tempname d1 d2 ...
        mlvecsum 'lnf' 'd1' = formula for ∂ln ℓ_j/∂θ_{1j}, eq(1)
        mlvecsum 'lnf' 'd2' = formula for ∂ln ℓ_j/∂θ_{2j}, eq(2)
        ...

        matrix 'g' = ('d1', 'd2', ... )
        if ('todo'==1 | 'lnf'==.) exit

        tempname d11 d12 d22 ...
        mlmatsum 'lnf' 'd11' = formula for ∂²ln ℓ_j/∂θ²_{1j}, eq(1)
        mlmatsum 'lnf' 'd12' = formula for ∂²ln ℓ_j/∂θ_{1j}∂θ_{2j}, eq(1,2)
        mlmatsum 'lnf' 'd22' = formula for ∂²ln ℓ_j/∂θ²_{2j}, eq(2)
        ...

        matrix 'H' = ('d11','d12',... \ 'd12'','d22',... )
end
```

Argument 'todo' will contain 0, 1, or 2. You are to fill in the log likelihood if 'todo'==0; the log likelihood and gradient if 'todo'==1; or the log likelihood, gradient, and negative Hessian otherwise.

You are passed the parameter row vector in 'b'. Use mleval to obtain the linear-equation values $\theta_{ij} = \mathbf{x}_{ij}\boldsymbol{\beta}_i$ from it.

Access the dependent variables (if any) by referring to $ML_y1, $ML_y2, ... Use $ML_y1, $ML_y2, ... just as you would any existing variable in the data.

The overall log-likelihood value is to be stored in scalar 'lnf'. Use mlsum to produce it.

The gradient vector is to be stored in vector 'g'. Use mlvecsum to produce it. You issue one mlvecsum per theta and then use matrix *name* = ... to put the results together.

The Hessian is to be stored in matrix 'H'. Use mlmatsum to produce it. You issue one mlmatsum per equation and equation pair and then use matrix *name* = ... to put the results together.

You must create any temporary variables for intermediate results as doubles: type gen double *name* = ...; do not omit the word double.

The estimation subsample is $ML_samp==1. You may safely ignore this if you do not specify ml model's nopreserve option. When your program is called, only relevant data will be in memory. If you do specify nopreserve, understand that mleval, mlsum, mlvecsum, and mlmatsum automatically restrict themselves to the estimation subsample; it is not necessary to code `if $ML_samp==1` on these commands. You merely need to restrict other commands to the `$ML_samp==1` subsample.

The weights are stored in variable `$ML_w`, which contains 1 if no weights are specified. If you use `mlsum` to produce the log likelihood and `mlvecsum` to produce the gradient vector, you may ignore this because these commands handle the weights themselves.

3. Issue the `ml model` statement as

    . ml model d2debug *myprog* (*equation for θ_1*) (*equation for θ_2*) ...

Note: Begin with method **d2debug** and not method **d2**. This allows testing whether the program produces correct first and second derivatives.

4. Verify that your program works; type

    . ml check

5. Find starting values; type

    . ml search

or use `ml init` or `ml plot`. We suggest that you use `ml search`. In most cases, `ml search` can be omitted because `ml maximize` calls `ml search`, but in that case `ml search` will not do as thorough a job.

6. Fit the model; type

    . ml maximize

Review the reported `mreldif()`s in the iteration log. Even if you have programmed the derivatives correctly, it is typical to observe the `mreldif()`s for the gradient vector falling and then rising and even observe the `mreldif()` being largest in the final iteration, when the gradient vector is near zero. This should not concern you. The reported `mreldif()`s should be 1e–3 or smaller in the middle iterations.

For the negative Hessian, early iterations often report large `mreldif()`s but eventually fall. The reported `mreldif()` should be 1e–5 or smaller in the last iteration.

If you suspect problems, reissue the `ml model` statement and then run `ml maximize` with the `gradient` option or the `hessian` option.

7. Now that you have verified that your program is calculating derivatives correctly, switch to using method **d2** by typing

    . ml model d2 *myprog* (*equation for θ_1*) (*equation for θ_2*) ...

Repeat steps 4, 5, and 6. You should obtain approximately the same results (relative differences are 1e–8 for coefficients and smaller than 1e–5 for standard errors). The iteration log will probably differ; method **d2** may require fewer or more iterations than method **d2debug**. This does not indicate problems. Method **d2debug** used the numerically calculated derivatives, and maximization is a chaotic process.

8. If later you wish to redisplay results, type

```
. ml display
```

D.5   Method lf0

1. To use method `lf0`, the likelihood function must meet the linear-form restrictions.

2. Write a program to evaluate $\ln \ell_j$, the contribution to the overall likelihood function for an observation. The outline for the evaluator is

```
program myprog
        version 18
        args todo b lnfj

        tempvar theta1 theta2 ...
        mleval 'theta1' = 'b'
        mleval 'theta2' = 'b', eq(2) // if there is a θ₂
        ...

        // if you need to create any intermediate results:
        tempvar tmp1 tmp2 ...
        generate double 'tmp1' = ...
        ...

        quietly replace 'lnfj' = ...
end
```

Argument '`todo`' will always contain 0; it can be ignored.

You are passed the parameter row vector in '`b`'. Use `mleval` to obtain the linear equation values $\theta_{ij} = \mathbf{x}_{ij}\boldsymbol{\beta}_i$ from it.

Access the dependent variables (if any) by typing `$ML_y1`, `$ML_y2`, ... Use `$ML_y1`, `$ML_y2`, ... just as you would any existing variable in the data.

Final results are to be stored in '`lnfj`'. ('`lnfj`' is just a temporary double-precision variable `ml` created for you. When `ml` calls your program, '`lnfj`' contains missing values.)

You must create any temporary variables for intermediate results as `doubles`: type `gen double` *name* = ...; do not omit the word `double`.

The estimation subsample is `$ML_samp==1`. You may safely ignore this if you do not specify `ml model`'s `nopreserve` option. When your program is called, only relevant data will be in memory. If you do specify `nopreserve`, understand that `mleval` automatically restricts itself to the estimation subsample; it is not necessary to code `if $ML_samp==1` on these commands. You merely need to restrict other commands to the `$ML_samp==1` subsample.

The weights are stored in variable `$ML_w`, which contains 1 in every observation if no weights are specified.

3. Issue the `ml model` statement as

>   . ml model lf0 *myprog* (*equation for θ_1*) (*equation for θ_2*) ...

Specify any `if` *exp* or weights at this point. Although method-d0 evaluators can be used with `fweights`, `aweights`, and `iweights`, `pweights` may not be specified.

Specify `if`, `in`, weights, and the `vce(opg)`, `vce(robust)`, `vce(cluster` *varname*), ... options at this point.

4. Verify that your program works; type

>   . ml check

5. Find starting values; type

>   . ml search

or use `ml init` or `ml plot`. We suggest that you use `ml search`. In most cases, `ml search` can be omitted because `ml maximize` calls `ml search`, but in that case `ml search` will not do as thorough a job.

6. Fit the model; type

>   . ml maximize

7. If later you wish to redisplay results, type

>   . ml display

D.6   Method lf1

1. To use method `lf1`, the likelihood function must meet the linear-form restrictions.

2. Write a program to evaluate $\ln \ell_j$, the contribution to the overall likelihood function for an observation, and to calculate the equation-level score variables. The outline for the evaluator is

```
program myprog
        version 18
        args todo b lnfj g1 g2 ...
        tempvar theta1 theta2 ...
        mleval 'theta1' = 'b'
        mleval 'theta2' = 'b', eq(2) // if there is a θ2
        ...
        // if you need to create any intermediate results:
        tempvar tmp1 tmp2 ...
        generate double 'tmp1' = ...
        ...
        quietly replace 'lnfj' = ...
        if ('todo'==0) exit
        quietly replace 'g1' = formula for ∂ln ℓj/∂θ1j, eq(1)
        quietly replace 'g2' = formula for ∂ln ℓj/∂θ2j, eq(2)
        ...
end
```

Argument 'todo' will contain 0 or 1. You are to fill in the log likelihood if 'todo'==0 or fill in the log likelihood and scores otherwise.

You are passed the parameter row vector in 'b'. Use mleval to obtain the linear equation values $\theta_{ij} = \mathbf{x}_{ij}\boldsymbol{\beta}_i$ from it.

Access the dependent variables (if any) by referring to $ML_y1, $ML_y2, ... Use $ML_y1, $ML_y2, ... just as you would any existing variable in the data.

The observation-level log-likelihood values are to be stored in 'lnfj'.

The scores are to be stored in g1, g2, ...

You must create any temporary variables for intermediate results as doubles: type gen double *name* = ...; do not omit the word double.

The estimation subsample is $ML_samp==1. You may safely ignore this if you do not specify ml model's nopreserve option. When your program is called, only relevant data will be in memory. If you do specify nopreserve, understand that mleval automatically restricts itself to the estimation subsample; it is not necessary to code if $ML_samp==1 on these commands. You merely need to restrict other commands to the $ML_samp==1 subsample.

The weights are stored in variable $ML_w, which contains 1 if no weights are specified.

3. Issue the ml model statement as

    . ml model lf1debug *myprog* (*equation for* θ_1) (*equation for* θ_2) ...

Note: Begin with method lf1debug and not method lf1. This allows testing whether the program produces correct scores.

4. Verify that your program works; type

    . ml check

5. Find starting values; type

    . ml search

or use ml init or ml plot. We suggest that you use ml search. In most cases, ml search can be omitted because ml maximize calls ml search, but in that case ml search will not do as thorough a job.

6. Fit the model; type

    . ml maximize

Review the reported mreldif()s in the iteration log. Even if you have programmed the derivatives correctly, it is typical to observe the mreldif()s falling and then rising and even observe the mreldif() being largest in the final iteration, when the gradient vector is near zero. This result should not concern you. The reported mreldif()s should be 1e–3 or smaller in the middle iterations. If you suspect problems, reissue the ml model statement and then run ml maximize with the gradient option.

7. Now that you have verified that your program is calculating derivatives correctly, switch to using method `lf1` by typing

    . ml model lf1 *myprog* (*equation for* θ_1) (*equation for* θ_2) ...

Repeat steps 4, 5, and 6. You should obtain approximately the same results. The iteration log will probably differ; method `lf1` may require fewer or more iterations than method `lf1debug`. This does not indicate problems. Method `lf1debug` used the numerically calculated derivatives, and maximization is a chaotic process.

8. If later you wish to redisplay results, type

    . ml display

D.7   Method lf2

1. To use method `lf2`, the likelihood function must meet the linear-form restrictions.

2. Write a program to evaluate $\ln \ell_j$, the contribution to the overall likelihood function for an observation, to calculate the equation-level score variables, and to calculate the Hessian of the overall likelihood function. The outline for the evaluator is

```
program myprog
        version 18
        args todo b lnfj g1 g2 ... H
        tempvar theta1 theta2 ...
        mleval 'theta1' = 'b'
        mleval 'theta2' = 'b', eq(2) // if there is a θ₂
        ...
        // if you need to create any intermediate results:
        tempvar tmp1 tmp2 ...
        generate double 'tmp1' = ...
        ...
        quietly replace 'lnfj' = ...
        if ('todo'==0) exit
        quietly replace 'g1' = formula for ∂ln ℓⱼ/∂θ₁ⱼ, eq(1)
        quietly replace 'g2' = formula for ∂ln ℓⱼ/∂θ₂ⱼ, eq(2)
        ...
        if ('todo'==1) exit
        tempname d11 d12 d22 ...
        mlmatsum 'lnf' 'd11' = formula for ∂²ln ℓⱼ/∂θ₁ⱼ², eq(1)
        mlmatsum 'lnf' 'd12' = formula for ∂²ln ℓⱼ/∂θ₁ⱼ∂θ₂ⱼ, eq(1,2)
        mlmatsum 'lnf' 'd22' = formula for ∂²ln ℓⱼ/∂θ₂ⱼ², eq(2)
        ...
        matrix 'H' = ('d11','d12',... \ 'd12','d22',... )
    end
```

Argument `'todo'` will contain 0, 1, or 2. You are to fill in the log likelihood if `'todo'==0`; the log likelihood and scores if `'todo'==1`; or the log likelihood, scores, and Hessian if `'todo'==2`.

You are passed the parameter row vector in 'b'. Use `mleval` to obtain the linear-equation values $\theta_{ij} = \mathbf{x}_{ij}\boldsymbol{\beta}_i$ from it.

Access the dependent variables (if any) by referring to $ML_y1, $ML_y2, ... Use $ML_y1, $ML_y2, ... just as you would any existing variable in the data.

The observation-level log-likelihood values are to be stored in 'lnfj'.

The scores are to be stored in g1, g2, ...

The Hessian is to be stored in H. Use `mlmatsum` to produce it. You issue one `mlmatsum` per equation and equation pair and then use `matrix` *name* = ... to put the results together.

You must create any temporary variables for intermediate results as `doubles`: type `gen double` *name* = ...; do not omit the word `double`.

The estimation subsample is $ML_samp==1. You may safely ignore this if you do not specify `ml model`'s `nopreserve` option. When your program is called, only relevant data will be in memory. If you do specify `nopreserve`, understand that `mleval` and `mlmatsum` automatically restrict themselves to the estimation subsample; it is not necessary to code `if $ML_samp==1` on these commands. You merely need to restrict other commands to the $ML_samp==1 subsample.

The weights are stored in variable $ML_w, which contains 1 if no weights are specified.

3. Issue the `ml model` statement as

    . ml model lf2debug *myprog* (*equation for* θ_1) (*equation for* θ_2) ...

    Note: Begin with method `lf2debug` and not method `lf2`. This allows testing whether the program produces correct first and second derivatives.

4. Verify that your program works; type

    . ml check

5. Find starting values; type

    . ml search

    or use `ml init` or `ml plot`. We suggest that you use `ml search`. In most cases, `ml search` can be omitted because `ml maximize` calls `ml search`, but in that case `ml search` will not do as thorough a job.

6. Fit the model; type

    . ml maximize

    Review the reported `mreldif()`s in the iteration log. Even if you have programmed the derivatives correctly, it is typical to observe the `mreldif()`s for the gradient vector falling and then rising and even observe the `mreldif()` being largest in the final iteration, when the gradient vector is near zero. This should not concern you. The reported `mreldif()`s should be 1e–3 or smaller in the middle iterations.

For the negative Hessian, early iterations often report large `mreldif()`s but eventually fall. The reported `mreldif()` should be 1e–5 or smaller in the last iteration.

If you suspect problems, reissue the `ml model` statement and then run `ml maximize` with the `gradient` option or the `hessian` option.

7. Now that you have verified that your program is calculating derivatives correctly, switch to using method `lf2` by typing

> . `ml model lf2` *myprog* (*equation for* θ_1) (*equation for* θ_2) ...

Repeat steps 4, 5, and 6. You should obtain approximately the same results (relative differences are 1e–8 for coefficients and smaller than 1e–5 for standard errors). The iteration log will probably differ; method `lf2` may require fewer or more iterations than method `lf2debug`. This does not indicate problems. Method `lf2debug` used the numerically calculated derivatives, and maximization is a chaotic process.

8. If later you wish to redisplay results, type

> . `ml display`

E Listing of estimation commands

E.1 The logit model

begin `mylogit.ado`

```
program mylogit, properties(or svyb svyj svyr swml)
        version 18
        if replay() {
                if (`"`e(cmd)'"' != "mylogit") error 301
                Replay `0'
        }
        else    Estimate `0'
end

program Estimate, eclass sortpreserve
        syntax varlist(fv) [if] [in]            ///
                [fweight pweight iweight] [,     ///
                noLOg noCONStant                 /// -ml model- options
                noLRTEST                         ///
                vce(passthru)                    ///
                OFFset(varname numeric)          ///
                EXPosure(varname numeric)        ///
                Level(cilevel)                   /// -Replay- options
                OR                               ///
                *                                /// -mlopts- options
                ]

        // mark the estimation sample
        marksample touse

        // check syntax
        mlopts mlopts, `options'
        local cns `s(constraints)'
        gettoken lhs rhs : varlist
        if "`weight'" != "" {
                local wgt "[`weight'`exp']"
                // for initial value calculations
                if "`weight'" == "pweight" {
                        local awgt "[aw`exp']"
                }
                else    local awgt "`wgt'"
        }
        if "`log'" != "" {
                local qui quietly
        }
        if "`offset'" != "" {
                local offopt "offset(`offset')"
        }
        if "`exposure'" != "" {
                local expopt "exposure(`exposure')"
        }
```

```
        // markout missing values from the estimation sample
        markout 'touse' 'offset' 'exposure'
        _vce_parse 'touse', opt(Robust oim opg) argopt(CLuster): 'wgt', 'vce'
        if "'constant'" == "" {
                // initial value
                sum 'lhs' 'awgt' if 'touse', mean
                local b0 = logit(r(mean))
                if "'weight'" == "iweight" {
                        local n = r(sum_w)
                }
                else    local n = r(N)
                local initopt init(_cons='b0') search(quietly)
                if "'lrtest'" == "" {
                        local n1 = r(sum)
                        local lf0 = 'n1'*ln(invlogit('b0'))     ///
                                +('n'-'n1')*ln(invlogit(-'b0'))
                        local initopt 'initopt' lf0(1 'lf0')
                }
        }
        // fit the full model
        'qui' di as txt _n "Fitting full model:"
        ml model lf2 mylogit_lf2                         ///
                (xb: 'lhs' = 'rhs',                      ///
                        'constant' 'offopt' 'expopt'     ///
                )                                        ///
                'wgt' if 'touse',                        ///
                'log' 'mlopts' 'vce'                     ///
                'modopts' 'initopt'                      ///
                missing maximize

        // save a title for -Replay- and the name of this command
        ereturn local title "My logit estimates"
        ereturn local cmd mylogit

        Replay , level('level') 'or'
end

program Replay
        syntax [, Level(cilevel) OR ]
        ml display , level('level') 'or'
end
```

—— end `mylogit.ado` ————————

```
                                         ————————————— begin mylogit_lf2.ado ——————————
program mylogit_lf2
        version 18
        args todo b lnfj g1 H
        tempvar xb lj
        mleval 'xb' = 'b'
        quietly {
                generate double 'lj' = invlogit( 'xb')  if $ML_y1 == 1
                replace    'lj' = invlogit(-'xb')  if $ML_y1 == 0
                replace 'lnfj' = ln('lj')
                if ('todo'==0) exit

                replace 'g1' =  invlogit(-'xb')  if $ML_y1 == 1
                replace 'g1' = -invlogit( 'xb')  if $ML_y1 == 0
                if ('todo'==1) exit

                mlmatsum 'lnfj' 'H' = -1*abs('g1')*'lj', eq(1,1)
        }
end
                                         ————————————— end mylogit_lf2.ado ——————————
```

E.2   The probit model

```
                                         ————————————— begin myprobit.ado ——————————
program myprobit, properties(svyb svyj svyr swml)
        version 18
        if replay() {
                if ('"'e(cmd)'"' != "myprobit") error 301
                Replay '0'
        }
        else    Estimate '0'
end

program Estimate, eclass sortpreserve
        syntax varlist [if] [in]                ///
                [fweight pweight iweight] [,     ///
                noLOg noCONStant                 /// -ml model- options
                noLRTEST                         ///
                vce(passthru)                    ///
                OFFset(varname numeric)          ///
                EXPosure(varname numeric)        ///
                Level(cilevel)                   /// -Replay- options
                *                                /// -mlopts- options
        ]
        // check syntax
        mlopts mlopts, 'options'
        local cns 's(constraints)'
        gettoken lhs rhs : varlist
        if "'weight'" != "" {
                local wgt "['weight''exp']"
                // for initial value calculations
                if "'weight'" == "pweight" {
                        local awgt "[aw'exp']"
                }
                else    local awgt "'wgt'"
        }
```

```
        if "`log'" != "" {
                local qui quietly
        }
        if "`offset'" != "" {
                local offopt "offset(`offset')"
        }
        if "`exposure'" != "" {
                local expopt "exposure(`exposure')"
        }
        // mark the estimation sample
        marksample touse
        markout `touse' `offset' `exposure'
        _vce_parse `touse', opt(Robust oim opg) argopt(CLuster): `wgt', `vce'
        if "`constant'" == "" {
                // initial value
                sum `lhs' `awgt' if `touse', mean
                local b0 = invnormal(r(mean))
                if "`weight'" == "iweight" {
                        local n = r(sum_w)
                }
                else    local n = r(N)
                local initopt init(_cons=`b0') search(quietly)
                if "`lrtest'" == "" {
                        local n1 = r(sum)
                        local lf0 = `n1'*ln(normal(`b0'))          ///
                                +(`n'-`n1')*ln(normal(-`b0'))
                        local initopt `initopt' lf0(1 `lf0')
                }
        }
        // fit the full model
        `qui' di as txt _n "Fitting full model:"
        ml model lf2 myprobit_lf2                        ///
                (xb: `lhs' = `rhs',                      ///
                        `constant' `offopt' `expopt'     ///
                )                                        ///
                `wgt' if `touse',                        ///
                `log' `mlopts' `vce'                     ///
                `modopts' `initopt' missing maximize
        // save a title for -Replay- and the name of this command
        ereturn local title "My probit estimates"
        ereturn local cmd myprobit
        Replay , level(`level')
end
program Replay
        syntax [, Level(cilevel) ]
        ml display , level(`level')
end
```

———————————————————————————————————— end myprobit.ado ————————

———————————————————————— begin `myprobit_lf2.ado` ————————————

```
program myprobit_lf2
        version 18
        args todo b lnfj g1 H
        tempvar xb lj
        mleval `xb' = `b'
        quietly {
                // Create temporary variable used in both likelihood
                // and scores
                generate double `lj' = normal( `xb') if $ML_y1 == 1
                replace      `lj' = normal(-`xb') if $ML_y1 == 0
                replace      `lnfj' = log(`lj')
                if (`todo'==0) exit

                replace `g1' =  normalden(`xb')/`lj'  if $ML_y1 == 1
                replace `g1' = -normalden(`xb')/`lj'  if $ML_y1 == 0
                if (`todo'==1) exit

                mlmatsum `lnfj' `H' = -`g1'*(`g1'+`xb'), eq(1,1)
        }
end
```

————————————————————————— end `myprobit_lf2.ado` ————————————

E.3   The normal model

———————————————————————————— begin `mynormal.ado` ————————————

```
program mynormal, properties(svyb svyj svyr swml)
        version 18
        if replay() {
                if (`"`e(cmd)'"' != "mynormal") error 301
                Replay `0'
        }
        else    Estimate `0'
end

program Estimate, eclass sortpreserve
        syntax varlist [if] [in]            ///
                [fweight pweight iweight] [,  ///
                noLOg noCONStant              /// -ml model- options
                HETero(varlist)               ///
                noLRTEST                      ///
                vce(passthru)                 ///
                OFFset(varname numeric)       ///
                EXPosure(varname numeric)     ///
                Level(cilevel)                /// -Replay- options
                EForm                         ///
                *                             /// -mlopts- options
        ]
        // check syntax
        mlopts mlopts, `options'
        local cns `s(constraints)'
        gettoken lhs rhs : varlist
        if "`weight'" != "" {
                local wgt "[`weight'`exp']"
                // for initial value calculations
                if "`weight'" == "pweight" {
                        local awgt "[aw`exp']"
```

```
            }
            else    local awgt "'wgt'"
    }
    if "'log'" != "" {
            local qui quietly
    }
    if "'offset'" != "" {
            local offopt "offset('offset')"
    }
    if "'exposure'" != "" {
            local expopt "exposure('exposure')"
    }

    // mark the estimation sample
    marksample touse
    markout 'touse' 'offset' 'exposure'
    _vce_parse 'touse', opt(Robust oim opg) argopt(CLuster): 'wgt', 'vce'
    if "'constant'" == "" {
            // initial values
            quietly sum 'lhs' 'awgt' if 'touse'
            local mean = r(mean)
            local lnsd = ln(r(sd))+ln((r(N)-1)/r(N))/2
            local initopt init(/mu='mean' /lnsigma='lnsd') search(off)

            'qui' di as txt _n "Fitting constant-only model:"
            ml model lf2 mynormal_lf2                         ///
                    (mu: 'lhs' =, 'offopt' 'expopt' )         ///
                    (lnsigma: 'hetero')                        ///
                    'wgt' if 'touse',                          ///
                    'log' 'mlopts' 'initopt'                   ///
                    nocnsnotes missing maximize
            if "'lrtest'" == "" {
                    local contin continue search(off)
            }
            else {
                    tempname b0
                    mat 'b0' = e(b)
                    local contin init('b0') search(off)
            }
    }

    // fit the full model
    'qui' di as txt _n "Fitting full model:"
    ml model lf2 mynormal_lf2                                 ///
            (mu: 'lhs' = 'rhs',                                ///
                    'constant' 'offopt' 'expopt'              ///
            )                                                  ///
            (lnsigma: 'hetero')                                ///
            'wgt' if 'touse',                                  ///
            'log' 'mlopts' 'vce'                               ///
            'contin' missing maximize

    if "'hetero'" == "" {
            ereturn scalar k_aux = 1
    }
    else    ereturn scalar k_aux = 0
    // save a title for -Replay- and the name of this command
    ereturn local title "My normal estimates"
    ereturn local predict mynormal_p
    ereturn local cmd mynormal

    Replay , level('level') 'eform'
```

```
        end
program Replay
        syntax [, Level(cilevel) EForm ]
        if `e(k_aux)' {
                local sigma diparm(lnsigma, exp label("sigma"))
        }
        ml display , level(`level') `sigma' `eform'
end
```
──────────────────────────────────── end `mynormal.ado` ───────

──────────────────────────────────── begin `mynormal_lf2.ado` ───────
```
program mynormal_lf2
        version 18
        args todo b lnfj g1 g2 H
        tempvar mu lnsigma sigma
        mleval `mu' = `b', eq(1)
        mleval `lnsigma' = `b', eq(2)
        quietly {
                generate double `sigma' = exp(`lnsigma')
                replace `lnfj' = ln( normalden($ML_y1,`mu',`sigma') )
                if (`todo'==0) exit

                tempvar z
                generate double `z' = ($ML_y1-`mu')/`sigma'
                replace `g1' = `z'/`sigma'
                replace `g2' = `z'*`z'-1
                if (`todo'==1) exit

                tempname d11 d12 d22
                mlmatsum `lnfj' `d11' = -1/`sigma'^2        , eq(1)
                mlmatsum `lnfj' `d12' = -2*`z'/`sigma'      , eq(1,2)
                mlmatsum `lnfj' `d22' = -2*`z'*`z'          , eq(2)
                matrix `H' = (`d11', `d12' \ `d12'', `d22')
        }
end
```
──────────────────────────────────── end `mynormal_lf2.ado` ───────

E.4   The Weibull model

── begin `myweibull.ado` ────────

```
program myweibull, properties(hr svyb svyj svyr swml)
        version 18
        if replay() {
                if ('"'e(cmd)'"' != "myweibull") error 301
                Replay '0'
        }
        else    Estimate '0'
end

program Estimate, eclass sortpreserve
        syntax varlist(fv) [if] [in]            ///
                [fweight pweight iweight] [,     ///
                FAILure(varname numeric)         ///
                noLOg noCONStant                 /// -ml model- options
                noLRTEST                         ///
                vce(passthru)                    ///
                OFFset(varname numeric)          ///
                EXPosure(varname numeric)        ///
                Level(cilevel)                   /// -Replay- options
                HR                               ///
                *                                /// -mlopts- options
        ]
        // check syntax
        mlopts mlopts, 'options'
        local cns 's(constraints)'
        gettoken time rhs : varlist
        if "'failure'" == "" {
                tempvar failure
                generate byte 'failure' = 1
        }
        if "'weight'" != "" {
                local wgt "['weight''exp']"
                // for initial value calculations
                if "'weight'" == "pweight" {
                        local awgt "[aw'exp']"
                }
                else    local awgt "'wgt'"
        }
        if "'log'" != "" {
                local qui quietly
        }
        if "'offset'" != "" {
                local offopt "offset('offset')"
        }
        if "'exposure'" != "" {
                local expopt "exposure('exposure')"
        }
        // mark the estimation sample
        marksample touse
        markout 'touse' 'failure' 'offset' 'exposure'
        _vce_parse 'touse', opt(Robust oim opg) argopt(CLuster): 'wgt', 'vce'
        if "'constant'" == "" {
                // initial value
                sum 'time' 'awgt' if 'touse', mean
                local b0 = r(mean)
```

```
              sum 'failure' 'awgt' if 'touse', mean
              local b0 = ln('b0'/r(mean))
              local initopt init(_cons='b0') search(quietly)

              'qui' di as txt _n "Fitting constant-only model:"
              ml model lf2 myweibull_lf2                        ///
                      (lneta: 'time' 'failure' = ,              ///
                              'offopt' 'expopt'                 ///
                      )                                         ///
                      /lngamma                                  ///
                      'wgt' if 'touse',                         ///
                      'log' 'mlopts' 'initopt'                  ///
                      nocnsnotes missing maximize
              if "'lrtest'" == "" {
                      local contin continue search(off)
              }
              else {
                      tempname b0
                      mat 'b0' = e(b)
                      local contin init('b0') search(off)
              }
      }
      // fit the full model
      'qui' di as txt _n "Fitting full model:"
      ml model lf2 myweibull_lf2                        ///
              (lneta: 'time' 'failure'= 'rhs',          ///
                      'constant' 'offopt' 'expopt'      ///
              )                                         ///
              /lngamma                                  ///
              'wgt' if 'touse',                         ///
              'log' 'mlopts' 'vce'                      ///
              'contin' missing maximize                 ///
              diparm(lngamma, exp label("gamma"))       ///
              diparm(lngamma,                           ///
                      function(exp(-@))                 ///
                      derivative(exp(-@))               ///
                      label("1/gamma")                  ///
              )

      ereturn scalar k_aux = 1
      ereturn local failure 'failure'
      ereturn local depvar 'time' 'failure'
      // save a title for -Replay- and the name of this command
      ereturn local title "My Weibull estimates"
      ereturn local cmd myweibull

      Replay , level('level') 'hr'
end

program Replay
      syntax [, Level(cilevel) HR ]
      ml display , level('level') 'hr'
end
```
――――――――――――――――――――――――――― end myweibull.ado ―――――

─────────────────────── begin `myweibull_lf2.ado` ───────────────────────
```
program myweibull_lf2
        version 18
        args todo b lnf g1 g2 H
        tempvar leta lgam p M R
        mleval `leta' = `b', eq(1)
        mleval `lgam' = `b', eq(2)
        local t "$ML_y1"
        local d "$ML_y2"
        quietly {
                generate double `p' = exp(`lgam')
                generate double `M' = (`t'*exp(-`leta'))^`p'
                generate double `R' = ln(`t')-`leta'
                replace `lnf' = -`M' + `d'*(`lgam'-`leta' + (`p'-1)*`R')
                if (`todo'==0) exit

                replace `g1' = `p'*(`M'-`d')
                replace `g2' = `d' - `R'*`p'*(`M'-`d')
                if (`todo'==1) exit

                tempname d11 d12 d22
                mlmatsum `lnf' `d11' = -`p'^2 * `M'                    , eq(1)
                mlmatsum `lnf' `d12' = `p'*(`M'-`d'+`R'*`p'*`M')     , eq(1,2)
                mlmatsum `lnf' `d22' = -`p'*`R'*(`R'*`p'*`M'+`M'-`d') , eq(2)
                matrix `H' = (`d11',`d12' \ `d12',`d22')
        }
end
```
─────────────────────── end `myweibull_lf2.ado` ───────────────────────

E.5   The Cox proportional hazards model

─────────────────────── begin `mycox.ado` ───────────────────────
```
program mycox, properties(hr swml)
        version 18
        if replay() {
                if (`"`e(cmd)'"' != "mycox") error 301
                Replay `0'
        }
        else    Estimate `0'
end

program Estimate, eclass sortpreserve
        syntax varlist(fv min=2) [if] [in]      ///
                [fweight] [,                     ///
                FAILure(varname numeric)         ///
                noLOg                            /// -ml model- options
                noLRTEST                         ///
                vce(passthru)                    ///
                Level(cilevel)                   /// -Replay- options
                HR                               ///
                *                                /// -mlopts- options
        ]
        // check syntax
        local diopts level(`level') `hr'
        mlopts mlopts , `options'
        local cns `s(constraints)'
        gettoken time rhs : varlist
```

```
        tempvar negt
        quietly generate double 'negt' = -'time'
        sort 'negt' 'failure'
        if "'failure'" == "" {
                tempvar failure
                generate byte 'failure' = 1
        }
        if "'weight'" != "" {
                local wgt "['weight''exp']"
        }
        if "'cns'" != "" {
                local lrtest nolrtest
        }

        // mark the estimation sample
        marksample touse
        markout 'touse' 'failure'
        _vce_parse, optlist(OIM) :, 'vce'

        // fit constant-only model
        if "'lrtest'" == "" {
                quietly ml model d0 mycox2_d0                       ///
                        (xb: 'negt' 'failure' = 'rhs',             ///
                                noconstant                          ///
                        )                                           ///
                        'wgt' if 'touse',                          ///
                        iterate(0)                                  ///
                        nocnsnotes missing maximize
                local initopt lf0(0 'e(ll)')
        }

        // fit the full model
        'qui' di as txt _n "Fitting full model:"
        ml model d0 mycox2_d0                                       ///
                (xb: 'negt' 'failure' = 'rhs',                     ///
                        noconstant                                  ///
                )                                                   ///
                'wgt' if 'touse',                                  ///
                'log' 'mlopts' 'vce' 'initopt'                     ///
                missing maximize

        // dependent variables
        ereturn local depvar 'time' 'failure'
        // save a title for -Replay- and the name of this command
        ereturn local title "My Cox estimates"
        ereturn local cmd mycox

        Replay , 'diopts'
end

program Replay
        syntax [, Level(cilevel) HR ]
        ml display , level('level') 'hr'
end
                                                    ──── end mycox.ado ────
```

```
                                          ─────── begin mycox2_d0.ado ───────
program mycox2_d0
        version 18
        args todo b lnf
        local negt "$ML_y1"              // $ML_y1 is -t
        local d "$ML_y2"                 // $ML_y2 is d
        tempvar xb B A sumd last L
        mleval 'xb' = 'b'
        // data assumed already sorted by 'negt' and 'd'
        local byby "by 'negt' 'd'"
        local wxb "$ML_w*'xb'"
        quietly {
                generate double 'B' = sum($ML_w*exp('xb'))
                'byby': generate double 'A' = cond(_n==_N, sum('wxb'), .) if 'd'==1
                'byby': generate 'sumd' = cond(_n==_N, sum($ML_w*'d'), .)
                'byby': generate byte 'last' = (_n==_N & 'd' == 1)
                generate double 'L' = 'A' - 'sumd'*ln('B') if 'last'
                mlsum 'lnf' = 'L' if 'last', noweight
        }
end
                                          ─────── end mycox2_d0.ado ───────
```

E.6   The random-effects regression model

```
                                          ─────── begin myrereg.ado ───────
program myrereg
        version 18
        if replay() {
                if ('"'e(cmd)'"' != "myrereg") error 301
                Replay '0'
        }
        else    Estimate '0'
end

program Estimate, eclass sortpreserve
        syntax varlist(fv) [if] [in] ,          ///
                panel(varname)                  ///
        [                                       ///
                noLOg noCONStant                /// -ml model- options
                noLRTEST                        ///
                vce(passthru)                   ///
                Level(cilevel)                  /// -Replay- options
                *                               /// -mlopts- options
        ]
        // check syntax
        local diopts level('level')
        mlopts mlopts , 'options'
        local cns 's(constraints)'
        gettoken lhs rhs : varlist
        if "'cns'" != "" {
                local lrtest nolrtest
        }
        if "'log'" != "" {
                local qui quietly
        }
```

```
        // mark the estimation sample
        marksample touse
        markout 'touse' 'panel', strok
        _vce_parse, optlist(OIM) :, 'vce'

        // capture block to ensure removal of global macro
capture noisily {
        // identify the panel variable for the evaluator
        global MY_panel 'panel'
        sort 'panel'

        if "'constant'" == "" {
                // initial values: variance components from one-way ANOVA
                sum 'lhs' if 'touse', mean
                local mean = r(mean)
                quietly oneway 'lhs' 'panel' if 'touse'
                local np  = r(df_m) + 1
                local N   = r(N)
                local bms = r(mss)/r(df_m)       // between mean squares
                local wms = r(rss)/r(df_r)       // within mean squares
                local lns_u = log( ('bms'-'wms')*'np'/'N' )/2
                if missing('lns_u') {
                        local lns_u = 0
                }
                local lns_e = log('wms')/2
                local   initopt search(off)             ///
                        init(/xb='mean' /lns_u='lns_u' /lns_e='lns_e')

                'qui' di as txt _n "Fitting constant-only model:"
                ml model d2 myrereg2_d2                          ///
                        (xb: 'lhs' = , 'offopt' 'expopt' )      ///
                        /lns_u                                  ///
                        /lns_e                                  ///
                        if 'touse',                             ///
                        'log' 'mlopts' 'initopt'                ///
                        nocnsnotes missing maximize
                if "'lrtest'" == "" {
                        local contin continue search(off)
                }
                else {
                        tempname b0
                        mat 'b0' = e(b)
                        local contin init('b0') search(off)
                }
        }

        // fit the full model
        'qui' di as txt _n "Fitting full model:"
        ml model d2 myrereg2_d2                          ///
                (xb: 'lhs' = 'rhs',                      ///
                        'constant' 'offopt' 'expopt'    ///
                )                                       ///
                /lns_u                                  ///
                /lns_e                                  ///
                if 'touse',                             ///
                'log' 'mlopts' 'contin'                 ///
                missing maximize

        // clear MY global
        global MY_panel
} // capture noisily
```

```
        // exit in case of error
        if c(rc) exit 'c(rc)'

        ereturn scalar k_aux = 2
        // save the panel variable
        ereturn local panel 'panel'
        // save a title for -Replay- and the name of this command
        ereturn local title "My rereg estimates"
        ereturn local cmd myrereg

        Replay , 'diopts'
end

program Replay
        syntax [, Level(cilevel) ]
        ml display , level('level')                     ///
                diparm(lns_u, exp label("sigma_u"))     ///
                diparm(lns_e, exp label("sigma_e"))
end
```

———————————————————————————————— end `myrereg.ado` ————————

———————————————————————————— begin `myrereg2_d2.ado` ————————

```
program myrereg2_d2
        version 18
        args todo b lnf g H
        tempvar xb z T S_z2 Sz_2 a last
        tempname s_u s_e
        mleval 'xb' = 'b', eq(1)
        mleval 's_u' = 'b', eq(2) scalar
        mleval 's_e' = 'b', eq(3) scalar
        scalar 's_u' = exp('s_u')
        scalar 's_e' = exp('s_e')
        // MY_panel contains the panel ID
        local by $MY_panel
        local y $ML_y1
        quietly {
                generate double 'z' = 'y' - 'xb'
                by 'by': generate 'last' = _n==_N
                by 'by': generate 'T' = _N
                by 'by': generate double 'S_z2' = sum('z'^2)
                by 'by': replace     'S_z2' = 'S_z2'[_N]
                by 'by': generate double 'Sz_2' = sum('z')^2
                by 'by': replace     'Sz_2' = 'Sz_2'[_N]
                generate double 'a' = 's_u'^2 / ('T'*'s_u'^2 + 's_e'^2)
                mlsum 'lnf'= -.5 *                              ///
                        (                                       ///
                                ('S_z2'-'a'*'Sz_2')/'s_e'^2 +   ///
                                ln('T'*'s_u'^2/'s_e'^2 + 1) +   ///
                                'T'*ln(2*c(pi)*'s_e'^2)         ///
                        )                                       ///
                                if 'last' == 1
                if ('todo'==0 | 'lnf'>=.) exit

                // compute the gradient
                tempvar S_z
                tempname dxb du de
                by 'by': generate double 'S_z' = sum('z')
                by 'by': replace 'S_z' = 'S_z'[_N]
                mlvecsum 'lnf' 'dxb' = ('z'-'a'*'S_z')/'s_e'^2   , eq(1)
                mlvecsum 'lnf' 'du' = 'a'^2*'Sz_2'/'s_u'^2       ///
                        -'T'*'a'                                 ///
```

```
                                      if `last'==1                          , eq(2)
                   mlvecsum `lnf' `de' = `S_z2'/`s_e'^2 -         ///
                           `a'*`Sz_2'/`s_e'^2 -                   ///
                           `a'^2*`Sz_2'/`s_u'^2 -                 ///
                           `T'+1-`a'*`s_e'^2/`s_u'^2              ///
                                      if `last'==1                          , eq(3)
                   mat `g' = (`dxb',`du',`de')
                   if (`todo'==1 | `lnf'>=.) exit

                   // compute the Hessian
                   tempname d2xb1 d2xb2 d2xb d2u d2e dxbdu dxbde dude one
                   mlmatsum `lnf' `d2u' = 2*`a'^2*`Sz_2'/`s_u'^2 -   ///
                           4*`s_e'^2*`a'^3*`Sz_2'/`s_u'^4 +         ///
                           2*`T'*`a'^2*`s_e'^2/`s_u'^2              ///
                                      if `last'==1                          , eq(2)
                   mlmatsum `lnf' `d2e' =                           ///
                           2*(`S_z2'-`a'*`Sz_2')/`s_e'^2 -         ///
                           2*`a'^2*`Sz_2'/`s_u'^2 -                ///
                           4*`a'^3*`Sz_2'*`s_e'^2/`s_u'^4 +        ///
                           2*`a'*`s_e'^2/`s_u'^2 -                 ///
                           2*`a'^2*`s_e'^4/`s_u'^4                 ///
                                      if `last'==1                          , eq(3)
                   mlmatsum `lnf' `dude' =                          ///
                           4*`a'^3*`Sz_2'*`s_e'^2/`s_u'^4 -        ///
                           2*`T'*`a'^2*`s_e'^2/`s_u'^2             ///
                                      if `last'==1                          , eq(2,3)
                   mlmatsum `lnf' `dxbdu' = 2*`a'^2*`S_z'/`s_u'^2  , eq(1,2)
                   mlmatsum `lnf' `dxbde' =                         ///
                           2*(`z'-`a'*`S_z')/`s_e'^2 -             ///
                           2*`a'^2*`S_z'/`s_u'^2                   , eq(1,3)
                   // `a' is constant within panel; and
                   // -mlmatbysum- treats missing as 0 for `a'
                   by `by': replace `a' = . if !`last'
                   mlmatsum `lnf' `d2xb2' = 1                                , eq(1)
                   generate double `one' = 1
                   mlmatbysum `lnf' `d2xb1' `a' `one', by($MY_panel) eq(1)
                   mat `d2xb' = (`d2xb2'-`d2xb1')/`s_e'^2
                   mat `H' = -(                                     ///
                           `d2xb',   `dxbdu', `dxbde'        \      ///
                           `dxbdu'', `d2u',   `dude'         \      ///
                           `dxbde'', `dude',  `d2e'                 ///
                   )
           }
   end
```

────────────────────────────── end `myrereg2_d2.ado` ──────────

E.7   The seemingly unrelated regression model

```
                                                    ———— begin mysureg.ado ————
program mysureg, properties(svyb svyj svyr)
        version 18
        if replay() {
                if (`""`e(cmd)'"' != "mysureg") error 301
                Replay `0'
        }
        else    Estimate `0'
end

program Estimate, eclass sortpreserve
        syntax anything(id="equations" equalok) ///
                [if] [in]                        ///
                [fweight pweight iweight] [,     ///
                noLOg                            /// -ml model- options
                noLRTEST                         ///
                vce(passthru)                    ///
                Level(cilevel)                   /// -Replay- options
                corr                             ///
                *                                /// -mlopts- options
        ]
        // parse equations
        ParseEqns `anything'
        local p `s(p)'
        local k_eq    `s(k_eq)'
        local yvars   `s(yvars)'
        local xvars   `s(xvars)'
        local eqns    `s(eqns)'
        local eqns0   `s(eqns0)'
        local ceqns   `s(ceqns)'
        local myeqns `anything'
        local nocons 0
        forval i = 1/`p' {
                local xvars`i' `s(xvars`i')'
                local xvars`i'_cns `s(xvars`i'_cns)'
                if "`xvars`i'_cns'" != "" {
                        local nocons 1
                }
        }

        // check syntax
        mlopts mlopts, `options'
        local cns `s(constraints)'
        if "`weight'" != "" {
                local wgt "[`weight'`exp']"
                // for initial value calculations
                local iwgt "[iw`exp']"
        }
        if "`log'" != "" {
                local qui quietly
        }

        // mark the estimation sample
        marksample touse
        markout `touse' `yvars' `xvars'
        markout `touse' `offset' `exposure'
```

```
_vce_parse 'touse', opt(Robust oim opg) argopt(CLuster): 'wgt', 'vce'
// initial values
tempname b0
InitialValues 'yvars' 'iwgt' if 'touse' , init('b0')
local initopt init('b0', skip) search(quietly)

'qui' di as txt _n "Fitting constant-only model:"
ml model d2 mysuregc_d2 'eqns0'          ///
        'iwgt' if 'touse',               ///
        'log' 'mlopts' 'initopt'         ///
        nocnsnotes missing maximize
if 'nocons' {
        mat 'b0' = e(b)
        local initopt init('b0', skip) search(off)
}
else    local initopt continue search(off)
if "'lrtest'" == "" & 'nocons' == 0 {
        local lf0 lf0('k_eq' 'e(ll)')
}

// iterate to solution using concentrated likelihood
'qui' di as txt _n "Fitting concentrated model:"
ml model d2 mysuregc_d2 'ceqns'          ///
        'iwgt' if 'touse',               ///
        'log' 'mlopts'                   ///
        'initopt'                        ///
        missing maximize
UnConcentrate 'yvars' 'iwgt' if 'touse', b('b0')
local initopt init('b0', copy) search(quietly)

// fit the full unconcentrated likelihood
'qui' di as txt _n "Fitting full model:"
ml model lf0 mysureg_lf0 'eqns'          ///
        'wgt' if 'touse',                ///
        'log' 'mlopts' 'vce'             ///
        'initopt' 'lf0' missing maximize waldtest(-'p')

// build stored results
tempname Sigma sd
matrix 'Sigma' = I('p')
matrix 'sd'   = I('p')
matrix colnames 'Sigma' = 'yvars'
matrix rownames 'Sigma' = 'yvars'
forval i = 'p'(-1)1 {
        mat 'Sigma'['i','i'] = _b[/sigma'i'_'i']
        forval j = '='i'+1'/'p' {
                mat 'Sigma'['i','j'] = _b[/sigma'i'_'j']
                mat 'Sigma'['j','i'] = 'Sigma'['i','j']
        }
        ereturn local xvars'i' 'xvars'i''
        ereturn local xvars'i'_cns 'xvars'i'_cns'
}
ereturn local xvars 'xvars'
ereturn local mleqns 'ceqns'
ereturn local myeqns 'myeqns'
ereturn matrix Sigma = 'Sigma'
ereturn matrix Vars = 'sd'
ereturn scalar k_aux = 'k_eq'-'p'
// save a title for -Replay- and the name of this command
ereturn local title "My sureg estimates"
ereturn local cmd mysureg
```

```
        Replay , 'diopts'
end

program Replay
        syntax [, Level(cilevel) notable corr ]
        if "'table'" == "" {
                ml display , level('level')
        }
        if "'corr'" != "" {
                di
                di as txt "Correlation matrix of residuals:"
                tempname corr
                matrix 'corr' = corr(e(Sigma))
                matrix list 'corr', noheader format(%6.4g)
        }
end

// Syntax:
//      ParseEqns (y1 xvars1) (y2 xvars2) ...
//
// Build the -ml- equivalent equations, including the equations for the
// variance matrix.
//
// Stored results:
//      s(p)            number of dependent variables
//      s(k_eq)         number of -ml- equations : unconcentrated model
//      s(yvars)        the dependent variables
//      s(xvars)        the independent variables from all the equations
//      s(xvars'i')     the independent variables from equation 'i'
//      s(xvars'i'_cns) "noconstant" or ""
//      s(eqns)         -ml- equations : unconcentrated model
//      s(ceqns)        -ml- equations : concentrated model
//      s(eqns0)        -ml- equations : constant-only concentrated model
program ParseEqns, sclass
        gettoken eq eqns : 0 , match(paren) parse("()") bind
        local p 0
        while "'eq'" != "" {
                local ++p
                local 0 'eq'
                syntax varlist [, noCONStant]
                gettoken yi xvarsi : varlist
                local xvars'p' 'xvarsi'
                local xvars'p'_cns 'constant'
                local meqns 'meqns' ('yi' : 'yi' = 'xvarsi', 'constant')
                local meqns0 'meqns0' ('yi' : 'yi' = )
                local yvars 'yvars' 'yi'
                local xvars 'xvars' 'xvarsi'
                gettoken eq eqns : eqns , match(paren) parse("()") bind
        }
        forval i = 1/'p' {
                forval j = 'i'/'p' {
                        local seqns 'seqns' /sigma'i'_'j'
                }
        }
        sreturn clear
        sreturn local p 'p'
        sreturn local k_eq = 'p'*('p'+3)/2
        sreturn local yvars    'yvars'
        sreturn local xvars    ': list uniq xvars'
        sreturn local eqns     'meqns' 'seqns'
        sreturn local ceqns    'meqns'
```

```
                sreturn local myeqns    `myeqns'
                sreturn local eqns0     `meqns0'
                forval i = 1/`p' {
                        sreturn local xvars`i' `xvars`i''
                        sreturn local xvars`i'_cns `xvars`i'_cns'
                }
        end
        // Take e(b) from the concentrated model and return the equivalent
        // unconcentrated coefficents vector.
        program UnConcentrate
                syntax varlist if [iw] , b(name)
                local wgt [`weight'`exp']
                local p : word count `varlist'
                tempname b0 s0 sv
                mat `b0' = e(b)
                forval i = 1/`p' {
                        tempvar r`i'
                        local y : word `i' of `varlist'
                        matrix score `r`i'' = `b0' `if', eq(`y')
                        quietly replace `r`i'' = `y' - `r`i''
                        local resids `resids' `r`i''
                }
                quietly matrix accum `s0' = `resids' `wgt' `if', noconstant
                matrix `s0' = `s0'/r(N)
                forval i = 1/`p' {
                        matrix `sv' = nullmat(`sv') , `s0'[`i',`i'...]
                }
                matrix `b' = e(b) , `sv'
        end
        // Compute the initial values for the concentrated constant-only model.
        program InitialValues
                syntax varlist if [iw] , init(name)
                tempname m
                tempvar one
                quietly generate double `one' = 1
                quietly matrix vecaccum `m' = `one' `varlist'    ///
                        [`weight'`exp'] `if', noconstant
                matrix `init' = `m'/r(N)
                matrix colnames `init' = _cons
                matrix coleq    `init' = `varlist'
        end
```
————————————————————————————————— end mysureg.ado ——————————

```
───────────────────────────── begin mysureg_d1.ado ─────────
program mysureg_d1
        version 18
        args todo b lnf g H
        local p : word count $ML_y
        local k = `p'*(`p'+1) / 2 + `p'
        tempname S iS sij isij isi ip
        matrix `S' = J(`p',`p',0)
quietly {
        // get residuals and build variance matrix
        local e 1
        forval i = 1/`p' {
                tempvar r`i'
                mleval `r`i'' = `b', eq(`i')
                replace `r`i'' = ${ML_y`i'} - `r`i''
                local resids `resids' `r`i''
                mleval `sij' = `b', eq(`=`p'+`e'') scalar
                matrix `S'[`i',`i'] = `sij'
                local ++e
                forval j = `=`i'+1'/`p' {
                        mleval `sij' = `b', eq(`=`p'+`e'') scalar
                        matrix `S'[`i',`j'] = `sij'
                        matrix `S'[`j',`i'] = `sij'
                        local ++e
                }
        }
        matrix `iS' = invsym(`S')
        // get score variables
        forval i = 1/`k' {
                tempvar g`i'
                quietly generate double `g`i'' = .
        }
        generate double `ip' = 0
        forval i = 1/`p' {
                matrix `isi' = `iS'[`i',1...]
                matrix colnames `isi' = `resids'
                matrix score `g`i'' = `isi', replace
                replace `ip' = `ip' + `r`i''*`g`i''
        }
        mlsum `lnf' = -0.5*(`p'*ln(2*c(pi))+ln(det(`S'))+`ip')
} // quietly
        if (`todo' == 0 | missing(`lnf')) exit
        // compute the scores and gradient
        tempname gi gs
        capture matrix drop `g'
quietly {
        local e = `p'+1
        forval i = 1/`p' {
                mlvecsum `lnf' `gi' = `g`i'', eq(`i')
                matrix `g' = nullmat(`g'), `gi'
                replace `g`e'' = 0.5*(`g`i''*`g`i''-`iS'[`i',`i'])
                mlvecsum `lnf' `gi' = `g`e'' , eq(`e')
                matrix `gs' = nullmat(`gs'), `gi'
                local ++e
                forval j = `=`i'+1'/`p' {
```

```
                                   replace `g`e'' = `g`i''*`g`j'' - `iS'[`i',`j']
                                   mlvecsum `lnf' `gi' = `g`e'' , eq(`e')
                                   matrix `gs' = nullmat(`gs'), `gi'
                                   local ++e
                        }
             }
             matrix `g' = `g' , `gs'
   } // quietly
   end
```
── end mysureg_d1.ado ────────────

── begin mysuregc_d2.ado ────────────
```
program mysuregc_d2
        // concentrated likelihood for the SUR model
        version 18
        args todo b lnf g H
        local p : word count $ML_y
        tempname W sumw

        // get residuals and build variance matrix
        forval i = 1/`p' {
                tempvar r`i'
                mleval `r`i'' = `b', eq(`i')
                quietly replace `r`i'' = ${ML_y`i'} - `r`i''
                local resids `resids' `r`i''
        }
        quietly matrix accum `W' = `resids' [iw=$ML_w] if $ML_samp, noconstant
        scalar `sumw' = r(N)
        matrix `W' = `W'/`sumw'
        scalar `lnf' = -0.5*`sumw'*(`p'*(ln(2*c(pi))+1)+ln(det(`W')))

        if (`todo' == 0 | missing(`lnf')) exit

        // compute gradient
        tempname gi iW iWi
        tempvar  scorei
        quietly generate double `scorei' = .
        capture matrix drop `g'
        matrix `iW' = invsym(`W')
        forval i = 1/`p' {
                matrix `iWi' = `iW'[`i',1...]
                matrix colnames `iWi' = `resids'
                matrix score `scorei' = `iWi', replace
                mlvecsum `lnf' `gi' = `scorei', eq(`i')
                matrix `g' = nullmat(`g'), `gi'
        }

        if (`todo' == 1 | missing(`lnf')) exit

        // compute the Hessian, as if we were near the solution
        tempname hij
        local k = colsof(`b')
        matrix `H' = J(`k',`k',0)
        local r 1
        forval i = 1/`p' {
                local c `r'
                mlmatsum `lnf' `hij' = -1*`iW'[`i',`i'], eq(`i')
                matrix `H'[`r',`c'] = `hij'
                local c = `c' + colsof(`hij')
                forval j = `=`i'+1'/`p' {
                        mlmatsum `lnf' `hij' = -1*`iW'[`i',`j'], eq(`i',`j')
```

```
                        matrix 'H'['r','c'] = 'hij'
                        matrix 'H'['c','r'] = 'hij''
                        local c = 'c' + colsof('hij')
                }
                local r = 'r' + rowsof('hij')
        }
end
```

─────────────────────────────────── end mysuregc_d2.ado ───────────

─────────────────────────────────── begin mysureg_lf0.ado ───────────
```
program mysureg_lf0
        version 18
        args todo b lnfj
        local p : word count $ML_y
        local k = 'p'*('p'+1) / 2 + 'p'

        tempname S iS sij isij isi
        tempvar ip
        matrix 'S' = J('p','p',0)
quietly {
        // get residuals and build variance matrix
        local e 1
        forval i = 1/'p' {
                tempvar r'i'
                mleval 'r'i'' = 'b', eq('i')
                replace 'r'i'' = ${ML_y'i'} - 'r'i''
                local resids 'resids' 'r'i''
                mleval 'sij' = 'b', eq('='p'+'e'') scalar
                matrix 'S'['i','i'] = 'sij'
                local ++e
                forval j = '='i'+1'/'p' {
                        mleval 'sij' = 'b', eq('='p'+'e'') scalar
                        matrix 'S'['i','j'] = 'sij'
                        matrix 'S'['j','i'] = 'sij'
                        local ++e
                }
        }
        matrix 'iS' = invsym('S')
        // get score variables
        tempvar scorei
        generate double 'scorei' = .
        generate double 'ip' = 0
        forval i = 1/'p' {
                matrix 'isi' = 'iS'['i',1...]
                matrix colnames 'isi' = 'resids'
                matrix score 'scorei' = 'isi', replace
                replace 'ip' = 'ip' + 'r'i''*'scorei'
        }
        replace 'lnfj' = -0.5*('p'*ln(2*c(pi))+ln(det('S'))+'ip')
} // quietly
end
```

─────────────────────────────────── end mysureg_lf0.ado ───────────

E.8   A bivariate Poisson regression model

```
─────────────────────────────────── begin make_lmybipoisson_mlib.do ───────────
// Likelihood evaluator and other functions for -mybipoisson-.

clear all

mata:
mata set matastrict on

// This is the method-lf0 evaluator() -- no derivatives.
void mybipoisson_lf0(transmorphic scalar ML, real scalar    todo,
                     real rowvector      b , real colvector lnfj,
                     real matrix         S , real matrix    H)
{
        real colvector   y1, y2, xi1, xi2, xi0

        y1  = moptimize_util_depvar(ML, 1)
        y2  = moptimize_util_depvar(ML, 2)
        xi1 = exp(moptimize_util_xb(ML, b, 1))
        xi2 = exp(moptimize_util_xb(ML, b, 2))
        xi0 = J(rows(y1), 1, exp(moptimize_util_xb(ML, b, 3)))

        lnfj = ln( bipoissonp_xi(y1, y2, xi1, xi2, xi0) )
}

// This is the method-lf2() evaluator. You can also use it with
// methods lf0 and lf1.
void mybipoisson_lf2(transmorphic scalar ML, real scalar    todo,
                     real rowvector      b , real colvector lnfj,
                     real matrix         S , real matrix    H)
{
        real colvector   y1, y2, xi1, xi2, xi0
        real matrix      G, Hraw, H11, H12, H13, H22, H23, H33

        y1  = moptimize_util_depvar(ML, 1)
        y2  = moptimize_util_depvar(ML, 2)
        xi1 = exp(moptimize_util_xb(ML, b, 1))
        xi2 = exp(moptimize_util_xb(ML, b, 2))
        xi0 = J(rows(y1), 1, exp(moptimize_util_xb(ML, b, 3)))

        // The following lines are necessary to handle constant-
        // only models. All of our functions assume xi1 and xi2
        // are colvectors; -ml- turns constant-only equations into
        // scalars.
        if (rows(xi1) == 1) {
                xi1 = J(rows(y1), 1, xi1)
                xi2 = J(rows(y1), 1, xi2)
        }

        lnfj = ln( bipoissonp_xi(y1, y2, xi1, xi2, xi0) )
        if (todo==0) return           // No derivatives

        G = ln_bip_xi_grad(y1, y2, xi1, xi2, xi0)
        S = J(rows(G), 3, 0)
        S[., 1] = G[.,1] :/ G[.,4]
        S[., 2] = G[.,2] :/ G[.,4]
        S[., 3] = G[.,3] :/ G[.,4]
        if (todo==1) return           // Just gradient

        Hraw = ln_bip_xi_hessian(y1, y2, xi1, xi2, xi0, G)
        H11 = moptimize_util_matsum(ML, 1, 1, Hraw[.,1], lnfj[1])
        H12 = moptimize_util_matsum(ML, 1, 2, Hraw[.,2], lnfj[1])
        H13 = moptimize_util_matsum(ML, 1, 3, Hraw[.,3], lnfj[1])
        H22 = moptimize_util_matsum(ML, 2, 2, Hraw[.,4], lnfj[1])
        H23 = moptimize_util_matsum(ML, 2, 3, Hraw[.,5], lnfj[1])
```

```
        H33  = moptimize_util_matsum(ML, 3, 3, Hraw[.,6], lnfj[1])
        H    = ( H11 , H12 , H13 \
                 H12', H22 , H23 \
                 H13', H23', H33 )
}

// This version of the bivariate Poisson PMF takes inputs in the
// (mu1, mu2, xi0)-space
real colvector bipoissonp(real colvector y1,
                          real colvector y2,
                          real colvector mu1,
                          real colvector mu2,
                          real colvector xi0) {
        real colvector  xi1, xi2

        // Go from (mu1, mu2, xi0)-space to (xi1, xi2, xi0)-space
        xi1 = mu1 - xi0
        xi2 = mu2 - xi0

        return( bipoissonp_xi(y1, y2, xi1, xi2, xi0) )
}

// This version of the bivariate Poisson PMF takes inputs in the
// (xi1, xi2, xi0)-space
real colvector bipoissonp_xi(real colvector y1,
                             real colvector y2,
                             real colvector xi1,
                             real colvector xi2,
                             real colvector xi0)
{
        real scalar     i, lim, p, r
        real colvector  pr

        pr = J(rows(y1), 1, .)
        for(i=1; i<=rows(y1); ++i) {
                lim = min( (y1[i], y2[i]) )
                p = 0
                for(r=0; r<=lim; ++r) {
                        p = p + poissonp(xi0[i], r) *
                                poissonp(xi1[i], y1[i]-r) *
                                poissonp(xi2[i], y2[i]-r)
                }
                pr[i] = p
        }
        return(pr)
}

// In the book we define Srj = Pr(y1-r)*Pr(y2-r)*Pr(r)
//                and    Sj = Sum_r Srj
// We don't do the final division by S so we can use these
// ingredients in our Hessian calculations.
real matrix ln_bip_xi_grad(real colvector y1,
                           real colvector y2,
                           real colvector xi1,
                           real colvector xi2,
                           real colvector xi0)
{
        real scalar     j, r, Rj, S, Sr, dSdt1, dSdt2, dSdt3
        real matrix     G

        G = J(rows(y1), 4, 0)   // sic
        for(j=1; j<=rows(y1); ++j) {
                Rj      = min( (y1[j], y2[j]) )
```

```
            S     = 0
            dSdt1 = 0
            dSdt2 = 0
            dSdt3 = 0
            for(r=0; r<=Rj; ++r) {
                    Sr = poissonp(xi0[j], r) *
                         poissonp(xi1[j], y1[j]-r) *
                         poissonp(xi2[j], y2[j]-r)
                    S  = S + Sr
                    dSdt1 = dSdt1 + Sr*((y1[j]-r) - xi1[j])
                    dSdt2 = dSdt2 + Sr*((y2[j]-r) - xi2[j])
                    dSdt3 = dSdt3 + Sr*(r - xi0[j])     // sic
            }
            G[j, 1] = dSdt1
            G[j, 2] = dSdt2
            G[j, 3] = dSdt3
            G[j, 4] = S
        }
        return(G)
}

// Returns observation-level hessian of log-likelihood
// [ d2l/dt1dt1 d2l/dt1dt2 d2l/dt1dt3 d2l/dt2dt2 d2l/dt2dt2 d2l/dt3dt3 ]
// In the book we define Srj = Pr(y1-r)*Pr(y2-r)*Pr(r)
//                 and   Sj  = Sum_r Srj
// The matrix G passed in has in the first three columns
// sum_r dSrj / dtk (k = 1,2,3).
// The fourth column has sum_r Srj = Sj.
real matrix ln_bip_xi_hessian(real colvector y1,
                              real colvector y2,
                              real colvector xi1,
                              real colvector xi2,
                              real colvector xi0,
                              real matrix    G)
{
        real scalar    j, k, m, r, Rj, Srj, u1, u2, u3,
                       S11, S12, S13, S22, S23, S33
        real matrix    Hraw

        Hraw = J(rows(y1), 6, 0)
        // j indexes observations
        for(j=1; j<=rows(y1); ++j) {
                // Get the cross-partial term
                Rj = min( (y1[j], y2[j]) )
                Srj = 0
                S11 = 0
                S12 = 0
                S13 = 0
                S22 = 0
                S23 = 0
                S33 = 0
                for(r=0; r<=Rj; ++r) {
                        Srj = poissonp(xi0[j], r) *
                              poissonp(xi1[j], y1[j]-r) *
                              poissonp(xi2[j], y2[j]-r)
                        u1  = (y1[j] - r - xi1[j])
                        u2  = (y2[j] - r - xi2[j])
                        u3  = (r - xi0[j])
                        S11 = S11 + Srj*(u1^2 - xi1[j])
                        S12 = S12 + Srj*u1*u2
                        S13 = S13 + Srj*u1*u3
```

```
                              S22 = S22 + Srj*(u2^2 - xi2[j])
                              S23 = S23 + Srj*u2*u3
                              S33 = S33 + Srj*(u3^2 - xi0[j])
                    }
                    Hraw[j, 1] = G[j,4]*S11 - G[j,1]*G[j,1]
                    Hraw[j, 2] = G[j,4]*S12 - G[j,1]*G[j,2]
                    Hraw[j, 3] = G[j,4]*S13 - G[j,1]*G[j,3]
                    Hraw[j, 4] = G[j,4]*S22 - G[j,2]*G[j,2]
                    Hraw[j, 5] = G[j,4]*S23 - G[j,2]*G[j,3]
                    Hraw[j, 6] = G[j,4]*S33 - G[j,3]*G[j,3]
                    Hraw[j, .] = Hraw[j, .] :/ ( G[j,4]^2 )
               }
               return(Hraw)
     }

     mata mlib create lmybipoisson, replace
     mata mlib add lmybipoisson *()
     mata mlib index

     end
     exit
                                      ———————— end make_lmybipoisson_mlib.do ————————



                                      ———————— begin mybipoisson.ado ————————
     program mybipoisson, properties(swml svyj svyr swml)
               version 18
               if replay() {
                         if (`"`e(cmd)'"' != "mybipoisson") error 301
                         Replay `0'
               }
               else    Estimate `0'
     end

     program Estimate, eclass sortpreserve
               syntax anything(id="equations" equalok) ///
                         [if] [in]                       ///
                         [fweight pweight iweight] [,    ///
                         noLOg                           /// -ml model- options
                         noLRTEST                        ///
                         vce(passthru)                   ///
                         TRACE                           ///
                         *                               ///
               ]

               // Check syntax
               mlopts mlopts other, `options'
               // Parse Replay options
               _get_diopts diopts, `other'

               // Parse equations
               ParseEqns `anything'
               foreach j in 1 2 {
                         local yvar`j'     `s(yvar`j')'
                         local xvars`j'    `s(xvars`j')'
                         if "`s(exposure`j')'" != "" {
                                   local expvar`j' `s(exposure`j')'
                                   local expopt`j' exposure(`expvar`j'')
                         }
                         if "`s(offset`j')'" != "" {
```

```
                          local offvar`j' `s(offset`j')'
                          local offopt`j' offset(`offvar`j'')
               }
               if "`s(noconst`j')'" != "" {
                          local nocon`j'  noconstant
               }
    }
    if "`weight'" != "" {
               local wgt "[`weight'`exp']"
    }
    if "`log'" != "" {
               local qui quietly
    }
    // Mark the estimation sample
    marksample touse
    markout `touse' `yvar1' `xvars1' `exposure1' `offset1'
    markout `touse' `yvar2' `xvars2' `exposure2' `offset2'

    // Check LHS variables to be nonnegative counts
    CheckLHS `yvar1' `yvar2' `touse'

    _vce_parse `touse', opt(Robust oim opg) argopt(CLuster): `wgt', `vce'
    // Initial values
    if "`trace'" != "" {
               local noise noisily
    }
    foreach j in 1 2 {
               tempname b0`j'
               capture `noise' poisson                              ///
                          `yvar`j'' `xvars`j'' `wgt' if `touse' ,   ///
                          `offopt`j'' `expopt`j'' `nocon`j''
               if _rc {
                          di in smcl as error                       ///
                             "Could not find initial values for " _c  ///
                             "equation `j'."
                          exit _rc
               }
               mat `b0`j'' = e(b)
               mat coleq `b0`j'' = "`yvar`j'':"
    }
    tempname b0
    mat `b0' = `b01', `b02', 0
    local initopt init(`b0', copy)

    local diparm1 diparm(ln_xi0, exp label("xi0"))

    // fit the full model
    ml model lf2 mybipoisson_lf2()                      ///
               (`yvar1' : `yvar1' = `xvars1',           ///
                          `nocon1' `offopt1' `expopt1'  ///
               )                                        ///
               (`yvar2' : `yvar2' = `xvars2',           ///
                          `nocon2' `offopt2' `expopt2'  ///
               )                                        ///
               /ln_xi0                                  ///
               `wgt' if `touse',                        ///
               `vce'                                    ///
               `log'                                    ///
               `mlopts'                                 ///
               `diparm'                                 ///
               `initopt'                                ///
```

```
                    missing                              ///
                    maximize
        ereturn scalar k_aux    = 1
        ereturn local  predict  mybipoisson_p
        ereturn local  cmd      mybipoisson

        Replay , `diopts'
end
program Replay
        syntax [, *]
        ml display , `options'
end
//      ParseEqns anything
//      Modified from mysureg.ado to fit our needs.
//
//      Takes "anything" and parses as
//              (y1 xvars1 [, EXPosure(varname) OFFset(varname) noCONStant])
//              (y2 xvars2 [, EXPosure(varname) OFFset(varname) noCONStant])
//
//      Returns, for j in {1,2}
//              s(yvar`j')
//              s(xvars`j')
//              s(exposure`j')
//              s(offset`j')
//              s(noconst`j')
//
program ParseEqns, sclass
        gettoken eq eqns : 0 , match(paren) parse("()") bind
        local p 0
        while "`eq'" != "" {
                local ++p
                local 0 `eq'
                syntax varlist(fv) [, EXPosure(varname) OFFset(varname) ///
                                  noCONStant ]
                gettoken lhs rhs : varlist
                local yvar`p'        `lhs'
                local xvars`p'       `rhs'
                local exposure`p'    `exposure'
                local offset`p'      `exposure'
                local noconst`p'     `constant'
                gettoken eq eqns : eqns , match(paren) parse("()") bind
        }
        if `p' != 2 {
                di in smcl as error "You must specify two, and only two, " _c
>
                di in smcl as error "equations."
                exit 198
        }

        sreturn clear
        foreach p in 1 2 {
                sreturn local yvar`p'      `yvar`p''
                sreturn local xvars`p'     `xvars`p''
                sreturn local exposure`p'  `exposure`p''
                sreturn local offset`p'    `offset`p''
                sreturn local noconst`p'   `nconst`p''
        }
end
```

```
program CheckLHS
        args yvar1 yvar2 touse
        local isae in smcl as error
        foreach j in 1 2 {
                summ 'yvar'j'' if 'touse', meanonly
                if r(min) < 0 {
                        di 'isae' "'yvar'j'' cannot be negative."
                        exit 398
                }
                if r(max) == r(min) {
                        di 'isae' "'yvar'j'' has no variance."
                }
                capture assert 'yvar'j'' == floor('yvar'j'')
                if _rc {
                        di 'isae' "'yvar'j'' has non-integer values."
                        exit 398
                }
        }
end

exit
```
——————————————————————————————————————— end `mybipoisson.ado` ——————————

——————————————————————————————————————— begin `mybipoisson_p.ado` ————————
```
// Predict handler for -mybipoisson-
program define mybipoisson_p
        version 18
        syntax [anything] [if] [in] [, SCores xb1 xb2 xi1 xi2 mu1 mu2 ]
        if "'scores'" != "" {
                // Let ml_p figure it out
                ml_p '0'
                exit
        }

        local n : word count 'anything'
        if 'n' < 1 | 'n' > 2 {
                exit 198
        }
        if 'n' == 1 {
                local vnam 'anything'
        }
        else {
                local vtyp ':word 1 of 'anything''
                local vnam ':word 2 of 'anything''
        }

        if "'xb1''xb2''xi1''xi2''mu1''mu2'" == "" {
                di in smcl "(option {bf:xb1} assumed)"
                local xb1 xb
        }
        tempname b
        mat 'b' = e(b)
        tempvar theta
        if "'xb1''xi1''mu1'" != "" {
                mat score 'theta' = 'b' 'if' 'in', eq(#1)
```

```
      }
      else {
             mat score 'theta' = 'b' 'if' 'in', eq(#2)
      }
      if "'xb1'" != "" {
             generate 'vtyp' 'vnam' = 'theta'
             lab var 'vnam' "Linear combination for equation 1"
      }
      else if "'xb2'" != "" {
             generate 'vtyp' 'vnam' = 'theta'
             lab var 'vnam' "Linear combination for equation 2"
      }
      else if "'xi1'" != "" {
             generate 'vtyp' 'vnam' = exp('theta')
             lab var 'vnam' "Latent mean for equation 1"
      }
      else if "'xi2'" != "" {
             generate 'vtyp' 'vnam' = exp('theta')
             lab var 'vnam' "Latent mean for equation 2"
      }
      else {
             tempname xi0
             sca 'xi0' = exp(_b[/ln_xi0])
             if "'mu1'" != "" {
                    generate 'vtyp' 'vnam' = exp('theta') + 'xi0'
                    lab var 'vnam' ///
                           "Expected count for ':word 1 of 'e(depvar)''"
             }
             else {
                    generate 'vtyp' 'vnam' = exp('theta') + 'xi0'
                    lab var 'vnam' ///
                           "Expected count for ':word 2 of 'e(depvar)''"
             }
      }
end
```

——————————————————————————————— end `mybipoisson_p.ado` ———————

References

AlMuhayfith, F. E., A. A. Alzaid, and M. A. Omair. 2016. On bivariate poisson regression models. *Journal of King Saud University – Science* 28: 178–189. https://doi.org/10.1016/j.jksus.2015.09.003.

Berndt, E. K., B. H. Hall, R. E. Hall, and J. A. Hausman. 1974. Estimation and inference in nonlinear structural models. *Annals of Economic and Social Measurement* 3/4: 653–665.

Binder, D. A. 1983. On the variances of asymptotically normal estimators from complex surveys. *International Statistical Review* 51: 279–292. https://doi.org/10.2307/1402588.

Breslow, N. E. 1974. Covariance analysis of censored survival data. *Biometrics* 30: 89–99. https://doi.org/10.2307/2529620.

Broyden, C. G. 1967. Quasi-Newton methods and their application to function minimization. *Mathematics of Computation* 21: 368–381. https://doi.org/10.1090/S0025-5718-1967-0224273-2.

Campbell, J. T. 1934. The Poisson correlation function. *Proceedings of the Edinburgh Mathematical Society* 4(1): 18–26. https://doi.org/10.1017/S0013091500024135.

Cleves, M., W. W. Gould, and Y. V. Marchenko. 2016. *An Introduction to Survival Analysis Using Stata*. Rev. 3rd ed. College Station, TX: Stata Press.

Davidon, W. C. 1959. Variable metric method for minimization. Technical Report ANL-5990, Argonne National Laboratory, U.S. Department of Energy, Argonne, IL. https://doi.org/10.2172/4252678.

Davidson, R., and J. G. MacKinnon. 1993. *Estimation and Inference in Econometrics*. New York: Oxford University Press.

Fletcher, R. 1970. A new approach to variable metric algorithms. *Computer Journal* 13: 317–322. https://doi.org/10.1093/comjnl/13.3.317.

———. 1987. *Practical Methods of Optimization*. 2nd ed. New York: Wiley.

Fletcher, R., and M. J. D. Powell. 1963. A rapidly convergent descent method for minimization. *Computer Journal* 6: 163–168. https://doi.org/10.1093/comjnl/6.2.163.

Fuller, W. A. 1975. Regression analysis for sample survey. *Sankhyā, Series C* 37: 117–132.

Gail, M. H., W. Y. Tan, and S. Piantadosi. 1988. Tests for no treatment effect in randomized clinical trials. *Biometrika* 75: 57–64. https://doi.org/10.1093/biomet/75.1.57.

Ghosh, I., F. Marques, and S. Chakraborty. 2021. A new bivariate Poisson distribution via conditional specification: properties and applications. *Journal of Applied Statistics* 48(16): 3025–3047. https://doi.org/10.1080/02664763.2020.1793307.

Goldberger, A. S. 1991. *A Course in Econometrics*. Cambridge, MA: Harvard University Press.

Goldfarb, D. 1970. A family of variable-metric methods derived by variational means. *Mathematics of Computation* 24: 23–26. https://doi.org/10.2307/2004873.

Gould, W. 2001. Statistical software certification. *Stata Journal* 1: 29–50. https://doi.org/10.1177/1536867X0100100102.

———. 2018. *The Mata Book: A book for Serious Programmers and Those Who Want to Be*. College Station, TX: Stata Press.

Greene, W. H. 2018. *Econometric Analysis*. 8th ed. New York: Pearson.

Grunfeld, Y., and Z. Griliches. 1960. Is aggregation necessarily bad? *Review of Economics and Statistics* 42: 1–13. https://doi.org/10.2307/1926089.

Harvey, A. C. 1976. Estimating regression models with multiplicative heteroskedasticity. *Econometrica* 44: 461–465. https://doi.org/10.2307/1913974.

Huber, P. J. 1967. The behavior of maximum likelihood estimates under nonstandard conditions. In Vol. 1 of *Proceedings of the Fifth Berkeley Symposium on Mathematical Statistics and Probability*, 221–233. Berkeley: University of California Press.

Kent, J. T. 1982. Robust properties of likelihood ratio tests. *Biometrika* 69: 19–27. https://doi.org/10.2307/2335849.

King, G. 1989. A seemingly unrelated poisson regression model. *Sociological Methods & Research* 17: 235–255. https://doi.org/10.1177/0049124189017003001.

Kish, L., and M. R. Frankel. 1974. Inference from complex samples. *Journal of the Royal Statistical Society, Series B* 36: 1–37. https://doi.org/10.1111/j.2517-6161.1974.tb00981.x.

Korn, E. L., and B. I. Graubard. 1990. Simultaneous testing of regression coefficients with complex survey data: Use of Bonferroni *t* statistics. *American Statistician* 44: 270–276. https://doi.org/10.1080/00031305.1990.10475737.

Lin, D. Y., and L. J. Wei. 1989. The robust inference for the Cox proportional hazards model. *Journal of the American Statistical Association* 84: 1074–1078. https://doi.org/10.1080/01621459.1989.10478874.

Loukas, S., and C. D. Kemp. 1986. The index of dispersion test for the bivariate Poisson distribution. *Biometrics* 42: 941–948. https://doi.org/10.2307/2530708.

Lütkepohl, H. 1996. *Handbook of Matrices.* New York: Wiley.

Marquardt, D. W. 1963. An algorithm for least-squares estimation of nonlinear parameters. *Journal of the Society for Industrial and Applied Mathematics* 11: 431–441. https://doi.org/10.1137/0111030.

Nash, J. C. 1990. *Compact Numerical Methods for Computers: Linear Algebra and Function Minimization.* 2nd ed. New York: Adam Hilger.

Ozuna, T., Jr., and I. A. Gomez. 1994. Estimating a system of recreation demand functions using a seemingly unrelated Poisson regression approach. *Review of Economics and Statistics* 76: 356–360. https://doi.org/10.2307/2109892.

Peto, R. 1972. Discussion on Professor Cox's Paper. *Journal of the Royal Statistical Society, Series B* 34: 205–207. https://doi.org/10.1111/j.2517-6161.1972.tb00900.x.

Press, W. H., S. A. Teukolsky, W. T. Vetterling, and B. P. Flannery. 2007. *Numerical Recipes in C: The Art of Scientific Computing.* 3rd ed. Cambridge: Cambridge University Press.

Rogers, W. H. 1993. sg17: Regression standard errors in clustered samples. *Stata Technical Bulletin* 13: 19–23. Reprinted in *Stata Technical Bulletin Reprints*, vol. 3, pp. 88–94. College Station, TX: Stata Press.

Royall, R. M. 1986. Model robust confidence intervals using maximum likelihood estimators. *International Statistical Review* 54: 221–226.

Shanno, D. F. 1970. Conditioning of quasi-Newton methods for function minimization. *Mathematics of Computation* 24: 647–656. https://doi.org/10.2307/1403146.

Stuart, A., and J. K. Ord. 1991. *Kendall's Advanced Theory of Statistics. Volume 2: Classical Inference and Relationship.* 5th ed. London: Arnold.

Welsh, A. H. 1996. *Aspects of Statistical Inference.* New York: Wiley.

White, H. 1980. A heteroskedasticity-consistent covariance matrix estimator and a direct test for heteroskedasticity. *Econometrica* 48: 817–838. https://doi.org/10.2307/1912934.

———. 1982. Maximum likelihood estimation of misspecified models. *Econometrica* 50: 1–25. https://doi.org/10.2307/1912526.

Williams, R. L. 2000. A note on robust variance estimation for cluster-correlated data. *Biometrics* 56: 645–646. https://doi.org/10.1111/j.0006-341X.2000.00645.x.

Wooldridge, J. M. 2010. *Econometric Analysis of Cross Section and Panel Data*. 2nd ed. Cambridge, MA: MIT Press.

Xu, X., and J. W. Hardin. 2016. Regression models for bivariate count outcomes. *Stata Journal* 16: 301–315. https://doi.org/10.1177/1536867X1601600203.

Author index

Subject index